Star Maps
History, Artistry, and Cartography
(Second Edition)

Nick Kanas

Star Maps

History, Artistry, and Cartography

(Second Edition)

 Springer

Published in association with
Praxis Publishing
Chichester, UK

Professor Emeritus Nick Kanas M.D.
University of California
San Francisco
U.S.A.

SPRINGER–PRAXIS BOOKS IN POPULAR ASTRONOMY

ISBN 978-1-4614-0916-8   ISBN 978-1-4614-0917-5 (eBook)
DOI 10.1007/ 978-1-4614-0917-5
Springer New York Heidelberg Dordrecht London

Library of Congress Control Number: 2012931281

Cover design: Jim Wilkie
Project management: OPS Ltd., Gt. Yarmouth, Norfolk, U.K.

Printed on acid-free paper

Springer is part of Springer Science+Business Media (www.springer.com)

Contents

APPENDICES

Preface to Second Edition

Star Maps: History, Artistry, and Cartography was first published in 2007. A second printing, which was similar to the first except for a few minor editorial changes, came out in 2009. When faced with the need to print more books in 2011, the question was whether or not to publish a third printing or issue a brand new edition. After all, how much new material could there be in a book that was mainly historical and focused on a fairly discrete and specialized topic?

As it turned out, plenty! This second edition, published in 2012, contains over 50 new pages, has 44 new images (16 in color), and lists over 60 additional references, most of which come from primary sources. These changes are in part due to new material that has appeared in scholarly map and astronomy journals and specialty magazines. But in addition, I have responded to helpful recommendations from the reviewers of the first edition, not to mention comments made by people attending my lectures on the history of celestial cartography. And to be honest, I have purchased new prints for my collection and developed new sub-interests in the past few years that I wanted to include.

As a result, I have reread the entire first edition and made a number of editorial changes and updates. In Chapter 3, mention is made of the Neoplatonist Hypatia. In Chapter 4, Caspar Vopel rather than Gerard Mercator is credited with introducing the constellations Antinous and Coma Berenices on his celestial globe, and a figure showing Antinous has been added. In Chapter 6, a new Hevelius map replaces the former Figure 6.6. Significant additions have been made to Chapter 8, including new figures and text in the celestial globe, volvelle, telescope, planets, and Moon sections; mention of the Nebra sky disk and astronomer Thomas Harriot; and entirely new sections on asteroids, deep sky objects, playing card maps, and frontispieces and title pages. In Chapter 9, there is new material on colonial America and on astronomy writer Hannah Mary Bouvier. In Chapter 10, there are several new figures and new sections on astronomer Johann von Littrow and computerized star maps.

A new Chapter 11 collects all the color figures into a single Color Plate Gallery. Corrections and updates also appear in the appendices.

Although much has been added to this second edition, not every recommended change was made. For example, some reviewers of the first edition suggested that I add material on prehistoric and indigenous sky maps. However, as I stated in the Preface to the first edition, the main focus of this book is the European sky mapping tradition, and to go beyond the cultures mentioned in Chapter 2 would take us too far afield. Similarly, the historical information in Chapters 3 and 4 was selected to be a primer for the reader to better understand the meaning of the sky maps described in subsequent chapters, not to be an exhaustive history of astronomy. For this, the reader is referred to the texts referenced in the chapter bibliographies.

Nick Kanas
January 2012

Foreword to First Edition

The representation of celestial bodies (stars, planets, comets and other extraterrestrial phenomena) has been an important part of cartography for millennia. Star maps of early Chinese, Indian, Mesopotamian, Egyptian and other cultures are significant in their own right but also because of their influence on Greek, Roman, Islamic and, later, European celestial cartography. The history of changing images and practices in this field up to the present is discussed and illustrated in admirable detail in *Star Maps*, by Nick Kanas.

As with a number of others interested in cartography, the author is a medical doctor who has devoted a great deal of time and resources to studying and writing about his long-term avocation. However, most collectors have an interest in a limited time period or geographical area (e.g., the eighteenth century or Jamaica). By contrast, Dr. Kanas has a longer and larger agenda: the representation of the heavens from antiquity to the present, as illustrated by maps and charts.

In order to cover his wide-ranging subject, Nick Kanas' book features over 200 images, 76 in color. Each of these images is discussed in the text in appropriate detail. Most of the illustrations are miniaturized, but they are generally well reproduced and remarkably readable. Similarly, the text is lucid and can be appreciated by specialists, but also by collectors and others. In addition to the maps and their descriptions, the volume contains a glossary and other reading aids.

Undoubtedly, *Star Maps* will become an important reference work in civic, institutional, college and university libraries, but individuals who are interested in the topics it covers will want to have their own copy. The distinguished scientific publishing house of Springer, in association with Praxis of Chichester, U.K., is to be commended for undertaking to make such a valuable body of knowledge available to a potentially larger audience. Because of its wide coverage, its many illustrations and other features, and because of the importance of the subjects it treats, *Star Maps* will become an indispensable library accession. It should also be acquired for their own private collections by all of those who are interested in the large and important body of knowledge it covers.

Norman J. W. Thrower
Professor Emeritus
University of California, Los Angeles

Preface to First Edition

In the 17th and 18th Centuries, a number of beautiful sky atlases were produced in Europe that showed the constellations as allegorical representations of classical Greek heroes, heroines, and monsters. But these constellation images also had a scientific purpose, in that they were placed in coordinate systems of celestial latitude and longitude that allowed the stars to be mapped in the sky. In addition, many of these atlases depicted diagrams of the solar system that reflected both contemporary and ancient cosmological systems, thus tracing the development of our view of the heavens over time.

Such images have generally disappeared from modern-day celestial charts, which instead focus on showing thousands of stars and deep-sky objects such as galaxies and nebulae that are not visible to the naked eye. With the discovery of ever more wonders in space and with plans to finish the International Space Station and then move on to explore the Moon and Mars, there is a renewed interest in the heavens. Increasing numbers of people are buying telescopes and becoming amateur astronomers, and they are using star charts to help them navigate in the sky. But, at the same time, the beauty and awe generated by the celestial void has captured our imagination and delighted our aesthetic sense, and there is a longing for the old images. For example, antiquarian map societies are prospering, and celestial maps are now viewed as a specialty of map collecting.

Up until now, discussions of star maps have been found in either general histories of astronomy or in catalogs of celestial atlases that have failed to trace their development over time. What is needed is a more integrated book that discusses celestial cartography in terms of constellation development, changing views of the universe, and advances in mapmaking techniques, while at the same time capturing the beauty of the heavens using images from antiquarian celestial prints and atlases. This, in a nutshell, is the intention of this book.

My decision to write this book is the culmination of a number of factors having to do with my long-standing interests in amateur astronomy (from childhood) and antiquarian map collecting (from young adulthood). How I got here from there may serve as an example of the appeal of these two activities.

Since the launch of Sputnik I, the world's first artificial satellite, on October 4, 1957, I have been hooked on space. Although I read science fiction novels and joined the space cadet secret decoder club before then, trying to find Sputnik moving through the sky one evening on a bluff overlooking the Willamette River in Portland, Oregon, made me realize that the night sky was pretty interesting. This notion was reinforced by viewing Saturn and its rings through a telescope that someone had set up the same night. Shortly thereafter, I received a 6-cm (2.4-inch) refracting telescope as a holiday present, and I began my 50 + year avocation as an amateur astronomer. Subscribing to *Sky & Telescope* magazine, I looked forward to the monthly star charts in order to see what I could see in the heavens. Although my range was limited, my imagination wasn't, and I dreamed of viewing more of the planets and deep-sky objects that I had been reading about.

This was realized when I finished my schooling and took my first professional job in 1977 as a psychiatry professor at the University of California in San Francisco, a position I still hold. Now, I could afford to buy a larger 20-cm (8-inch) reflecting telescope, and I joined the San Francisco Amateur Astronomers. After my first club "star party" one cold November night on a mountain north of the city, I realized two things: I needed a warmer jacket (even in California), and I needed a good star atlas. I bought a copy of *Norton's Star Atlas*, and I was on my way. As my interests expanded, I bought additional star atlases to help me locate the dim objects I wanted to see. I found these star maps to be fascinating, especially those that showed actual constellation images. I read up on some of the mythology behind the images (mostly from the Ancient Greeks) and imagined being a part of this folklore as I scanned the heavens looking for my deep-sky prey.

While visiting relatives in Rhode Island one summer in 1982, my wife and I chanced upon an antique store in Newport that was displaying two antiquarian constellation prints on a wall. One featured Sagittarius and the other showed a number of constellations around the southern celestial pole. I bought them for a whopping $24 (total!). I subsequently found out that these prints were from the 1776 French edition of a sky atlas written by John Flamsteed, the first English Astronomer Royal. Several years later in 1989, while on a sabbatical in London, I visited a special exhibit on celestial cartography at the British Museum, and I really became hooked at the beauty and sense of history of these old star maps. Since then, I have continued to acquire antiquarian celestial prints and books. I have joined the California Map Society, the Washington Map Society, and the International Map Collectors' Society (based in London), and I have learned a great deal about collecting old maps through these associations.

But I was frustrated by the fact that there was not a single book on celestial cartography that could inform me about the various aspects of my collecting, such as the meaning behind the squiggly lines that accompanied the epicycles of my cosmologically oriented charts, who some of the people were who produced the great classical star atlases, how celestial maps evolved and changed over time, ways to protect and preserve them, etc. What I needed was a book that not only was a primer for the collector but also had sufficient reference detail to allow me to identify and understand my maps. Nothing like this appeared, so I decided to write such a book some day.

And here it is, "only" 25 years after I acquired my first antiquarian star maps! This book has been written for three of my alter egos: (1) amateur astronomers who want to know more about star maps and their development, (2) antiquarian map collectors who want to expand their horizons from terrestrial to celestial maps, and (3) people out there who can appreciate the beauty and history behind these wonderful works of art and science and who want to know more about them. I have tried to tell the story of how star maps came into being and evolved over time, as well as to illustrate their artistry through the figures. Except where indicated, all of the images in the book are from the pieces that my wife and I have acquired over the years from here and there. I have digitally photographed nearly all of the illustrations in this book from actual celestial prints, in part to show their diversity, and in part to provide actual examples for the collector who may want to compare his or her prints with someone else's.

The story of star map development is written and illustrated in the 10 chapters of the book, which may be summarized as follows. Chapter 1 defines the two types of celestial maps: those illustrating cosmological systems, which generally include members of our solar system, and those showing constellation images and the locations of stars in coordinate grids. In addition, this chapter discusses basic orienting concepts that are necessary for the understanding of the maps. Chapter 2 deals with the cosmologies and constellations of four non-European areas that either influenced or were influenced by European star map development: China, Mesopotamia, Egypt, and India. Also in this chapter is a section on ancient astrology. Chapter 3 presents a review of European cosmology from the pre-Socratic philosophers to the time of Newton, with an emphasis on topics relevant to understanding what is shown in the cosmological type of star map (e.g., geocentric versus heliocentric cosmologies, deferents and epicycles). Included are sections on printing and medieval astrology. Chapter 4 traces the development of constellations in Europe, with an emphasis on topics relevant to understanding what is shown in the constellation type of star map (e.g., constellation images depicted at different times in history, currently obsolete constellations). The astronomically sophisticated reader may wish to skip Chapters 3 and 4. However, as in all the chapters, the figures in Chapters 3 and 4 are taken from antiquarian celestial books and prints, so even skimming readers may wish to take a look at them in order to enhance their familiarity with the range of such images.

The text of the remaining chapters deals more specifically with antiquarian celestial maps. Chapter 5 reviews early star maps found in manuscripts and printed works up to 1600. Chapter 6 discusses the Golden Age of pictorial star maps from 1603 to 1801 in Europe, with a special focus on four of the most influential cartographers: Bayer, Hevelius, Flamsteed, and Bode. Chapter 7 continues the discussion of the Golden Age in terms of other important contributors in Europe. Chapter 8 discusses special topics that are relevant to mapping the heavens: celestial globes and gores, volvelles, astronomical instruments before and including the telescope, and members of our solar system. Chapter 9 focuses on the history of star mapping in early America. Finally, Chapter 10 deals with the transition to star maps without constellation images in the 1800s, and the importance of astrophotography, along with improvements in the graphic arts and computer technology, in producing the star-rich and precise atlases of Norton, Becvar, Tirion, and others up to the present day.

The first four appendices are designed for advanced amateur astronomers, map collectors, and collector-wannabees who wish to know more about the process of collecting and who want a general reference for specific maps that go beyond the material found in the chapters. A glossary and index round out the book

The legends to the figures contain dimensions in centimeters (cm). This not only gives an indication of the size of the print or page for the general reader, but it also allows collectors to compare their maps with my images to help in establishing authenticity and state of printing. For maps and other images with a border, I have followed the convention of giving the vertical by horizontal dimensions as measured from the inner borderlines. If these don't exist, I have given the height and width of the block mark (for woodblocks) or plate mark (for engravings), or indicated the page dimensions. Where there are hemispheres and planispheres, I have usually given the diameter, again in cm.

It is my hope that this book will stimulate *you* to take a look at the heavens with a new eye, appreciating their scientific wonders for sure, but at the same time seeing the sky as a haven for beauties and beasts of old. You can become a direct participant: economically priced telescopes are available, antiquarian celestial maps can still be found, and there are many amateur astronomy and map-collecting organizations that are ready to help you along the way. Have a pleasant journey!

Nick Kanas
April 2007

To Carolynn, who has joined me in my collecting passion and who has lovingly encouraged me to write this book, even into the wee hours and on sunny days.

Acknowledgments

A book of this type cannot be written in a vacuum, and I would like to thank a number of people for their help and support. First and foremost is my wife Carolynn, who has spent hours looking at celestial prints with me and who has encouraged me to write this book despite its intrusion into our personal time. Andrew and Peter have also been patient over the years hearing about dad's esoteric hobby and listening to explanations about epicycles and cosmologies.

Norman Thrower, Professor Emeritus of Geography (Cartography) at the University of California, Los Angeles, has provided helpful cartographic and editorial advice to an earlier draft of this book, as well as moral support through our mutual association in the California Map Society. Owen Gingerich, Professor Emeritus of Astronomy and History of Science at the Harvard–Smithsonian Center for Astrophysics, has likewise provided helpful editorial suggestions and valuable astronomical advice. Although I have included the suggestions of Professors Thrower and Gingerich in the final version of this book, the responsibility for its content and accuracy remain mine.

Others have helped me in different ways. My friend and fellow celestial map collector, Bob Gordon, has given me moral support, confirmatory information on some questions I had on some of the maps, and digital images of pieces from his wonderful celestial map collection. He has responded graciously to my myriad requests for print dimensions and exact labeling content on some of his maps, and I have appreciated his patience and responsiveness. Peter Barber, Head of Map Collections at the British Library, and his staff have been enormously helpful during my trips to London to research celestial maps and atlases from their vast holdings. Specifically, I would like to thank Nicola Beech, Sue Young, Geoff Amitage, and especially Carlos Corbin for their patience and assistance.

I would also like to mention several people who provided me special assistance in obtaining images for this book: Sandra Powlette at the British Library; Doris Nicholson at the Bodleian Library at Oxford University; Devon Pyle-Vowles at the Adler Planetarium & Astronomy Museum in Chicago; Stacey McCarroll Cutshaw at the Boston University Art Gallery; Linda Schubert at Rand McNally & Company in Skokie, Illinois; and Jonathan Potter, who kindly provided me with two digital images from his map shop in London and gave me his input on a draft of Appendix A from a dealer's perspective.

I also would like to thank Professor Gingerich, Mr. John Ventre at the Cincinnati Observatory Center, and Mr. Paul Luther at AstronomyBooks in Bernardston, Massachusetts, for providing me with information on O.M. Mitchel's edition of Burritt's book and atlas (see Section 9.6.2 in Chapter 9).

Several of my friends and colleagues in the International Map Collectors Society, the California Map Society, the San Francisco Amateur Astronomers, and the amateur astronomy publication world have also provided me with support and encouragement in writing this book. Some have served as referees of my book proposal and provided helpful advice. I would especially like to thank the following: Robert Clancy, Susan Gole, David Kalifon, Hans Kok, Joshua Roth, Phil Simon, and James C. White.

Clive Horwood, my publisher at Praxis Publishing Ltd. in Chichester, England, has been very supportive during the process of producing this book, and as an author juggling a number of competing projects and struggling with deadlines, I have appreciated his patience. Neil Shuttlewood, whose company Originator Publishing Services in Norfolk, England, has been responsible for the copyediting, has been very responsive to my questions about digital photography and page proof production. Finally, Rhea Siegel, Esq., at Springer Publications in New York, has given me helpful advice about permissions and copyrights.

Unless otherwise indicated, the figures in this book have been produced from digital photographs taken from antiquarian books and prints that are part of the Nick and Carolynn Kanas Collection. Permissions to use and photograph the images from other sources have been obtained, and these sources are acknowledged in the legends to the figures. Every effort has been made to source the original copyright holders, and I apologize to any that I may have missed through oversight or inability to contact via e-mail or phone.

Figures

(boldface indicates page numbers of color versions in the Color Plate Gallery)

Tables

Abbreviations and acronyms

AAVSO	American Association of Variable Star Observers
HDC	Henry Draper's Catalog
IAU	International Astronomical Union
IC	Index Catalogue
NGC	New General Catalogue
SAO	Smithsonian Astrophysical Observatory
SDUK	Society for the Diffusion of Useful Knowledge

1

What is a star map?

From 1600 to 1800, a number of beautiful star atlases were printed that depicted the constellations according to ancient myths and tales. In Europe, where the quality of celestial atlases was unmatched, classical Greek traditions prevailed, and the constellations were given allegorical visual representations that consisted of heroes and heroines, real and imaginary animals, scientific instruments, and artistic tools. These images were placed in celestial coordinate systems that allowed the positions of the stars to be mapped in the sky and formed the backdrop for predictions of the location of the planets and other heavenly bodies throughout the year. But there was a second kind of image that was found in these celestial atlases as well. These images consisted of diagrams of heavenly bodies or of the entire solar system that reflected both contemporary and ancient cosmological systems. The components of these systems were shown with reference to each other in the sky and in some cases to the background stars. Let's look at these two types of star map in more detail.

1.1 CONSTELLATION MAPS

An example of a constellation type of star map is shown in Figure 1.1. This plate is from a celestial atlas first published by Fortin in 1776 and shows the sky around the central constellation of Cygnus the swan. Our eye is first drawn to the beautiful constellation images, here shown in vivid color. In some star atlases, the color was original, but in most cases (such as this one) color was added later to enhance the beauty and decorative quality of the plates. Further perusal of this plate reveals that the names of the constellations are in the French language, indicative of the French origins of the atlas from which the plate comes. According to the title, four con-stellations are featured: la Lyre (Lyra the lyre), le Cygne (Cygnus the swan), le Lezard

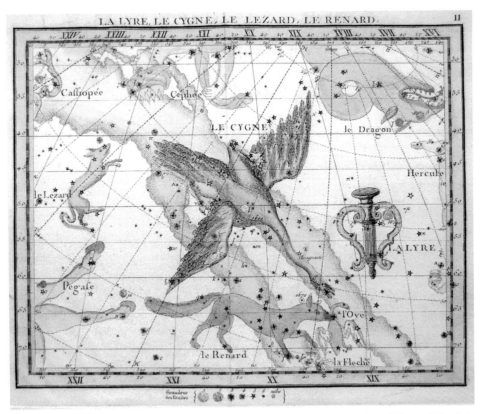

Figure 1.1. View of several northern constellations, taken from the 1795 edition of Fortin's *Atlas Céleste de Flamsteed*. 15.7 × 20.6 cm. Note the constellations of Lacerta ("le Lezard", which looks more like a dog than a lizard); Cygnus (flying in the Milky Way); Lyra (with the bright star "Wega"); and Vulpecula ("le Renard") clutching in its jaws Anser ("l'Oye", which no longer exists). *See also* Color Plate Gallery.

(Lacerta the lizard, looking here more mammalian that reptilian), and le Renard (Vulpecula the fox). Clenched in the jaws of Vulpecula is a separate constellation, l'Oye (Anser the goose). Lyra and Cygnus are among the 48 traditional constellations that have been with us for some 2,000 years, since ancient Greek times. In contrast, Lacerta and Vulpecula are "new" constellations that were introduced by the famous Polish astronomer Johannes Hevelius in 1687. Although these two constellations are still with us today, Anser is not, being a victim of the actions of the International Astronomical Union in 1922 when it purged the sky of many constellations in an effort to standardize their number and to discourage the actions of astronomers eager to honor patrons or celebrate events by inventing constellations for their star atlases (the I.A.U. settled on 88 constellations, which are now considered official). So, throughout history, constellations have come and gone, and what one sees in a given star atlas reflects what was current in the mind of its creator.

But Figure 1.1 shows more than constellations. Note the presence of two grid-like coordinate patterns, one using dotted lines and one solid lines. The first was based on an imaginary line in the sky representing the path of the Sun called the ecliptic. The second was based on an imaginary line in the sky that was a projection of the Earth's equator called the celestial equator. These will be described more fully later on, but for now the point to note is that these two systems allowed any object in the sky to be "mapped" according to the coordinate system being used. This was a major advance over pre-Renaissance systems, where the location of heavenly bodies was in reference to their location in a constellation (e.g., "the star at the end of the right foreleg" or "the planet at the tip of the tail").

Plates such as this one were often used to accompany a star catalog, which gave detailed information about stars such as their location in the sky, brightness (or magnitude), etc. Some of this information could be incorporated in the constellation map. For example, note that according to the scale at the bottom, Vega is a magnitude 1 star since it is indicated on the plate by a large gold symbol. In contrast, the star near the tip of Lacerta's nose is much dimmer at magnitude 5. Also shown in the plate are the locations of various nebulae, the supernova of 1670 (near Vulpecula's upraised ear), and the cloud-like Milky Way running diagonally from upper left to lower right.

1.2   COSMOLOGICAL MAPS

A particularly rich example of a cosmological star map is shown in Figure 1.2, which was produced by Johann Doppelmayr and appeared in an atlas from the famous Homann publishing house around 1720. In a sense, this beautiful plate depicts the state of astronomical knowledge in the early 1700s. In the center is a spectacular representation of the Sun-centered heliocentric world introduced by the great Polish astronomer, Nicholas Copernicus (1473–1543 AD). The view here includes descriptions from the Dutch astronomer Christiaan Huygens (1629–1695 AD), complete with sunburst and showing the orbits of the planets and their moons revolving concentrically around the Sun. Throughout, there is written and numerical information on the proportionate diameters of the planets, and the rest of the universe surrounding our solar system is shown in the form of the twelve constellations of the zodiac.

In the upper left corner, we see representations of the then-known planets along with the Sun. There is an attempt to show these heavenly objects to scale in terms of diameter, and some surface features are indicated on some of the planets. In the upper right corner, we see heavenly clouds and diagrams of other solar systems that were thought to exist around the stars. In the lower left corner we see a depiction of the solar eclipse of May 12, 1706. On the miniature map of the Earth, the state of California was represented as an island, which was thought to be the case throughout much of the 17th Century and early 18th Century.

Figure 1.2. A plate produced by Doppelmayr for Homann Publications, ca. 1720, which also appeared in Homann's 1742 *Atlas Coelestis*. 48.2 × 56.8 cm, 43.6 cm dia. hemisphere. It depicts the state of astronomical knowledge in the early 1700s. Note the Copernican cosmological system in the center, complete with the planets and their moons and textual and numerical information on the proportionate diameters of the planets. See text for a description of the other images. *See also* Color Plate Gallery.

In the lower right corner below the illustration of a lunar eclipse, we see representations of three cosmological systems that are being introduced to us by Urania, the goddess of astronomy. Going left to right, we first see the system of the ancient Greek astronomer, Claudius Ptolemy (ca. 100–ca. 178 AD), which is partially obliterated by contemporary astronomical instruments (allegorically implying that modern science has dispatched this view). In this system, the Earth is in the center, followed by the orbits of our Moon, Mercury, Venus, the Sun, Mars, Jupiter, Saturn, and the fixed stars. A hybrid system developed by the Danish astronomer Tycho Brahe (1546–1601 AD) is next. In this model, the Earth is in the center and is orbited by the Moon and the Sun. However, Mercury, Venus, Mars, Jupiter, and Saturn orbit the Sun, whereas the fixed stars continue to surround the central Earth. The final model is labeled *sic ratione* (i.e., according to reason) and is the Copernican system. This labeling and the central dominance of Copernicus' ideas in this print clearly indicate the opinions of Doppelmayr and Homann, even though alternative models are included in the corner for historical purposes.

1.3   WHAT MAKES THESE IMAGES MAPS?

Images such as these were found in atlases devoted exclusively to the heavens. But celestial images also found their way into geographical atlases, which often included a plate or two of the constellations or of the solar system to accompany the terrestrial maps that were the focus of the book. This led credence to the notion that these celestial representations were a type of map. This notion was supported by the fact that celestial globes depicting the constellations had been known since classical Greek times, and beginning with the Renaissance, celestial globes often were paired with terrestrial globes for sale to people interested in geography and exploration. More recently, celestial images have been included in standard texts of cartography.

But is it accurate to call these celestial representations maps, or is the juxtaposition of celestial images with terrestrial images simply "guilt by association"? To gain some perspective on this question, let's consider one definition of a map: "A (flat) representation of the earth's surface or a part of it, showing physical, geographical, or political features ... a similar representation of the sky showing the relative positions of stars etc., or of the surface of a planet etc ... A diagram representing the spatial distribution of anything or the relative positions of its components ..." (O.U.P., 2002, p. 1697). Professor Norman Thrower, the distinguished geographer from the University of California, Los Angeles, further characterizes maps as follows: "Viewed in its development through time, the map details the changing thought of the human race, and few works seem to be such an excellent indicator of culture and civilization" (Thrower, 1999, p. 1). For example, in terrestrial maps, different countries appear and disappear over time depending on politics and world events. From the above quotes, it would seem that to be called a map a celestial image needs to illustrate the physical location of something in space, to accurately show the relative position of its components, and to reflect the assumptions and ideas of the times. How well do celestial maps do this?

Based on the examples shown above, pretty well. The images give us information regarding the location of heavenly bodies and the relative position of their component parts. The stars and constellations are accurately plotted in a coordinate system referenced to the ecliptic or celestial equator. The surfaces of heavenly bodies such as the planets may be represented by showing important physical features. Finally, the images reflect the state of knowledge and the politics of the times in which they were created. Thus, star maps such as these meet the essential criteria of being true maps, albeit maps that are truly out of this world.

1.4   CIRCLES IN THE SKY

1.4.1   The Sun–Earth orientation

Since ancient times, people have identified regions on the Earth based on the apparent location of the Sun in the heavens. When looking at the sky throughout the year at a given time each day (say, noon), the Sun appears to gradually increase in elevation day by day to a certain height, then decrease day by day to a certain depth, and so on. We now know that this apparent rising and falling is due to the fact that the Earth revolves around the Sun in a year, and because the Earth's axis is tilted some $23\frac{1}{2}$ degrees to the plane of its orbit, this affects the height of the Sun in the sky. In the summer, the Sun most directly beams its rays to people in the Northern Hemisphere; hence, it appears higher in the sky (left image in Figure 1.3), and the hours of daylight are the longest. If we project a direct line from this highest elevation of the Sun onto the Earth's surface, a circle of latitude is defined which is called the Tropic of Cancer. Alternatively, in the winter the Sun most directly faces the Southern Hemisphere, so to people in the Northern Hemisphere it appears lowest in the sky (right image in Figure 1.3), and the hours of daylight are the shortest. A line projected onto the Earth when the Sun is at its lowest defines a circle called the Tropic of Capricorn. These two extremes of time when the Sun is at its highest or lowest at noon are called the

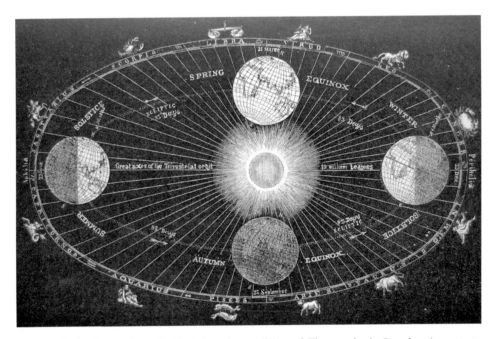

Figure 1.3. An image from the first American edition of Flammarion's *Popular Astronomy*, translated with his sanction by J. Ellard Gore and published in 1894. 23.2 × 15.5 cm (page size). Note the revolution of the Earth around the Sun against the backdrop of the zodiacal constellations.

solstices. The summer solstice occurs around June 22 every year, whereas the winter solstice occurs around December 22 every year. In between these two extremes, the Sun shines most directly on the equator. Because these times of equal day and night occur twice a year, we refer to them as the spring or vernal equinox (around March 21) and the fall or autumnal equinox (around September 23).

How did the tropics get their names? To comprehend this, it is important to understand that for millennia people have had an intimate relationship with the sky and often oriented themselves on Earth with the stars. In a similar manner, they oriented themselves in the sky according to conventions used on Earth. As people looked up at the heavens, they realized that the Sun, Moon, and planets (sometimes called "wandering stars") appeared to move in a region of the sky close to the ecliptic, and the 12 constellations in this region were given special significance. Because most of these constellations represented animals, they collectively were referred to as the constellations of the zodiac (like the word "zoo", this term comes from the Greek word for "animal"). Look again at Figure 1.3 and imagine yourself at the summer solstice looking at the Sun at noon. Although it is daylight and you cannot see the stars, you know that the summer constellations were high in the sky the evening before and that the Sun is currently in that part of the zodiac occupied by the constellation Cancer (if there was a sudden solar eclipse, you would in fact see this constellation behind the Sun). So, during the summer when the Sun is at its highest in the sky for the Northern Hemisphere and shines its direct rays at the northernmost latitude on the Earth, we call this latitude the Tropic of Cancer. In a similar manner, if you are looking at the Sun at noon at the time of the winter solstice, the Sun is in the part of the zodiac occupied by the constellation Capricornus, so the latitude on Earth of most direct southern rays is called the Tropic of Capricorn.

1.4.2   The armillary sphere

In order to better visualize in three dimensions the relationship between the Earth perceived as the center of the universe and the sky around it, the ancients created an instrument called the armillary sphere. There is evidence that Eratosthenes used an armillary sphere in the 3rd Century BC to determine the angle in the sky between the celestial equator and the ecliptic. Scholars at the great library in Alexandria also used this instrument to visualize the heavens, compute the coordinates of the stars on the celestial sphere, and make calculations involving angular distances between heavenly bodies. At about the same time during the Han dynasty (207 BC–220 AD) in China, the armillary sphere was developed and became an important astronomical tool. Its popularity continued in Islamic countries during the 8th to 15th Centuries AD, when several treatises were written on its construction and use. During the Renaissance, the armillary sphere continued to be used for education and calculation, and one of the first European makers of this instrument was Johannes Regiomontanus (1436–1476), who we shall meet in Section 3.8.2. During the Age of Exploration (1400 to 1800), the armillary sphere was used in training navigators; in fact, it became an important symbol of Portuguese exploration during the heyday of their activities in the 15th and

16th Centuries, often appearing figuratively on banners and paintings. Most armillary spheres were geocentric, with the Earth appearing in the center, and it was not until the 18th Century that Sun-centered armillaries were produced. But, by then, advances in spherical trigonometry and the development of the telescope and other instruments that could more accurately determine the positions of heavenly bodies in the sky led to the armillary sphere becoming more of a decorative piece than a tool for celestial calculations.

The armillary sphere consisted of a number of nested rings, or armillaries (which means "bracelet" in Latin). These were of varying diameters and represented the principal celestial spheres. One of these rings was fixed and was connected to the base of the instrument; this represented the observer's horizon. The other rings could be moved or pivoted around an axis to match the latitude and longitude of the area in the sky being observed. Usually, there were rings representing the meridian, the celestial equator, and the ecliptic. In some larger versions, there were also rings representing the orbits of the Sun and planets. Many of the rings were calibrated in degrees. In the center was a sphere representing the Earth. Most armillary spheres that were used for observation and calculation were large and made out of brass, but those that were primarily used for demonstrational purposes were smaller and often constructed of cheaper material such as wood rather than metal.

A schematic of this instrument is shown in Figure 1.4. Imagine yourself way out in space looking back on the rest of the universe. There in the center is the Earth ("Terra", in black), oriented with the north pole pointing directly up. At a $23\frac{1}{2}$ degree angle around the Earth is the ecliptic ("Ecliptica") with the 12 constellations of the zodiac. In parallel to the spherical Earth, the universe itself can be conceived of as a sphere, with many of the circles on the Earth being projected onto the universal sphere. Thus, the projection of the equator is the celestial equator (here the "Aequienoctoalis"). Note also the projections of the Tropics of Cancer and Capricornis, the Arctic and Antarctic Circles, and the Arctic and Antarctic Poles. Like the Earth, the sky may be divided into 360 degrees, going up or down from 0 to 90 degrees from the celestial equator. In this diagram, the orientation shows east ("Oriens") to the left and west ("Occidens") to the right. Just as on the Earth, many of the circumference lines on the celestial sphere whose center is the center of the universe/Earth are called the "great circles", and they include the ecliptic, the celestial equator, and the horizon (which here is shown parallel to the celestial equator).

Note that there is a place in the sky where the great circle of the celestial equator crosses the great circle of the ecliptic. For the ancients, this location occurred at the positions of the equinoxes, when the Sun was in the constellations of Libra and Aries (Figure 1.3). In fact, the place when the Sun first entered the constellation Aries was called "first point of Aries", and this had great significance for ancient astronomers and astrologers and was considered the 0-degree point of celestial longitude in the sky. In the last 2,000 years, a wobbling of the Earth's axis has caused a "precession" of the heavens so that the place in the sky where the celestial equator and ecliptic cross is now in the neighboring constellation of Pisces, as is shown in Figure 1.4. Nevertheless, the first point of Aries continues to have special meaning

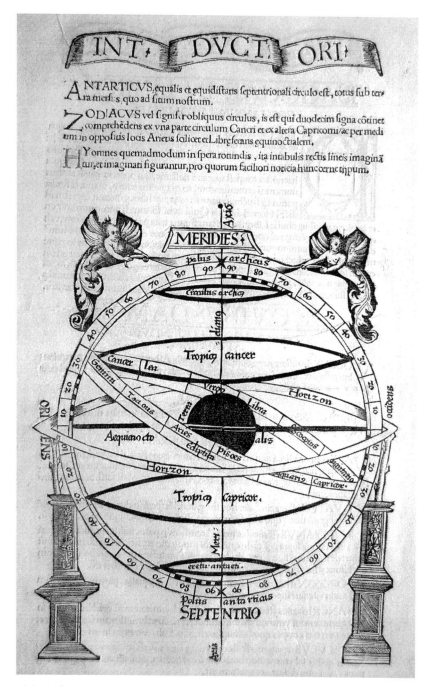

Figure 1.4. A figure of an armillary sphere, from Lorenz Fries' 1522 edition of Ptolemy's *Geographia*. 37.3 × 24.9 cm (page size). Note the central Earth and the projections of its great circles into the sky. See text for details.

for astrologers. In fact, the great north-to-south circle that goes through the two equinoctial points has a special significance and appears on some star charts as the "Equinoctial Colure". In a complementary manner, the great circle that goes through the two solstitial points is called the "Solstitial Colure". (For a view of the first point of Aries on a celestial globe, see Figure 8.2.)

1.5   DIRECTIONS IN THE SKY

In any map, it is important to orient yourself to direction before you can figure out what you are seeing. Typically, most terrestrial maps are oriented with north at the top and east to the right. However, this was not always the case. For example, some earlier terrestrial maps had east at the top, since this was the direction to Jerusalem as seen from Western Europe. Thus, the orientation of a map is a matter of convention.

A similar situation has existed with celestial maps, and two major systems have been dominant. The earliest star maps were oriented to the ecliptic in the age when the zodiac was prominent. In this coordinate system, this line was the 0 point of latitude, and 90 degrees to the north was the north ecliptic pole (and 90 degrees to the south was the south ecliptic pole). One could also measure degrees of longitude along the ecliptic from a point designated as 0 degrees longitude, which by convention was taken to be at the First Point of Aries. Thus, in the ecliptic system, one could locate any object in the sky in terms of its degree of celestial longitude and degree of celestial latitude.

Later on, the stellar coordinate system was oriented to the celestial equator. In this system, the night sky (for Europeans and others living in the Northern Hemisphere) appeared to rotate around Polaris, the pole star, which is close to the north equatorial pole. Since the sky appeared to move from east to west at 15 degrees per hour (due to the rotation of the Earth on its axis from west to east), an object in the sky that was being observed through a telescope or whose height was being measured using a fixed quadrant that was, say, 30 degrees due east would drift into the field of view by just waiting 2 hours. As clocks began to be perfected in the 17th Century, longitudinal positions in the sky could now be obtained by recording when a celestial object like a star crossed the meridian. Directions along the celestial equator were now measured in terms of hours of right ascension. That is, the sky was divided into 24 hour lines, each 15 degrees apart at the celestial equator. Areas above and below the celestial equator continued to be measured in degrees, which were referred to as degrees of declination. During the transition from the ecliptic to the celestial equatorial coordinate system, star maps often showed one superimposed upon the other, which made for a somewhat cluttered and confusing array of lines (Figure 1.5). In modern star atlases, the equatorial system with its right ascensions and declinations is almost always used exclusively.

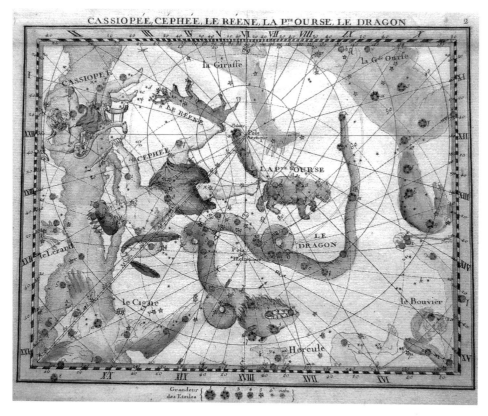

Figure 1.5. View of the north pole, taken from the 1776 edition of Fortin's *Atlas Céleste de Flamsteed*. 15.7 × 20.7 cm. Note the double-grid system centered on the poles of the ecliptic and the celestial equator. *See also* Color Plate Gallery.

Another directional issue relates to the way the constellation figures are presented. In some cases, these figures are shown facing us from the front; in other cases we see them from behind. Some figures are carefully oriented to the star patterns according to classical Greek descriptions. That is to say, the description might say that a bright star is located in the left shoulder of the figure, and that is the way it is shown on the star map. In other cases, there is no one-to-one relationship between the figure and the underlying star pattern. Specific examples of these variations in constellation orientation will be given when discussing specific star maps in later chapters.

A final directional issue concerns the orientation of the star patterns themselves. Some star maps use a "geocentric" perspective, where the patterns appear the way we see them when we look up into the night sky. But others show a left-to-right reversal. How can this be? The reason is that these maps are following an old convention going back to the time when stars were commonly painted or engraved on a spherical globe and were oriented as if the viewer was in Heaven outside of the celestial sphere looking back at the resulting constellations. This "external" orientation causes the star patterns we see in the sky to be reversed in the horizontal direction, as if reflected in a mirror. Celestial globes and their construction will be described in Section 8.1.

1.6   PROJECTIONS OF STAR MAPS

The Earth closely approximates a sphere in shape, which is why terrestrial globes produce accurate depictions of the relative shapes and locations of countries and other geographical features. Since maps printed in a book or a loose print are two-dimensional representations of such three-dimensional features, their accuracy depends upon the way they are projected onto the surface of the paper, vellum, or other flat medium. There are relative trade-offs in doing this. Some projection systems make the countries look more or less real in terms of their shapes, but the distances within and between these countries are not accurate. Other projections allow for more accurate measurements (and are therefore useful for navigation purposes), but they distort the appearance of the countries and other features. We all have had the experience of seeing a flat map with accurate representations of countries near the Equator but with larger than real representations of geographical features near the poles (such as a Greenland that is larger than the United States or Europe!).

Since the celestial vault has also been viewed as being spherical in shape, the same problems with projection have occurred in making star maps as well, especially those depicting constellations. In this case, the distortion involves star patterns and con-stellation images instead of countries. On a celestial globe oriented to the celestial equator, certain characteristics can be noted: (1) all east–west latitude lines are parallel but decrease in length as one goes from the equator to the poles; (2) all north–south meridian lines of longitude meet at the poles and are equal in length; (3) the distance along meridians (e.g., longitude lines) between any two parallels (e.g., latitude lines) is the same; and (4) all meridians and parallels meet at right angles. The coordinate grid system of a celestial map on a flat surface violates at least one of these characteristics, and this has ramifications in accurately plotting stellar positions and the locations of heavenly bodies (such as planets and comets) in the sky.

Map projections can be classified into three major groups: azimuthal, cylindrical, and conic. In azimuthal projections, the surface of the globe is projected onto a flat surface that touches it at a single point. In celestial maps, these projections are often used to show the stars and constellations in a celestial hemisphere, with the pole in the center (the so-called "polar" projection). In the "orthographic" version, the hemi-sphere is viewed as it might appear from deep space, but the distances between the parallels progressively decrease outward from the center pole, creating a "crushing" distortion in star patterns near the edges. However, distances are true along the parallels themselves. In the "stereographic" version, the parallels as well as the meridians project as circles. The distances between parallels increase as one goes outward from the center, but this projection is conformal, meaning that the shape of any small area of the mapped surface is unchanged, offering some advantages to navigation (unlike the orthographic projection, where the shape changes based on location). Another navigation-friendly projection is the "gnomonic", where all the great circles (meridians and equator) are shown as straight lines, and so the path of the shortest distance between any two points is also a straight line. Although useful for planning routes from one point to another, this projection results in a great deal of distortion of features, and for this reason these maps usually just include a small area of coverage.

Cylindrical projections are created as if a tube was rolled round a globe along its equator. The parallels and meridians are straight lines and are usually at right angles to each other. An example for terrestrial maps is the Mercator projection. In celestial maps, however, a version that was popular during the Renaissance and into the 17th Century owing to its simplicity and ease of construction was the "trapezoidal" projection. Examples include star maps found in the 17th-Century celestial atlases of Bayer and Hevelius. Here, the coordinate system shows straight, equally spaced parallel lines, but the straight meridians converge, not necessarily at the same point. As a result, a trapezoidal figure is created by the coordinate system. This projection allowed distances between stars and other features to be measured with a simple straight edge. But, because of the resulting distortion, these maps usually show just a small area of coverage. This projection is rarely used today in celestial maps.

Conic projections are made as if a cone were placed around the globe and touched it along a line of latitude. The resulting parallels are arcs of concentric circles, and they are spaced evenly along the meridian lines. The meridians are straight lines and, as on a globe, come closer together as they approach the pole. Conic projections tend to minimize distortion and are used frequently today. The famous German astronomer Johann Bode used this projection in his celestial atlas of 1801.

One unique projection that was popularized by John Flamsteed, the first Astronomer Royal of England, in his great celestial atlas of 1729 was the "sinusoidal" projection (sometimes called the "Sanson–Flamsteed" projection). In her book on celestial maps, astronomy historian Deborah Warner states: "In this modified conical projection all the parallels of declination are equidistant straight lines; on each map a central hour circle is projected as a straight vertical line and hour circles to either side are projected as sine curves …" (Warner, 1979, p. 81). Here, an attempt is made to reproduce the main features of the coordinate system found on a celestial globe. Although more closely simulating a sphere than its trapezoidal predecessors, the sinusoidal projection still showed distortion in its star patterns (especially the farther one looks from the central vertical meridian), and this quality led Bode to abandon this projection (which he had used previously) for the less distorted standard conic projection.

1.7 MANUSCRIPTS AND PRINTS

Prior to the mid-15th Century, star maps (and terrestrial maps) produced in Europe were largely drawn by hand. They tended to be used to illustrate the text in books, which were also written by hand as manuscripts, and free-standing celestial images were quite rare. These illustrations often were hand-colored and were quite beautiful, although accuracy was usually sacrificed for art. For example, early depictions of constellations in manuscripts typically emphasized the form of the constellation figure being discussed rather than accurately representing the location of the stars within that figure. Manuscripts often were written and illustrated by monks, and they usually took months to complete. In addition, the pool of potential readers was small in the Middle Ages, as was the pool of people who could afford to buy these labor-intensive books. Thus, the spread of star maps was quite limited in the manuscript era.

This all changed with the spread of printed maps and books in the Renaissance, as will be discussed in Section 3.7. For now, it is important to realize that printed star maps were produced in three forms, paralleling terrestrial maps and other printed works. The earliest form was relief printing. In this approach, a printing block was prepared, usually from wood (hence the term "woodblock") but nowadays from metal, by chiseling away all of the non-printing area from the original surface, leaving the part to be printed as a flat, raised area. Ink was then applied to the raised area, a sheet of vellum (parchment made of sheepskin) or paper was placed over it, and the inked image was pressed onto this medium by pressure from a printing press. The Chinese had been printing using this method since at least the 9th Century AD. In the West the method had been developed around 1400, and with the production of a sturdy printing press and movable type by Gutenberg in the 1450s, woodblocks became the preferred method of creating celestial and other printed images for the next 100 years. They persisted longer in books, which often used relief methods for the printed text and so continued this style for the accompanying illustrations. Woodblock images were characterized by relatively thick and unequal lines, a dark rim around the edges of the printed areas due to the build up of ink, and a raised ghost image on the reverse side due to the pressure of the printing process.

The second form of printing was intaglio, which is Italian for incising or engraving. In this form, the lines to be printed were cut into the plate by a burin or other sharp engraving tool. Originally made of wood, copper became the preferred medium for a plate because it was soft enough to be engraved but durable enough to survive many printings. Acid could also be used to eat away or etch lines into the metal by placing acid-resistant wax on top and then exposing the metal surface below by carving lines in the wax. After the image was made on the plate, ink was applied and collected into the grooves, and the rest of the surface was wiped clean. Then, the paper was placed over the plate and put through the printing press under great pressure. As the paper fibers pushed into the grooved areas, the image was produced. A visible plate mark was often produced on the paper that bordered the image. Engravings were developed in the early 1400s, and acid etchings in the early 1500s, but these intaglio processes did not replace woodblock printing as the preferred method for producing terrestrial and celestial maps until the late 1500s. Nevertheless, they offered many advantages. The image consisted of fine, crisp lines that were well suited for the degree of detail required in mapmaking. In addition, by varying the depth of the copper engraving or etching, more or less ink could collect, making the resulting printed lines darker or lighter. Deeper grooves also resulted in wider lines, giving a certain sculptured, three-dimensional quality to the image that was not found in the flatter relief maps. Finally, the metal plates were fairly durable and could go through a number of printing runs, despite the high pressures required by the process (especially in the early 1800s, when steel began to be used for the plates instead of copper). Even when some areas became worn away or when an update was required, the plates could be repaired or modified for the second state of the image.

In the early 1800s a third printing approach began to be used that did not depend upon cutting an image into a plate but instead used an image drawn onto its surface. The original and most common of these planographic approaches, lithography, was based upon the observation that oil and water did not mix. In this method, the image to be printed was drawn on a flat semi-absorbent stone plate using an oily, greasy pencil. The surface was then dampened with water, which avoided the oily image. Next, a roller covered with greasy printer's ink was passed over the surface. The ink would adhere to the oily image on the plate but not to the area covered with water. When paper was then pressed over the plate, it would then pick up the inky image. The result was a somewhat flat but evenly toned image, and the lower pressure of the printing process did not produce an impression on the reverse side of the paper or a plate mark. Later advances, such as offset procedures and the use of zinc or aluminum instead of stone, made planographic procedures cheaper, faster, and more reliable. Instead of laboriously cutting into a plate, the image could simply be drawn on its surface, which allowed for great detail and versatility. Planographic approaches continue to be used today.

All three printing forms—relief, intaglio, and planographic—lent themselves to color as well as to black and white. In the case of celestial prints, colorizing might double the cost to the consumer, so many star atlases and individual prints were purchased only in black-and-white versions. Some of these were hand-colored later by non-professionals (e.g., apprentices), which would be cheaper than if the print was originally bought in a colored version, but which of course could vary in quality. Many antiquarian star maps purchased today have been colorized by contemporary professionals, and the result is a beautiful, but more expensive, display print that may rival a painting when framed and hung on a wall.

Thus, star maps can be found in non-printed manuscripts and in printed relief, intaglio, and planographic versions. Often, this gives a clue as to when the map was produced and its ability to show ever more precise images and stellar positions. But as we shall see over the next chapters, the beauty of the maps was not necessarily dependent upon type of production, and in some cases even woodblocks produced striking images that enhanced the appeal of the image shown.

1.8   BIBLIOGRAPHY

Gascoigne, B. (2004) *How to Identify Prints: A Complete Guide to Manual and Mechanical Processes from Woodcut to Inkjet*, 2nd edn. New York: Thames & Hudson.

Kanas, N. (2002) Mapping the solar system: Depictions from antiquarian star atlases. *Mercator's World*, **7**, 40–46.

Kanas, N. (2005) Are celestial maps really maps? *Journal of the International Map Collectors' Society*, **101**, 19–29.

OUP (2002) *The Shorter Oxford English Dictionary*, 5th edition, Vol. 1, A–M. Oxford, U.K.: Oxford University Press.

Thrower, N.J.W. (1999) *Maps and Civilization: Cartography in Culture and Society*, 2nd edition. Chicago: University of Chicago Press.

Warner, D.J. (1979) *The Sky Explored: Celestial Cartography 1500–1800*. Amsterdam: Theatrum Orbis Terrarum.

2

Non-European cosmology and constellation development

The study of the structure and evolution of the universe belongs to the field of cosmology, as opposed to the description and representation of this structure, which is cosmography. Cosmographical images are depicted in many celestial charts, so it is important to have a sense of various cosmological systems. Since printed star maps are essentially a European development, I will go into much detail in the next chapter about the thread of cosmological development that began with the classical Greeks as background for an understanding of what is shown on these maps.

However, a number of cultures interacted with Europe and influenced the course of its development. In particular, Mesopotamia and Egypt played a role in Greek astronomy. In turn, Hellenistic astronomy influenced India after the conquests of Alexander the Great, and in turn India influenced Islamic (and then European) astronomy during the Middle Ages. Although China had less direct impact, the Chinese were great traders with India and Europe and had some scientific influences on the West as well.

An example of the fertility that ensued from such inter-cultural contact has been nicely summarized by the great historian of astronomy, Otto Neugebauer:

> Three different systems of astronomical reference were independently developed in early antiquity: the "zodiac" in Mesopotamia, the "lunar mansions" in India, and the "decans" in Egypt. The first system alone has survived to the present day because it was the only system which at an early date (probably in the fifth century BC) was associated with an accurate numerical scheme, the 360-division of the ecliptic. The lunar mansions, i.e., the twenty-seven or twenty-eight places occupied by the Moon during one sidereal rotation, were later absorbed into the zodiacal system which the Hindus adopted through Greek astronomy and astrology. With Islamic astronomy the mansions returned to the west but mainly as an astrological concept. A similar fate befell the decans. When Egypt became part of the Hellenistic world the zodiacal signs soon show a division into three

decans of 10° each. As "drekkana" they appear again prominently in Indian astrology, and return in oriental disguise to the west, forming an important element in the iconography of the late Middle Ages and the Renaissance (Neugebauer, 1983, p. 205).

For these reasons, this chapter will focus on cosmological developments in four ancient cultures: China, Mesopotamia, Egypt, and India. Although the Mesopotamian constellation system was on the direct path leading to Greek constellation development, Chinese, Egyptian, and Indian constellations for a time preserved their own unique constellation features that were separate from those in the West. Since some star maps depict these non-European models, I will present an overview of these as well. For information on other interesting cosmological development in places such as early Britain, Australia and Polynesia, Africa, and the Americas, the reader is referred to the standard texts mentioned in the Bibliography (e.g., Selin, 2000; Walker, 1996).

2.1   CHINA

2.1.1   Cosmology

The Chinese have been recording celestial events since at least the time of the Shang Dynasty (ca. 1600 BC–ca. 1046 BC). The evidence for this has been markings made on oracle bones, which were fragments of mammalian bones or tortoise carapaces that were subjected to heat. The paths made by the resulting cracks were read as answers to questions inscribed on the bone asking about current or future events. Some of these bones have been dated to ca. 1300 BC and recorded stars, solar eclipses, and even a nova that occurred near the star that we now call Antares. In subsequent centuries, the Chinese systematically recorded a number of celestial events, including the probable earliest sighting of Halley's comet in 611 BC, a nova noted by Hipparchus in 134 BC, sunspots (seen through smoky crystal or jade) from around 28 BC, the stellar explosion that created the Crab Nebula in 1054 AD, and the novae described by Tycho Brahe and Kepler in 1572 AD and 1604 AD, respectively.

The reason for this diligence has to do with the way the Chinese viewed themselves and their world. According to their perspective, there was a close association between the Earth and the heavens. In fact, events on the Earth mirrored those in the sky, and vice versa. For example, the appearance of a comet or nova near the area of the sky representing the emperor might mean that he was corrupt or serving his people poorly. Consequently, the Chinese emperors employed astronomers and astrologers to monitor celestial events and to look for signs that might portend the future. Careful astronomical records were kept for centuries, and patterns related to such things as eclipses and planetary orbits were noted, much as was the case for the Babylonians.

The Chinese believed that there were five elements: wood, fire, earth, metal, and water. These were related to each other in complicated ways. For example, wood could produce fire, which could produce earth, which could produce metal, which could produce water, which could produce wood again. However, wood could destroy earth, but fire would mask this process, and fire could destroy metal, but earth masked this process, and so on. Each element was also associated with certain numbers, parts of the body, grains, and animals, as well as to the planets (wood, Jupiter; fire, Mars; earth, Saturn; metal, Venus; and water, Mercury).

There were three main cosmological models in ancient Chinese thinking. The oldest, which was developed by the 3rd Century BC, conceived of the heavens as a large dome covering a similar dome-shaped Earth, which nevertheless had a square base. The highest point of the Earth's dome was the North Pole. The heavens rotated around the Earth to the west like a turning mill-stone. Although the Sun and Moon moved to the east, they were dragged along by the heavens and thus appeared to set in the west. The second model, which was developed by the 1st Century AD, viewed the heavens as resembling a round egg, with the Earth floating within it like the yolk. Both inside and outside the heavens was water, which allowed both the heavens and Earth to rotate. The Earth and the celestial bodies were supported by *qi*, a vaporous substance that was analogous to the Greek *pneuma* and the Hindu *prana*. The final model, which was developed by the 3rd Century AD, visualized the heavens as being infinite. Celestial bodies, including the Earth, floated within it at rare intervals, and the movement of these bodies was directed by *qi*. Despite the prescience of the infinite space model for today's cosmology, the egg model was most favored until the arrival of the Jesuits and their western ideas.

2.1.2   Time and the calendar

Issues involving time and the calendar also were important. Oracle bones suggest that even as far back as the Shang Dynasty the Chinese had been using a luni-solar calendar, whereby both the lunar and solar cycles were important. The lunar months (or lunations) alternated between 29 and 30 days, and an extra intercalary lunar month was added every three or four years to match up with the solar year, which the Chinese knew was $365\frac{1}{4}$ days. Through their careful record keeping, they also discovered what was later called the Metonic Cycle, whereby the number of days in 235 lunations were equivalent to 19 solar years that included 7 intercalary months, as well as other cyclical patterns that were even longer. They also had 24 fortnightly periods based on the solar cycle, so that each corresponded to the movement of the Sun by about 15 degrees in longitude on the ecliptic. Note that, like the Babylonians, the Chinese were interested in patterns of events that could be predicted algebraically through the keeping of records, rather than making predictions based on geometric models of nature, which was the strategy used by the ancient Greeks. New calendars have been published regularly in China since 104 BC. The year usually began with the winter solstice. There were minor modifications made to the calendrical system over the centuries until the adoption of the Western-style calendar in 1644.

2.1.3 Chinese constellations

The Chinese oriented themselves to the north celestial pole, around which all the stars revolved. Our current pole star, *Alpha Ursae Minoris* (i.e., the brightest star in the constellation of Ursa Minor, in which the Little Dipper is located) was not the pole star to the ancient Chinese due to precession, but other stars received this honor. For example, in the 2nd Millennium BC, *Alpha Draconis* was the pole star, and Ho *et al.* (2000) has calculated that a faint star in our constellation of Camelopardalus was the "pivot star" during the early Tang period (7th Century AD). Due to their philosophical orientation that events on the Earth and in the heavens mirrored each other, and to their belief that China was the center of the world, it was natural for them to think that the area around the north celestial pole represented the emperor and the imperial household. *Beta Ursae Minoris* was the brightest star in this area at the time, so it was thought to represent the emperor. The second brightest star, *Gamma Ursae Minoris*, stood for the crown prince, and a fainter star in the area represented the empress. Two long chains of stars represented the walls of the imperial palace, and other stars enclosed by these walls in the "Purple Forbidden Enclosure" stood for concubines, eunuchs, and other court officials. The Chinese saw our Big Dipper asterism as a bushel or plough, and it was thought to regulate the seasons as it moved around the pivot star (Figure 2.1).

Any abnormal occurrence in the sky, such as a nova, comet, meteor, or eclipse, might portend a (usually) negative repercussion for society, especially for that aspect represented by the area of the sky in which the occurrence took place. For example, a nova discovered in an agricultural-related area of the sky would likely signify poor crops, or a comet moving into the Purple Forbidden Enclosure would bode poorly for the emperor and his central government. The location of the planets and other celestial phenomena (e.g., zodiacal light, clouds) also had astrological ramifications. Such events might signify that a ruler was misconducting his government or following an immoral path, actions that would disturb the natural order and lead to famines, plague, and disturbances in the heavens. In general, predictable phenomena were good signs and unpredictable phenomena were bad signs. Royal astrologers were kept busy interpreting the meaning of unusual celestial events, and their prognostications were often treated as state secrets (especially if they were negative). It should be noted that Chinese astrology mainly involved areas of interest to society rather than to individuals, except in the case of the emperor and his court.

By the 5th Century BC, the Chinese had developed a system of dividing the broad area of the sky through which the Moon moved into 28 unequal parts called lunar mansions (Figure 2.2). Each was numbered and named for a constellation or asterism located more or less along the celestial equator. For example, the 18th lunar mansion was called *Mao* (representing a Stopping Place) and was formed by the stars of the Pleiades, and the 21st lunar mansion was called *Shen* (representing an Investigator) and was nearly identical to our modern Orion. The lunar mansions served as reference points, and by linking them with the north celestial pole the location of a heavenly body could be identified. For example, the location of a star could be described in terms of how many degrees south of the north celestial pole it was

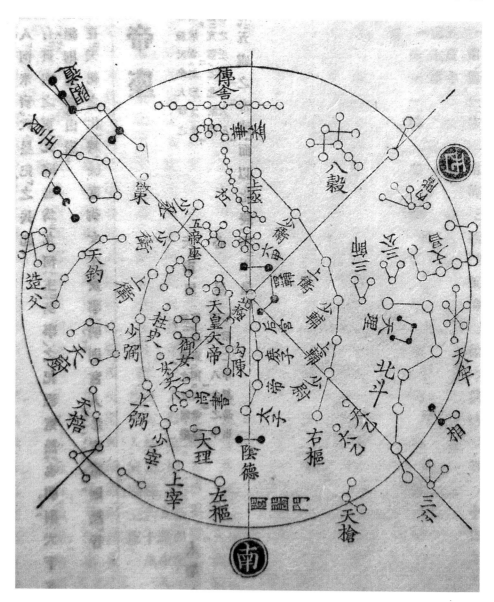

Figure 2.1. The Chinese northern circumpolar constellations, from the 1901 edition of a book first written in Japan in 1712 by Terashima Ryoan, a naturalist and physician at Osaka Castle. The title, *Wakan Sansai Zue*, states that this is a Japanese/Chinese picture book of the heavens, the Earth, and human beings. The Japanese adopted the Chinese view of the heavens. 26.2 × 17.5 cm (page size). Note the two vertical chains of stars, which represented walls around the Purple Forbidden Enclosure, and the Big Dipper beyond the right wall, which the Chinese viewed as a bushel or plough.

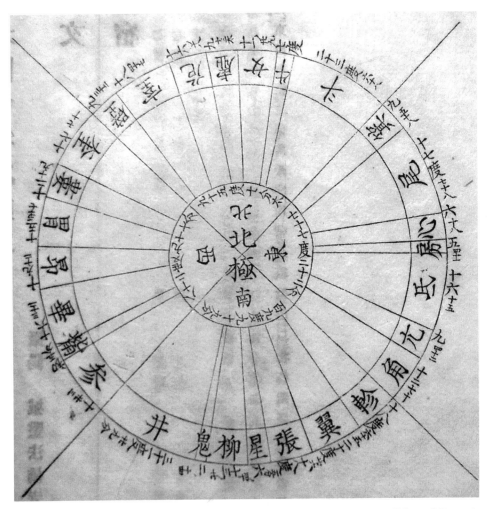

Figure 2.2. A diagram of the 28 Chinese lunar mansions, from the 1901 edition of Ryoan's *Wakan Sansai Zue*. 26.2 × 17.5 cm (page size). Note that, although the area of the sky represented by each mansion constellation was different in size, they were organized into four equal-sized "palaces" of seven mansions, indicated by the crossed lines.

and how many degrees it was from the edge of the nearest lunar mansion. Note that the Chinese celestial sphere contained $365\frac{1}{4}$ degrees, not the 360 degrees that we use today. This system was probably put in use by the 3rd or 4th Centuries BC. Thus, the Chinese employed an equatorial celestial coordinate system centuries before it was used in the West (which preferred an ecliptic-oriented system until the 18th Century). For this reason, many of their astronomical instruments used a mounting oriented to the equator and were the forerunners of our modern telescopic equatorial mounts.

The Chinese had been creating star maps and catalogs since at least the 5th Century BC. In the 4th and 3rd Centuries BC, three notable Chinese astronomers, Shi Shen, Gan De, and Wu Xian each created their own star map and catalog. Chinese author Deng Yinke states that the catalog of Shi Shen provided equatorial coordinates for 120 stars, and both he and Gan De observed the five known planets and noted that Jupiter's sidereal period was 12 years (close to the exact 11.86 years). The earliest existing book to systematically describe the Chinese constellations in the sky was the *Tianguan Shu* by Sima Qian (ca. 145 BC–ca. 87 BC). Some 90 constellations were mentioned, including the 28 lunar mansions. These were organized into five palaces. The Central (or Purple) Palace was the area surrounding the north celestial pole and has been alluded to earlier. The rest of the sky was divided into four equal segments that were called the palaces of the North (or Somber Warrior, represented by an entwined turtle and snake), East (or Azure Dragon), South (or Red Bird), and West (or White Tiger). Each of these palaces represented one of the four seasons, and each consisted of seven lunar mansions. Stars in these areas represented and were named for more mundane aspects of Chinese society, such as temples, philosophical concepts, shops and markets, farmers, soldiers, etc.

In the 3rd Century AD, astronomer Chen Zhuo integrated the records of Shi Shen, Gan De, and Wu Xian. The result was a star map and catalog of 1,464 stars grouped into 284 constellations. Early in the 4th Century AD, the imperial astronomer Qian Luozhi cast a bronze celestial globe with stars colored on it to distinguish the listings of these earlier astronomers. A similar range of stars and constellations is also reflected in the earliest existing printed star map, the Chinese *Tunhuang* manuscript, dating back to the later Tang Dynasty (618 AD–907 AD). Most of these constellations were different from those we are familiar with, although a few were patterned the same way. The great Chinese historian, Joseph Needham, mentions five: Great Bear, Orion, Auriga, Corona Australis, and Southern Cross.

As the Chinese had more contact with Indian and then Islamic astronomers, they became exposed to the Greek system of constellation development, and some of these ideas were incorporated into Chinese thought. This continued when the Jesuits entered China in the 16th Century, as we shall see below.

2.1.4   Chinese influences in Korea and Japan

In large part, the astronomical ideas expressed above were imported by both the Koreans and Japanese, in part as a reflection of the political dominance of China in the region. There were a few minor differences. For example, despite being forced to use the Chinese calendar system, Korean calendars were independently calculated after the early 11th Century, and the two systems were not successfully resolved until the early 15th Century. In Japan some of the mythology associated with the Sun goddess *Amaterasu* and with *Subaru* (the Pleiades), as well as the appearance of the three belt stars of Orion to govern times for the cultivation of rice and millet at the latitude of Japan, needed to be integrated with Chinese models. Other differences exist that are beyond the reach of this book—see Selin (2000) and Walker (1996) for more details. But, in essence, the Koreans and Japanese used Chinese methods to observe the skies, record celestial events, conceive of time and the calendar, and orient themselves to their universe. They also adopted the Chinese constellations.

2.1.5   Outside influences on China

Early China has a reputation of being closed to outside influences, but is this really true? It would appear that there was indeed limited early contact between the Chinese and the classical Greeks and that astronomy developed differently in each country. For sure, there were some similarities: both cultures developed star catalogs, both were interested in the calendar, and both tracked the movement of the planets in the heavens. However, the constellation systems were quite different, as will be shown in the next chapter. In addition, China seemed to echo the Babylonian algebraic traditions, where the positions of heavenly objects in the sky were determined by calculating from patterns identified through centuries of record keeping. The Greeks, by contrast, developed a new way of speculative thinking, where they created geometrical models to explain heavenly phenomena and applied methods of geometry and spherical trigonometry to these models to calculate the location of objects in the sky. In addition, the Chinese were interested in the celestial equator and the circumpolar region, whereas the Greeks were interested in the ecliptic, the location of the Sun, Moon, and planets.

But Needham (1970) has pointed out a number of factors that suggest there was contact between China and the West later on. For example, by the 1st Century AD, there were numerous trade routes from China to other places, such as India, the Middle East, and the scholarly city of Alexandria, Egypt. These included both land and sea routes, and Chinese navigators were adept at using the stars to guide them through the water of the Indian Ocean. We also know that paper, which was invented in China around 105 AD, and the printing of books, which was developed in western China around 870 AD, both made their way to Europe during the Middle Ages, along with other technologies. But what about scientific activities? According to Needham, Chinese pure science seemed to have been filtered out; it went into Arabic cultures but did not penetrate further west. However, the exchange between China and Islam was rich. For example, Needham points out that after al-Tusi's famous observatory and library were built at Maragha in the late 13th Century AD, astronomers were sent from China to collaborate.

Buddhism was introduced into China in the 1st Century AD, and this opened the door to Indian science and medicine. Traditional Indian astronomy had little impact initially, but in the 5th Century it underwent changes due to the influences of Greek astronomy, and these ideas began to be introduced into China. In subsequent centuries many Indian astronomers served at national observatories in China and had an impact, especially in the area of calendrical reform. By the 8th Century, astronomers from Persia also worked in China, and together with later Islamic astronomers they added additional input from the West.

Along with other sciences, Chinese astronomy declined during much of the Ming period (1368 AD–1644 AD), but it was revived in the early 1580s with the arrival of the Jesuit Matteo Ricci, who followed St. Francis Xavier's successful mission to Japan from 1549 to 1552. In addition to gifts, such as clocks, maps, armillary spheres, and sundials, Ricci also brought knowledge of Western mathematics and astronomy to China, and he translated several books in these areas into Chinese. After Ricci's death in 1610, other Jesuit missionaries followed, including Johann Adam Schall von Bell, whose knowledge of astronomy greatly impressed the Chinese, and his successor, Ferdinand Verbiest. As a result, a number of Western ideas infiltrated into the Chinese view of the heavens, such as the Earth-centered cosmology of Aristotle and Ptolemy, Tycho Brahe's hybrid model of the solar system and his ideas on astronomical instruments, the use of the ecliptic coordinate system, and classical Greek views of the constellations. Although some of the concepts were adopted, they were changed to suit Chinese sensibilities. For example, the Chinese zodiac constellations used a set of animals rather than the mixed god/animal representations from the Greek system (Figure 2.3). Although relations between the Chinese court and the Pope soured in the early 18th Century, with the result that all missionaries were expelled from China by the end of the century, a few Jesuits employed in the astronomy bureau were allowed to remain.

2.2   MESOPOTAMIA

2.2.1   Historical interlude

Although much of Mesopotamian cosmology reflected ideas initiated in the city of Babylon, there were additional contributions from other groups that lived in and around the Tigris and Euphrates Rivers (essentially modern-day Iraq). Thus, a brief review of the history and peoples of this area is in order.

The Sumerians, who lived in the southern part of the area near the delta of the two rivers, were the first group to emerge. They developed cuneiform writing on clay tablets, and ancient samples place their civilization as existing earlier than 3000 BC. Their city-states dominated the area, the most famous of which was Ur. In the mid-2300s BC, semitic Akkadians from the central part of the region invaded and conquered Sumer under their king, Sargon, and tablets were written in both the Sumerian and Akkadian languages. Regional power shifted from Ur to Babylon in the 19th Century BC, especially under Hammurabi, who unified the area into one empire and developed a law code that influenced Western thinking to the present day. For some 300 years, Babylon was the center of this empire, and it became a rich and powerful city. Its influence as a center of learning continued even after it fell to the Hittites around 1530 BC and then became part of the Cassite Empire from around 1500 to the mid-1100s BC.

In the 14th Century BC the Assyrians from the mountainous north began to assert themselves in the region, and within a few hundred years they had taken over much of Mesopotamia. By the mid-7th Century BC, they controlled large areas of modern Iraq, Syria, Israel, and Egypt from their capital, Nineveh, although Babylon continued to be viewed as a commercial and intellectual center. But later in this

Figure 2.3. The 12 Chinese constellations of the zodiac (left), from the 1894 American edition of Flammarion's *Popular Astronomy*. 23.2 × 15.5 cm (page size). Note that the depictions are different from those of the Greeks and include a rat, ox, tiger, rabbit, dragon, snake, horse, sheep, monkey, rooster, dog, and pig. To the right is a drawing of an ancient Chinese medal with the Big Dipper engraved on it.

century there was a revolt, Nineveh was destroyed in 612 BC, and Babylon again came into prominence under the Chaldean (or "new" Babylonian) Empire. Its great king, Nebuchadnezzar, expanded its influence and built a great palace with its hanging gardens. However, in 539 BC Cyrus the Great conquered Babylon, and the region became part of the Persian Empire.

In 331 BC Alexander the Great took Babylon from the Persians on his way to India, initiating the Hellenistic period. Upon Alexander's death in 323 BC, his successor in Mesopotamia and Iran, General Seleucus, began the Seleucid dynasty and continued the Greek influence in the region. After being conquered by the Parthinians from Iran in the mid-2nd Century BC, the area came under the control of the Romans in 64 BC, and the influence of Babylon declined.

2.2.2   Cosmology

The mythology of the Sumerians viewed the cosmos as being ruled by three primordial gods: An, the god of the remote heavens; En-Lil, the god of the sky and wind; and En-Ki, the god of the waters around and below the Earth (including the underworld). Earth itself was created from a primordial unity when En-Lil intervened to separate the heavens from the area below. The Sumerians worshiped these and lesser gods, and each city-state had its favorites.

With the unification of the empire under Hammurabi, the Babylonian sun-god Marduk was given supremacy, and a mythology was created that expanded his powers, which is summarized in the *Enuma Elish*, a mythology text dating from the late 2nd Millennium BC. Now the gods are created out of a watery chaos, the sweet sea being the primeval male Apsu, and the salty sea being the primeval female Tiamat. From them are descended the sky Anu and the Earth Ea, and Ea becomes the father of Marduk. When Apsu and Tiamet threaten to destroy all of their offspring, Ea succeeds in killing Apsu (taking over the sweet water domains for himself). Marduk subsequently kills the powerful Tiamat only after the other gods agree that by doing so he will become the supreme god. From her corpse, he rearranges the cosmos into areas governed by cosmic deities: the heavens (Anu), the sky/wind (Enlil—not to be confused with the Sumerian En-Lil), the subterranean waters (Ea), and the Earth itself, which Marduk took over. One story describes this system as consisting of three heavenly areas and three earthly areas (including the underworld). In either event, the Earth is seen as the center of the universe, with Marduk's temples in Babylon making this city its cosmic capital. This design gave mythological credence to the idea of a unified Heaven and Earth.

Astrological interests led the astronomer-astrologers in Mesopotamia to keep careful records of celestial events for centuries in tablets such as the *Enuma Anu Enlil*. This source consisted of some 70 clay tablets written early in the 1st Millennium BC and later excavated from the ruins of Ashurbanipal's library at Nineveh. Especially during the Chaldean period and in the years thereafter, a number of tables were produced that recorded data, especially from lunar and planetary events. From these records a number of celestial patterns were deduced, such as the orbits of the planets, the periodic appearances of comets, the times of solar and lunar eclipses, and the variable speeds of the heavenly bodies. Some of these patterns (e.g., the variable speed of the Moon in the sky) could be characterized mathematically in one of two ways developed by the Mesopotamians. One way, System A, assumed that the velocity was held constant over a period of time (say, several days), and then changed suddenly to another value during a second period. Plotted against time over several months, a crenulated pattern emerged, giving average approximations of the changes in velocity. The other way, System B, gave the measured positions of the Moon in celestial longitude for each day, thus tracking its actual speed in smaller increments. If plotted against time, a zig-zag pattern emerged, as the velocity of the Moon was seen to first increase for a while, then decrease. The second system was more complicated since it took actual incremental values rather than averages for its computations, but it was more accurate as well. Both of these mathematical systems were in use from about the 3rd Century BC, with System A being the first to be invented.

Calculations were assisted by the mathematical system originally started by the Sumerians. Rather than using a decimal system based on powers of 10, a sexagesimal system based on powers of 60 was used. For example, a vertical stroke made by a stylus stood for a "1", and a wedge mark like a ">" stood for "10". These marks were built up like Roman numerals up to a value of 60. But unlike the Roman system, Mesopotamian mathematicians used a place-value notation, whereby numbers larger than 60 were indicated by adding similar marks, but separated from the others by a space. This system allowed for the use of basic mathematical operations, such as addition, subtraction, multiplication, etc., provided that the person doing the calculations kept the base 60 in mind. The Sumerians, and the Babylonians after them, became quite proficient in mathematics and over time developed calculation tables for multiplication, reciprocals, square roots, etc. Elements of this sexagesimal system persist today in our 360-degree circle, 60-minute hour, and 60-second minute.

2.2.3   Time and the calendar

The Mesopotamian calendar was initially based on the lunar cycle alone, with the month beginning on the evening when the lunar crescent was first visible. As far back as 1800 BC, the Babylonians were recording the times of moonrise and the date of the new Moon. Under Hammurabi, the calendar was unified throughout the empire, and the months were given Babylonian names. But in time the calendar became luni-solar in its orientation, where the lunar months were integrated in with the solar year. Since the actual number of lunations in a year is a fraction over 12, and since each lunar cycle averaged about 29.5 days, rules had to be set up to govern whether in a given year an extra "intercalary" month should be added and which months would have 29 or 30 days. However, these rules were sometimes applied haphazardly. Around 400 BC, astronomers in Babylon began using the Metonic Cycle, proposed by the Athenian astronomer Meton, who realized that the number of days in 235 lunations were equivalent to 19 solar years, provided that the equivalent of seven months were added during this time span. Thereafter, the Babylonians followed a regular pattern of intercalating seven months in every 19 years.

Based on later Greek sources, we know that the Mesopotamians used water clocks, or clepsydras, to measure time during the day. In these devices, which could have been a simple bowl, time was indicated by marks on the inner wall showing the changing level as the water dripped out of a narrow opening at a constant rate.

2.2.4   Mesopotamian constellations and the zodiac

The names of constellations were being recorded on clay tablets as far back as the time of the Sumerians, around 3000 BC. Some of these names are familiar to us today: the bull (Taurus), the lion (Leo) and the scorpion (Scorpius). This interest in forming constellations may have reflected their desire to organize the sky in a mythologically meaningful manner, particularly the area through which traveled the Sun, Moon, and planets, which we now call the ecliptic. In this way, a reference point for describing the location of these heavenly bodies was made, and this information was useful in preparing calendars for agricultural and social purposes, improving navigation at sea, and making astrological predictions.

Since the calendar consisted of 12 months by the time of the first Babylonian period (around 1800 BC), it seemed reasonable to divide this area into a like number of parts. By 1100 BC, a system had been created where three groups of 12 stars were arranged in three paths across the sky, each of which was related to a creator god. These are described in the *Mul Apin* clay tablets, which were produced early in the 1st Millennium BC and contained a catalog of important stars and some 60 constellations, along with their rising and setting times. The middle path was roughly plus or minus 17 degrees from the ecliptic line and was related to Anu. The path north of this area was named for Enlil, and the path south for Ea. Based on his review of the appearances and locations of the constellations in the sky taken from the *Mul Apin* clay tables and other sources, astronomy historian Bradley Schaefer has concluded that the bulk of the Mesopotamian constellations were developed between 1300 and 1100 BC by Assyrian observers in the northern part of the region. The influence of this system spread widely into India, China, Egypt, and Greece.

Paralleling this development was the creation of 18 "constellations" that were easily observed at night to be in the path of the Moon. These included not only star groups more or less similar to our own, but also some asterisms that we do not recognize as constellations. Historian Nicholas Campion (2000) gives a list of these 18 groups from the *Mul Apin*, and this list includes a number of familiar names: the bull (Taurus), the twins (Gemini), the crab (Cancer), the lion (Leo), the scales (Libra), the scorpion (Scorpius), and the goat fish (Capricornus). Thus, a lunar zodiac was created that was based primarily on star groupings, and this soon took on astrological meaning. For example, the *Enuma Anu Enlil* contains omens whereby the positions of the planets are described in relation to some of these constellations. It was perhaps inevitable that this 18-constellation lunar zodiac would evolve into a 12-constellation solar zodiac more similar to our own by the 5th Century BC. Although the constellations differed in size, they were given $\frac{1}{12}$th of the ecliptic each, which transitioned into an astrological area of 30 degrees (Steele and Gray, 2007). This focus on the ecliptic and the zodiac (rather than on the celestial equator, such as happened in China) was transported to Greece, and this became the preferred orientation in the West for describing the positions of the heavenly bodies until the 1700s.

2.3   EGYPT

2.3.1   Cosmology

The ancient Egyptians have had a civilized society for millennia, and their mythological system and views of an afterlife were major components of this society. In fact, the sky figured prominently in this mythology. The sky goddess, Nut, was often described in papyrus texts and portrayed on coffin lids and temple ceilings as a naked woman, sometimes arching over her consort Geb, who was the Earth deity. Ra, the Sun god, was frequently shown as entering the mouth of Nut every sunset, traversing her body during the night, and finally being reborn from her every morning at sunrise (Figure 2.4). Associated with this image were figures representing the Moon, planets, and constellations, which will be discussed below. From the union of Nut and Geb came a number of Egyptian gods and goddesses, such as Isis, Osiris, Seth, and Nephthys, and their offspring. This rich mythology dates back to at least the 3rd Millennium BC and was often depicted in connection with descriptions and images of the afterlife. One common image was the weighing of the deceased's heart to see if it was lighter than the feather of truth, suggesting purity and the lack of sinful behavior

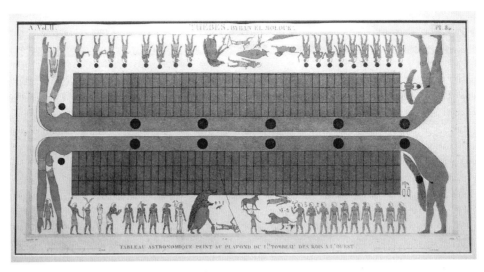

Figure 2.4. Drawing of a ceiling painting from a temple at Thebes, from *Description de l'Egypte,* ca. 1802. This book resulted from Napolean's military and scientific expedition to Egypt. 25.7×55 cm. Note the double depiction of Nut, the sky goddess, with the Sun shown entering her mouth at sunset, traversing her body, and flying out of her at sunrise. Note also a number of traditional Egyptian constellations in the center. See Section 2.3.4 for details.

"weighing it down" during the course of a life. In Figure 2.5 Maat, the goddess of truth, is presiding over this activity, with the deceased man shown on her left. On the other side of the scales is Thoth, the god of wisdom and writing who is shown in his baboon form, who is prepared to record the verdict. Below is Ammit the devourer, with crocodile head, lion body, and hippopotamus legs, waiting to destroy the heart of the deceased if there is an unfavorable outcome.

2.3.2   Time and the calendar

The practice of religion and its festivals, as well as other activities such as planting crops and regulating a complex society, created a need for an accurate method of determining time. As an agricultural people, the Egyptians noted that there was a yearly rise in the Nile River, which flooded their soil and prepared it for planting. By the beginning of the 3rd Millennium BC, they were recording an association between this flooding and the first pre-dawn or heliacal appearance of the star Sirius, which they called Sothis (which occurs in our mid-July). They set up a calendar system of three seasons: flooding, planting, and harvesting. Each consisted of four lunar months, which were named for important agricultural and religious festivals that occurred within the month. Unlike the ancient Mesopotamians and the Chinese, the Egyptian month began with the disappearance of the waning crescent Moon before sunrise, not with the appearance of the new crescent just after sunset. This interest in the dawn sky might have been related to their interest in the daily rebirth of Ra, as mentioned above. Like other luni-solar people who realized that there was not

Figure 2.5. Chromolithograph of an Egyptian papyrus "Judgment of the Dead", from Binion's (1887) *Ancient Egypt or Mizraim.* 21 × 43.6 cm. Note the goddess of truth, Maat, presiding over the weighing of the heart of the deceased man on the left. Thoth, the god of wisdom and writing, is ready to record the outcome, and the "devourer" is waiting to destroy the heart if there is an unfavorable outcome. Being a lunar deity, Thoth has a Moon over his head. *See also* Color Plate Gallery.

an even number of lunar months in a solar year, they added an intercalary month periodically. For the Egyptians, this occurred whenever Sothis rose late in the 12th month (about every three years), and this kept the lunar months synchronized with their New Year festival.

In addition to this agricultural/religious calendar, the Egyptians developed a parallel administrative civil calendar around the beginning of the 3rd Millennium BC that was based on 12 months of 30 days each, followed by five extra days. Since this 365-day calendar lost $\frac{1}{4}$ day each year, it soon lost step with the agricultural/ religious calendar, but a systematic correction of one day each four years (like our leap-year) was not instituted until late in the 1st Millennium BC.

In the civil calendar each 30-day month was divided into three 10-day periods and was associated with the heliacal rising of a star or group of stars called a decan. These were located in a band parallel to but slightly south of ecliptic. Discounting daybreak and nightfall, about 12 decans could be seen rising during the darkness of night, and it was logical to use these as time markers. In fact, priests could regulate the times of nightly temple services by watching the successive appearances of the decans. Another way of monitoring time at night was through a clepsydra, or water clock, which was used as far back as the 16th Century BC. The custom of dividing the night into 12 decanal "hours" may have led to the parallel division of the daytime into 12 hours. To tell the time during the day, sundials and shadow clocks were used, examples of which date back to about the 13th Century BC. Clepsydras of course could also be used during the daytime.

In the latter part of the 2nd Millennium BC, tomb paintings began to show images of a person sitting before a grid surrounded by stars. This is thought to be a method of telling time at night by recording the array of stars transiting the meridian behind an actual sitting person at a given time and comparing this array with standard tables. Although this probably was an attempt to increase the accuracy of measuring the time at night using the stars, it was not terribly successful since there was no attempt to standardize the size of the sitter. Thus, a star appearing near the left shoulder of one person might be documented as near the left ear of another.

2.3.3   Orientation of temples

Much has been written about the orientation of Egyptian pyramids and temples with reference to the sky. Popularized by the great British scientist J. Norman Lockyer in his book *The Dawn of Astronomy*, the idea was that the Egyptians oriented the axes of their religious structures in the direction of the cardinal compass points (especially the meridian line) or some important astronomical event, such as the rising or setting of the Sun or a star during a religious festival day or during an equinox or solstice (Figure 2.6). For example, the entrances for the three pyramids at Giza all face north, and the entrance corridors are angled such that one could see the northern circum-polar stars from them. Many structures located close to the Nile were oriented on an east–west axis, but this was probably because the Nile flows northward, and it was appropriate to align a rectangular building facing toward the river for aesthetic (not necessarily religious) reasons. In some cases, a temple was oriented so that the inner shrine was illuminated by the rays of the rising Sun during a certain festival day. In other cases, there seemed to be an intent to orient a building toward the rising or setting point of a bright star that had a special meaning, like Sirius. But there did not seem to be a universal pattern, and one gets the impression that a supporter of the orientation hypothesis can almost always find a good justification for a particular orientation. Work in this area continues, with scientists performing statistical analyses on a number of temples looking for specific orientation patterns and ways of explaining them—see Belmonte *et al.* (2006, 2008, 2010) and Shaltout *et al.* (2005, 2007a, b) in the Bibliography.

2.3.4   Egyptian constellations

The Egyptians developed their own constellation system based on important gods and animals in their mythology, although it was not as extensive as in other cultures. For example, historian and mathematician Hugh Thurston (1996) mentions an Egyptian catalog of the universe dating to about 1100 BC that lists only five con-stellations, two of which are similar to our Orion and Ursa Major. Of course, if one includes the decan star groups, then this number jumps by 36.

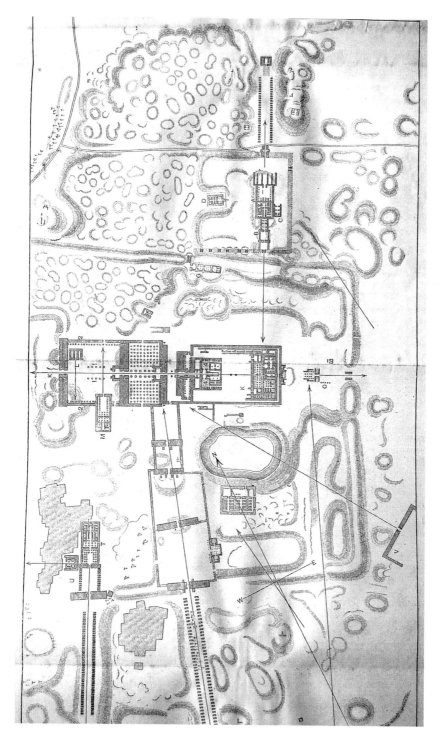

Figure 2.6. A pull-out plate of the plan of the temples at Karnak, from J. Norman Lockyer's (1894) *The Dawn of Astronomy.* 17.5 × 41.2 cm. Note the arrows indicating the orientations of the main buildings. For example, the text tells us that temple M at the upper center faces the direction where the star *Gamma Draconis* rises in the sky, and temple L faces where the star Canopus sets.

Like the Mesopotamians and Chinese, the circumpolar constellations were important to the Egyptians, not so much because they never set but because they never appeared before the rising Sun. Thus, they were often linked with the powers of darkness and with ferocious animals. For example, the circumpolar area around our Draco was often associated with a crocodile or hippopotamus, and the Plough (our Big Dipper) asterism of Ursa Major was viewed as the thigh or foreleg of an ox or bull (representing the evil god Seth). Thoth, a lunar deity, was usually shown with a Moon symbol above its head. It was also depicted as a baboon (an animal that shrieks before dawn), representing the transition from night to day. Both of these images could be combined, as is illustrated in Figure 2.5. Nut was frequently associated with the Milky Way as it arched across the sky.

Lull and Belmonte (2006) have recently analyzed a number of Egyptian images located in tombs and on temple ceilings to find parallels between the traditional Egyptian constellations and the constellations that we visualize today. For example, they viewed the female hippopotamus image in the lower left center of Figure 2.4 as representing the large area of the sky centering around the circumpolar constellation Draco, from Lyra to Bootes. The crocodile on her back represented the area around the head of Serpens. The reclining lion and crocodile images in the lower right center depicted Leo and Hydra, respectively. And, of course, the thigh/bull combination located in the upper center part of Figure 2.4 was Ursa Major.

Following the death of Alexander the Great, one of his generals, Ptolemy, took over the administration of Egypt, thus initiating the Ptolemaic Period (323–30 BC). During this time, Greek ideas involving the cosmos and astrology began to gain influence in Egypt. In addition, Greek constellations were intermingled with those native to Egypt in images on temple ceilings and other monuments. A case in point is the famous image that dates from just after this period that was once located on the ceiling from the Temple of Hathor at Dendera. This so-called "Dendera zodiac" depicts the circle of the heavens held up by 12 figures representing the constellations of the zodiac (Figure 2.7). The outer ring of figures in the circle represents the 36 traditional decanal stars and constellations. The innermost figures are Egyptian constellations, and these are surrounded by the Greek constellations of the zodiac mixed in with images representing the planets, depicted as gods holding staffs.

2.3.5   Differences from China and Mesopotamia

Egyptian cosmology differed in two important ways from that of China and Mesopotamia. First, it was less interested in perceiving omens from celestial events. Portent-based interpretations were not central to the Egyptian mythology, and a system of astrology did not appear until it was imported from Mespotamia and Greece. Without the stimulus of omens, the Egyptians consequently did not produce the kinds of regular records of eclipses, planetary movements, or other celestial events that we have seen in other ancient cultures.

Second, the Egyptians eschewed mathematical approaches to astronomical events. Although some celestial patterns were noted and used qualitatively when they related to religious notions or agricultural needs (e.g., the heliacal rising of

Sirius), a numbering system amenable to algebraic calculations did not develop along the Nile. Instead of a place–value system, the Egyptians had symbols for different numbers (like 1, 10, 100), and they simply repeated them as was necessary. Thus, although many early Greek astronomers and philosophers, like Thales of Miletus, Democritus, and Plato, spent time with scholars in Egypt, it is likely that they were more influenced by Babylonian imports than native Egyptian traditions, since the direction of Greek astronomy seemed to follow Mesopotamian models more than anyone else. Two exceptions were the 365-day civil calendar and the division of the day and night into 12 hours, both of which were homegrown Egyptian products that were taken up by the Greeks.

2.4   INDIA

2.4.1   Cosmology

Cosmology in India goes back several thousand years to Vedic times. Historian and engineer Subhash Kak has categorized early Indian astronomy into several periods. The first was Rgvedic astronomy (Kak: ca. 4000 BC–2000 BC; other scholars: several centuries later), which focused on the motions of the Sun and Moon, the observations of planetary periods, and the division of the sky into *naksatras* (which will be described in Section 2.4.3). Historian John North has cited creation myths from this period that included ideas that the universe was a building of wood made by the gods, with the heavens and the Earth supported by posts, or that it was created from the body of a primeval giant and inhabited by a world-soul. The Sun was sometimes seen as an astral god drawn in a chariot by seven horses.

The second period related to the texts of the Brahmanas (2000 BC–1000 BC), which described the non-uniform motions of the Sun and Moon in non-circular orbits, calculated the cycles of time that were related to the relative positions of the heavenly bodies, formalized the luni-solar calendar with its intercalations, presented cosmological ideas related to the "strings of wind" joining the Sun with the planets, and suggested that the Earth rotated on its axis (an idea later popularized by Aryabhata around 500 AD). This is known to us from later Indian texts and from the writings of Lagadha (ca. 1350 BC, or perhaps centuries later), whose *Vedanga Jyotisa* is the only extant astronomical text from the Vedic period, according to Kak. During this period, there was much symbolism in the field, with certain numbers having special meanings and altars being constructed to represent the geometry of the heavens. Kak relates this to the connection or equivalence between astronomical, terrestrial, physiological, and psychological realms that characterizes Indian thought and is often represented by similar numbers—for example, the 360 bones of the infant being related to the 360 days of the year.

The third period consisted of early Puranic and early Siddhantic writings (1000 BC–500 BC). Kak views the Siddhantas as being more mathematical and the Puranas as being more encyclopedic and empirical but also more cryptic and

Figure 2.7. Copper schematic engraving of the famous "Dendera zodiac" planisphere at the Temple of Hathor at Dendera, from Denon's (1808) *Viaggio nel Basso e Alto Egitto.* 29.1 × 28.8 cm. Note the traditional Egyptian constellations in the center: hippopotamus (area around Draco) and thigh of an ox (Big Dipper). These are surrounded by figures representing the Greek zodiac and the planets (depicted as gods holding staffs). On the rim of the circle are figures representing the 36 decans.

speculative. These sources provided information on the relative sizes and distances of the Sun, Moon, and planets; introduced the concept of kalpa (i.e., a day of Brahma, the creator of time, equaling 4.32 billion years); and described and further developed the great cycles of time that were of interest to early Indian astronomers. Some of these implied that the planets revolved around the Sun, which in turn went around the Earth. There were also hints of a primitive epicycle theory.

In addition, Kak cites later Indian sources as giving a figure for the speed of light that is much like our own and supposedly reflected an earlier Puranic tradition; however, he believes that the accuracy of this value was probably a lucky guess. Some early Indian sources described an atomic theory that consisted of four atoms (earth, water, fire, and air) that combined to form matter. Light rays were a stream of high-velocity fire atoms.

After 500 BC, there were additional Puranic and Siddhantic writings. Kak describes two models of the universe mentioned in the Puranas. One conceived of it as consisting of seven underground worlds below the orbital plane of the planets and seven regions that encircled the Earth. In the center of the flat, circular Earth is a large mountain, Meru, which represents the axis of the universe. In another model, there is a central Earth that is orbited by the Sun, beyond which are the orbits of the Moon, asterisms, planets (in order: Mercury to Saturn), and then Ursa Major followed by the pole star. Beyond this are four additional spheres. Surrounding our universe is the limitless space with countless other universes. This cosmology envisions cycles of creation and destruction of 8.64 billion years, or a day and night of Brahma. The universe itself was said to last for 100 Brahma years (each of which has 360 Brahma days and nights).

Of the later Siddhantic writings, those of the great Indian astronomer Aryabhata (born 476 AD) were influential in southern India and dealt with the size of the universe and distances to the Sun and Moon, as well as making refinements to Puranic ideas concerning the relative diameters of the Earth, Sun, and Moon (although the angular sizes of the planets were too large by a factor of four). He also presented epicyclic models of the orbits of the planets that were different from those presented by Greek astronomers. For example, to account for two anomalies in location of a planet in the sky using a geocentric framework (due to the ellipticity of its real orbit around the Sun and to the fact that it is observed from a moving Earth), he employed two concentric epicycles rather than using an equant like Ptolemy (see Section 3.1.12). His writings also included some spherical trigonometry, a procedure for calculating the duration of an eclipse, and a mathematics section that allowed one to calculate an accurate value for pi. Finally, although his cosmology included Mount Meru as the center of the Earth, he also made statements supporting the rotation of the Earth and the revolution of the planets around the Sun.

A competing Siddhantic system was put forth by Brahmagupta (born 598 AD) that made improvements to some of Aryabhata's ideas and calculations and was influential in northern and western India, as well as in the Islamic world through its Arabic translations. Later, Bhaskara II (ca. 1150 AD) produced a comprehensive Siddhanta that was based on Brahmagupta's work and further developed the epicyclic theories involving the motions of the planets. He also developed notions of trigonometry that probably reflected Islamic influences, as did later Indian astronomy as well.

Mention should be made of the great stone observatories created by Jai Singh in the early 1700s, which were modeled after that built by Ulugh Beg at Samarkand. Although out of date when they were built, this effort nevertheless demonstrated a valiant attempt at observational astronomy, which had generally been neglected in India in favor of mathematical astronomy until the late 14th Century. Their remains can still be seen at Delhi, Jaipur, and Ujjain.

2.4.2   Time and the calendar

Traditionally, the Indian calendar was based on lunar months, each of which began with the full Moon. In time, this was integrated with the solar year. According to Kak, the lunar and solar calendars were brought into harmony in a variety of ways depending on local traditions: adding 11 days each year to the 354-day lunar year (i.e., 12 months of 29.5 days each); adding five days to a year made up of 30-day months; adding an intercalary 13th month twice in every five years; etc.

Attention was paid to both equinoxes and solstices, with the ritual year starting with the winter solstice and the civil year starting with the spring equinox. The ritual year was divided into two halves: when the Sun moved north in the sky, and when it moved south. The summer solstice was the midpoint, and it was recognized as far back as the Brahmanas that the number of days in each half of the year was not equal (a fact also noted by the Greeks, but not until the 5th Century BC—see Section 3.1.9 and Figure 3.2). Ceremonies and festivals marked the time, such as the closing rite at the end of the year to celebrate the first ploughing. There were also sacrificial rituals every four months, ceremonies for the full and new Moons, and rites to mark the passage of the day.

2.4.3   Indian constellations

In keeping with the lunar calendar system, Indians astronomers during the Rgvedic period divided the sky along the Moon's path into 27 equal parts called *naksatras* (Figure 2.8). Specific stars or constellations associated with these areas were also called *naksatras*. In later literature the number of both the regions and associated stars was increased to 28, which better matched the Moon's progress in the sky. Other constellations were recognized that were similar to our own, such as the Bears (Ursa Major and Minor), the two divine Dogs (Canis Major and Minor), the Boat (Argo Navis), and the Pleiades in Taurus. According to Kak, in the Brahmanas period, Orion and the bright star Sirius were singled out, as well as possibly stars in our Gemini, Capricornus, and Cassiopeia.

The naked eye planets also were known and named since the Rgvedic period. In Vedic mythology, they were traditionally the offspring of other heavenly beings and were themselves equated with the gods: Mercury (Visnu), Venus (Indra), Mars

(Skanda, the son of Siva), Jupiter (Brahman), and Saturn (Yama). Venus was sometimes associated with the twins Asvins, reflecting its appearance as both a morning
and evening planet. The Sun was linked to Siva, and the Moon to Uma, Siva's
wife. The planets were also associated with colors (e.g., Mercury and Jupiter, yellow;
Venus, white; Mars, red; Saturn, black). They were also part of references that
alluded to the 34 lights in the sky, which were the 27 *naksatras*, the Sun, the Moon,
and the five planets.

2.4.4   Outside influences

There is evidence for contact between Indian and Mesopotamian cultures during the
Assyrian period. For example, North (1995) cites similarities with some of the
statements found in the *Mul Apin* clay tables, which were produced early in the

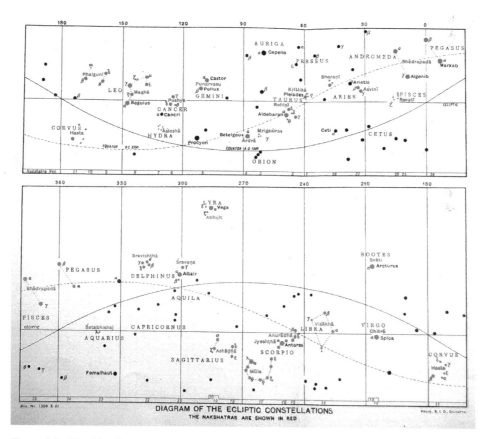

Figure 2.8. The 27 *naksatra* constellations from Vedic mythology, from G.R. Kaye's *Memoirs
of the Archaeological Survey of India, No. 18: Hindu Astronomy*, published in Calcutta in 1924.
22.7 × 27.4 cm. Note that they include both constellations and individual stars, some of which
are familiar (e.g., *Krittika*, the Pleiades; *Svati*, the star Arcturus). *See also* Color Plate Gallery.

1st Millennium BC, and later Vedic texts. In addition, the Persians moved into northwest India during the late 5th Century, bringing with them Babylonian ideas involving astronomy and astrology, including omens related to those found in the *Enuma Anu Enlil*.

During the Hellenistic period after Alexander the Great's conquests, Greek influences made their way into the region. Needham (1970) points out that settlers were left behind in India to form the Greek kingdoms of Bactria and Sogdia. In fact, an examination of Indian texts gives some insight into pre-Ptolemaic astronomy from Greece, especially since few Greek texts survive from this period as a result of being replaced by Ptolemy's great work. Gradually, the ideas of Aristotle and Ptolemy took hold, and Indians made refinements to epicyclical theory (such as the use of an oval-shaped epicycle) that they later shared with the Arabs. They also adopted the 7-day week and the dedication of each day to the deities of the Sun, Moon, and five known planets. In addition, they adopted the Greek constellation system. In later Indian manuscripts, one sees the zodiac represented, along with traditional Greek constellations. Often, the figures were dressed in traditional Indian clothing or were altered to match Indian prototypes, sometimes resulting in curious hybrids (Figure 2.9). But the influence is unmistakable. However, Kak (2000) has provided some evidence for reciprocal influence from India to the West, in that the Druids used a calendar system similar to that mentioned in the *Vedanga Jyotisa*, and they employed a 27-day lunar month suggesting a linkage of the lunar phases and the 27 *naksatras*. Also, some of the Venus mythologies of Mesopotamia and Greece seem to have been predated by Vedic texts, as well as images of elephants in ancient European artwork. Finally, some elements of Indian geometry and mathematics predated those in Babylonia and Greece.

2.5   ASTROLOGY IN ANCIENT TIMES

Although existing clay tablets tell us that the Sumerians were the first to record the names of constellations, regular observations of the Moon and planets began with the Babylonians. For example, the Venus tablet, which was made around 1600 BC, contains 59 omens grouped into 8-year cycles based on the first and last appearances of Venus in the sky. Thus, by this time the Babylonians were observing and recording events in the sky, recognizing the periodicity of some of these events and using this information to make predictions about the future. These predictions generally related to issues involving society at large, such as the weather, agricultural productivity, and politics, rather than to specific individuals (excepting, of course, the king and his court). Thus, there was a fusion between what we now call astronomy and astrology, representing a worldview that events in the sky resonated with events on Earth and that knowledge of these events could affect man's future. This preoccupation with knowing and influencing one's future occurred in other areas as well, such as the omens made from reading the entrails (especially the livers) of animals and from abnormal births.

Astrological issues continued into the Assyrian period. For example, clay tablets record that the 8th-Century emperor Sargon II used the advice of court astrologers in planning his military campaigns. In addition, the *Enuma Anu Enlil* was a comprehensive compendium of astronomical observations and some 7,000 astrological omens. Most of the material dealt with the appearances and movement of the Moon and Sun, although the planets and weather issues also were included. By now, the planets had taken on special meanings associated with personifications of gods. Mars was the "star" of Nergal, the god of pestilence, and was seen as an evil body, whereas Jupiter was associated with Marduk and was seen as being lucky.

Figure 2.9. Indian constellations, from the 1894 American edition of Flammarion's *Popular Astronomy*. 23.2 × 15.5 cm (page size). Note that the figures are stylized using an Indian perspective. The outer constellations represent the zodiac, and the inner ones represent the Sun, planets, and the Moon and its ascending and descending nodes.

After the fall of the Assyrian empire, there was a shift in astrological emphasis, particularly during the Persian occupation. Although astronomical observations continued to be made, the role of astrology in making political decisions declined, and a new form developed that related to predicting one's individual destiny based on the positions of the heavenly bodies in the sky at the time of conception or (more practically) at birth: natal astrology. The idea was that the configuration of heavenly bodies when one came into the world would influence a person's subsequent personality and destiny. Once again, the link between the heavens and the Earth was presupposed, but this time the effects of this relationship were personal and potentially available to everyone, although astronomy historian Nicholas Campion (2000) states that the small number of surviving birth charts compared with the large number of astronomical tables suggests that the former were mainly used by the social elite. Campion gives an example of a birth chart for a nameless child born on April 29, 410 BC, that gives the date, the names of its father and grandfather, astronomical details at the time of birth, and astrological predictions.

This change in favor of birth charts may have reflected the increased information on planetary periodicities and the meanings of birth defects that had been acquired over the previous centuries. Alternatively, it may have reflected the discomfort of the Persian rulers for making political decisions based on earlier Babylonian ideas. In addition, the main religion of Persia, Zoroastrianism, viewed a person's soul as coming from and being influenced by the heavens, particularly the planets, and this idea supported the value of having a personal horoscope. Finally, the shaping of the birth chart was influenced by ideas involving the zodiac, a Mesopotamian invention.

The Mesopotamian zodiac and constellation system were imported into Greece midway through the 1st Millennium BC, as were many astrological concepts. But, once in Greece, natal astrology became more rational and precise. In earlier times the heavens and Earth were unified and mutually affected each other. Thus, events on Earth could affect those in the heavens, and vice versa. In the Greek system, however, the celestial regions were seen as being purer and metaphysically superior to the sub-lunar regions. Consequently, events in the heavens could bring changes on Earth, but the reverse did not occur. In addition, astrology developed a more scientific emphasis, in keeping with cosmological developments that encouraged mathematics and geometrical model building. According to astronomy historian James Evans (2004), the area of the sky and its relationship to the horizon became important, being divided into four centers: ascendant or horoscopic point, mid-heaven, setting point, and under-earth. Pairs of zodiac signs were designated as the solar and lunar houses of a planet. The precise location of heavenly bodies in the zodiac at different points in time became part of the astrological forecast. As Campion puts it, Babylonian astrology relied on what was observed, whereas Greek astrology depended on accurate predictions that could be made of planetary positions.

According to Evans, the Greek system of horoscopic astrology really began to grow in the 1st Century BC and continued into the Roman period. It was during this time that it also picked up elements of eastern mysticism. This was especially true in Greco-Roman Egypt, where there was a syncretic fusion of Western concepts involving the constellations, especially those of the zodiac, with native Egyptian elements

Figure 2.10. "Planisphere Egyptien" representing the Egyptian sky according to Athanasius Kircher, from Charles-François Dupuis' *L'Origine de tous les Cultes ou Religion Universelle*, ca. 1795. 18.1 cm dia. Note the syncretic mixture of western and eastern zodiacal images around the periphery and the more traditional western constellations in the center. *See also* Color Plate Gallery.

(Figure 2.10). Not only were the constellation images altered to conform to both systems, but the use of the decans was included. Each of the zodiac signs was divided into three 10-degree areas, which were named and associated with a decan (Figure 2.11). Furthermore, each Egyptian constellation was linked to one of the zodiac signs. Based on archeological evidence, Evans has given us a picture of astrological practice in Greco-Roman Egypt. This included the use of papyrus horoscopes as well as marble or ivory astrologer's boards engraved with numbers and figures of zodiac signs and decans, on which were placed colored stones with images of planetary gods. By setting the stones on the board, the astrologer could illustrate for his client the

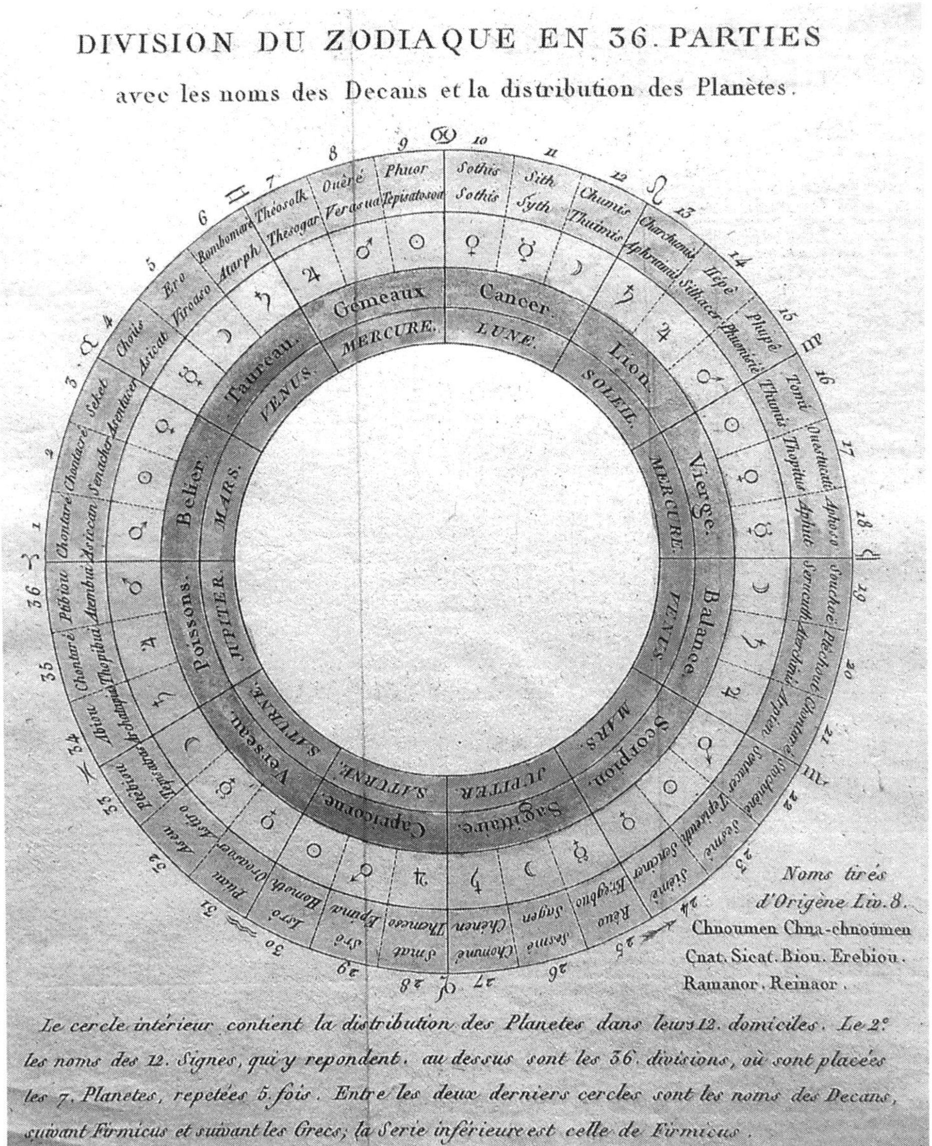

Figure 2.11. A figure from Charles-François Dupuis' *L'Origine de tous les Cultes ou Religion Universelle*, ca. 1795, which summarizes important astrological information and is reminiscent of a Greco-Roman astrological board from Egypt (see text). 16.2 cm dia. Note the concentric rings that represent (from the outside in) the names of the 36 decans and the symbols for their planetary associations, and the names of the 12 zodiac constellations and the names of their planetary associations. *See also* Color Plate Gallery.

basis for a prediction. Much of this astrological work occurred in temples dedicated to the cult of Serapis, who himself was a Hellenized fusion of Egyptian (e.g., Osiris and Apis) and Greek (e.g., Helios or Zeus) gods. A Hellenistic astrologer's board has also been found outside Egypt, in southern Croatia (Forenbaher and Jones, 2011).

The principles of ancient astrology were written down in texts, such as Marcus Manilius' *Astronomica* and Dorotheus of Sidon's *Carmen*, both from the 1st Century AD. However, the definitive text of astrology was Claudius Ptolemy's 2nd-Century-AD *Tetrabiblos*. We shall meet this famous scholar from Alexandria, Egypt, in Section 3.1.12 in the context of his astronomical works. But his well-known astrological treatise summed up the state of this subject to that time and became the foundation for modern astrology as it was to be practiced in the West. Originally known as the *Mathematical Treatise in Four Books*, it did indeed consist of four parts. Book 1 dealt with the basic principles of astrology, such as the characteristics of the heavenly bodies and signs of the zodiac (e.g., favorable/unfavorable, masculine/feminine), and the various alignments of the Sun, Moon, and planets. Book 2 dealt with astrological issues related to society at large, such as which planets ruled over which countries and the impact of the heavenly bodies on the weather. Book 3 dealt with the individual, such as the importance of the phase of the Moon or the sign that was rising in the sky at the time of conception in predicting future events in a person's life, and the astrological influences of issues antecedent to conception, such as those related to his or her parents and siblings. Book 4 dealt with additional astrological issues that Ptolemy considered as being more external, such as later occupation, marriage, children, and travel. In the *Tetrabiblos*, Ptolemy defended astrology as being scientific because it operated according to natural laws. This treatise was a textbook rather than a how-to-do-it manual, focusing more on a systematic presentation of general themes than on specific details of practice. For over 1,200 years, Islamic and European scholars regarded this book as the definitive reference in astrology (much as Ptolemy's other books on astronomy and geography were considered in their areas of specialty).

When the Roman Empire broke up, astrology in the West began to lose its influence, as Greek mathematical astronomy went into hibernation and as the Catholic Church began to condemn its practices. But as we shall see in Section 3.6, astrology continued on in the East and became even more complex and mystical during the Middle Ages.

2.6   BIBLIOGRAPHY

Belmonte, J.A. and Shaltout, M. (2006) On the orientation of ancient Egyptian temples: (2) New experiments at the oases of the western desert. *Journal for the History of Astronomy*, **37**, 173–192.

Belmonte, J.A., Shaltout, M., and Fekri, M. (2008) On the orientation of ancient Egyptian temples: (4) Epilogue in Serabit el Khadim and Overview. *Journal for the History of Astronomy*, **39**, 181–211.

Belmonte, J.A., Fekri, M., Abdel-Hadi, Y.A., Shaltout, M., and Garcia, A.C.G. (2010) On the orientation of ancient Egyptian temples: (5) Testing the theory in Middle Egypt and Sudan. *Journal for the History of Astronomy*, **41**, 65–93.

Britton, J. and Walker, C. (1996) Astronomy and astrology in Mesopotamia. In: C. Walker (ed.), *Astronomy before the Telescope*. New York: St. Martin's Press.

Campion, N. (2000) Babylonian astrology: Its origin and legacy in Europe. In: H. Selin (ed.), *Astronomy across Cultures: The History of Non-Western Astronomy*. Dordrecht, The Netherlands: Kluwer Academic.

Deng, Yinke (2005) *Ancient Chinese Inventions* (translated by Wang Pingxing). Beijing: China Intercontinental Press.

De Young, G. (2000) Astronomy in ancient Egypt. In: H. Selin (ed.), *Astronomy across Cultures: The History of Non-Western Astronomy*. Dordrecht, The Netherlands: Kluwer Academic.

Evans, J. (2004) Astrology in Greco-Roman Egypt. *Journal for the History of Astronomy*, **35**, 1–44.

Flammarion, C. (1894) *Popular Astronomy* (1st American edition, translated by J.E. Gore). New York: D. Appleton & Co.

Forenbaher, S. and Jones, A. (2011) The Nakovana Zodiac: Fragments of an astrologer's board from an Illyrian–Hellenistic cave sanctuary. *Journal for the History of Astronomy*, **42**, 425–438.

Gingerich, O. (1992) *The Great Copernicus Chase and Other Adventures in Astronomical History*. Cambridge, MA: Sky Publishing.

Ho, P.Y. (2000) *Li, Qi and Shu: An Introduction to Science and Civilization in China*. Mineola, NY: Dover Publications.

Hoskin, M. (1997) *Cambridge Illustrated History of Astronomy*. Cambridge, U.K.: Cambridge University Press.

Kak, S. (2000) Birth and development of Indian astronomy. In: H. Selin (ed.), *Astronomy across Cultures: The History of Non-Western Astronomy*. Dordrecht, The Netherlands: Kluwer Academic.

Kanas, N. (2002) Mapping the solar system: Depictions from antiquarian star atlases. *Mercator's World*, **7**, 40–46.

Kanas, N. (2005) Are celestial maps really maps? *Journal of the International Map Collectors' Society*, **101**, 19–29.

Lockyer, J.N. (1894) *The Dawn of Astronomy*. London: Cassell.

Lull, J. and Belmonte, J.A. (2006) A firmament above Thebes: Uncovering the constellations of ancient Egyptians. *Journal for the History of Astronomy*, **37**, 373–392.

Needham, J. (1970) *Clerks and Craftsmen in China and the West*. Cambridge, U.K.: Cambridge University Press.

Needham, J. and Ronan, C. (1993) Chinese cosmology. In: N.S. Hetherington (ed.), *Cosmology: Historical, Literary, Philosophical, Religious, and Scientific Perspectives*. New York: Garland Publishing.

Neugebauer, O. (1969) *The Exact Sciences in Antiquity*, 2nd edition. New York: Dover Publications.

Neugebauer, O. (1983) *Astronomy and History: Selected Essays*. New York: Springer-Verlag.

North, J. (1995) *The Norton History of Astronomy and Cosmology*. New York: W.W. Norton & Co.

Pannekoek, A. (1989) *A History of Astronomy*. New York: Dover Publications.

Park, S-R. (2000) History of astronomy in Korea. In: H. Selin (ed.), *Astronomy across Cultures: The History of Non-Western Astronomy*. Dordrecht, The Netherlands: Kluwer Academic.

Pingree, D. (1996) Astronomy in India. In: C. Walker (ed.), *Astronomy before the Telescope*. New York: St. Martin's Press.

Renshaw, S.L. and Ihara, S. (2000) A cultural history of astronomy in Japan. In: H. Selin (ed.), *Astronomy across Cultures: The History of Non-Western Astronomy*. Dordrecht, The Netherlands: Kluwer Academic.

Robbins, F.E. (transl. and ed.). *Ptolemy: Tetrabiblos*. Cambridge, MA: Harvard University Press (Loeb Classical Library).

Rochberg, F. (1993) Mesopotamian cosmology. In: N.S. Hetherington (ed.), *Cosmology: Historical, Literary, Philosophical, Religious, and Scientific Perspectives*. New York: Garland Publishing.

Ronan, C. (1996) Astronomy in China, Korea and Japan. In: C. Walker (ed.), *Astronomy before the Telescope*. New York: St. Martin's Press.

Schaefer, B.E. (2006) The origin of the Greek constellations. *Scientific American*, November, 96–101.

Selin, H. (2000) *Astronomy across Cultures: The History of Non-Western Astronomy*. Dordrecht, The Netherlands: Kluwer Academic.

Shaltout, M. and Belmonte, J.A. (2005) On the orientation of ancient Egyptian temples: (1) Upper Egypt and lower Nubia. *Journal for the History of Astronomy*, **36**, 273–298.

Shaltout, M., Belmonte, J.A., and Fekri, M. (2007a) On the orientation of ancient Egyptian temples: (3) Key points in Lower Egypt and Siwa Oasis, Part I. *Journal for the History of Astronomy*, **38**, 141–160.

Shaltout, M., Belmonte, J.A., and Fekri, M. (2007b) On the orientation of ancient Egyptian temples: (3) Key points in Lower Egypt and Siwa Oasis, Part II. *Journal for the History of Astronomy*, **38**, 413–442.

Steele, J.M. and Gray, J.M.K. (2007) A study of Babylonian observations involving the zodiac. *Journal for the History of Astronomy*, **38**, 443–458.

Sun, X. (2000) Crossing the boundaries between heaven and man: Astronomy in ancient China. In: H. Selin (ed.), *Astronomy across Cultures: The History of Non-Western Astronomy*. Dordrecht, The Netherlands: Kluwer Academic.

Thurston, H. (1996) *Early Astronomy*. New York: Springer-Verlag.

Walker, C. (1996) *Astronomy before the Telescope*. New York: St. Martin's Press.

Wells, R.A. (1996) Astronomy in Egypt. In: C. Walker (ed.), *Astronomy before the Telescope*. New York: St. Martin's Press.

Yano, M. (2005) Calendar, astrology, and astronomy. In: G. Flood (Ed.), *The Blackwell Companion to Hinduism*. Malden, MA: Blackwell Publishing.

3

European cosmology

Two cosmological models predominated in early European astronomy: the geocentric model, which placed the Earth in the center of the universe, and the heliocentric model, which placed the Sun in the center. Along the way there have been hybrids, but in all cases either our planet or our Sun has been in the center (or close to the center) of everything. This "everything" was skewed toward our solar system, in that the orbits of the Sun, Moon, and known planets were featured in individual concentric spheres, with all the stars lumped together in the next sphere, followed by a few additional spheres representing heaven. We now know that our Sun and its planets are located near the edge of a galaxy of billions of stars, and that our galaxy itself is just one of a plethora of other galaxies expanding from each other, with no clear center. But this understanding was not developed until the last century, so it is not relevant for most of the star maps we will be discussing in this book.

This chapter will not be a complete history of Western astronomy and cosmology. Instead, our review will begin with the ideas of the classical Greeks and their predecessors and continue on to the time of Newton, roughly spanning the period from 800 BC to 1700 AD. Along the way, certain aspects will be emphasized that will enhance our understanding of what is being depicted on star maps, especially those dealing with cosmography and the solar system. Readers wanting a more complete review of astronomy are referred to the Bibliography at the end of this chapter. The next chapter will focus on the development of constellations, which will set the tone for further chapters on the star maps themselves.

3.1 CLASSICAL GREEK ASTRONOMY

3.1.1 Precursors

Humans have always been interested in the sky. The ability to accurately locate the heavenly bodies and to predict celestial events had practical applications for many early cultures, as we saw in the last chapter. Examples of these applications included:

using the heavens to tell time and develop calendars; associating recurring celestial events with natural phenomena on Earth, such as the flooding of the Nile River at the time of the heliacal rising of Sirius, which had agricultural implications in terms of when to plant and sow crops; and using the sky to make omens or astrological predictions that affected the future of society in areas such as government and war.

In the areas around the Mediterranean, there was much cross-fertilization between several early cultures in the millennium before Christ. For the early Greeks, much of this fertilization was based on emigration (e.g., Greeks moving into the Black Sea area and Sicily), trade (e.g., between Greece and Egypt), and war (e.g., wars with Persia and Alexander the Great's conquests through the Middle East). Many early Greek astronomers and philosophers, like Thales, Pythagoras, and Plato, spent time with scholars in Egypt, where they were exposed to Egyptian and Babylonian traditions. From the Egyptians, they learned about the length of the year, its break-up into a 12-month calendar, the division of day and night into 12 hours each, and the use of water clocks for telling time at night. From the Mesopotamians, they learned a sophisticated system of constellations (especially involving the zodiac along the ecliptic), the use of sundials for telling time during the day, the application of algebraic concepts to long cycles of celestial events to make future predictions of eclipses and other phenomena, a sexagesimal system of mathematics that allowed for ease in calculations and for dividing the sky into degrees, and the early elements of astrology. For example, in his book entitled *Works and Days*, written about 800 BC, the Greek poet Hesiod described how the time of the year and celestial events such as the early twilight setting of the Pleiades star cluster could be used in determining the seasons and when to plant and pick crops.

However, the Greeks went beyond the simple recording of celestial phenomena and the use of repeating patterns to predict future events. They developed a new way of speculative thinking, where they created geometrical models to explain heavenly phenomena and applied methods of spherical geometry and trigonometry to these models to calculate the location of objects in the sky. Although the modeling process began relatively early among Greek philosophers, the application of mathematics to these models took several centuries to develop.

3.1.2   The cosmologies of early Greek philosophers

Writing around 800 BC, Homer pictured the Earth as a flat, circular disk that was surrounded by a great river, Oceanus. This river flowed back into itself and, via subterranean channels, produced the other rivers of the world. Over the Earth was the hemispheric vault of the heavens, whereas below it was the hemispheric vault of hell, Tartarus. According to this view, after the planets set, they floated around Oceanus back to the east, where they rose again the following evening.

The great classicist, Sir Thomas Heath (1991), documents excerpts from a variety of Greek commentators that trace the development of Greek cosmological systems. Many of the earliest models were proposed by Ionian philosophers from the town of Miletus, and, as we shall see, some of their speculations set the tone for later Greek thinking. The earliest of these philosophers was Thales (ca. 624 BC–ca. 547 BC), who excelled in several areas: astronomy, mathematics, politics, and business. He described a 365-day year, wrote about the solstices and equinoxes, and predicted a solar eclipse that took place in 585 BC. For him, the primary element was water, from which the other elements resulted: earth from condensation, air from rarefaction, and fire from heating. He viewed the Earth as a flat disk or cylinder floating on water, an idea he may have learned from the Egyptians and Babylonians.

In contrast was Thales' student, Anaximander (ca. 611 BC–ca. 546 BC), whose primordial substance (the Infinite) was not water or any other known element and from which arose, evolved, and passed away an infinite number of worlds. Our solar system developed when a hot sphere formed around the cold Earth and separated into rings of fire, representing the Sun, Moon, and stars. Each of these rings was surrounded by compressed and opaque air with a single circular vent, producing the appearance of a round heavenly body where the enclosed fire was able to shine through. These rings revolved around the central Earth, some of which were oblique to our planet's axis (hence, passing through the area later called the ecliptic). The Earth itself was viewed as a cylinder. Anaximander thought that the distance to the Moon was about 19 Earth radii and to the Sun about 27 Earth radii. He had three other claims to fame: he was said to be the first Greek to construct a sundial, he was the first to draw a map of the entire known inhabited world, and he hinted at a theory of evolution when he speculated that animals arose from slime in the sea and that humans were born inside fishes and later left the water and took to land.

Anaximenes (ca. 585 BC–ca. 528 BC), who was an associate of Anaximander, put air as the primary element, on which was supported a flat Earth. As moisture arose from the Earth and was rarefied, it became fire and produced the Sun, Moon, and stars. The fixed stars were attached to a crystalline sphere, whereas the wandering stars floated freely on the air. He postulated additional dark bodies floating among the heavens that sometimes came between us and the Sun or Moon, accounting for eclipses.

Heath goes on to describe the ideas of a number of other early Greek philosophers. Two of these warrant comment for their influence on later thinking. Empedocles of Acragas (ca. 490 BC–ca. 444 BC) postulated that there were four original elements (earth, water, air, and fire) and that all matter was made from their various combinations. He viewed the heavens as being a crystalline and somewhat egg-shaped sphere, with the fixed stars being attached to it. Within this crystalline body revolved two hemispheres, one with fire producing daytime and one with air producing nighttime. He also had the prescient idea that light traveled and took a finite amount of time to go from one place to another.

Another influential philosopher was Anaxagoras (ca. 500 BC–ca. 428 BC), who was born near what we now call Smyrna but later moved and worked in Athens. He deduced that the Moon shone by the reflected light from the Sun and proposed Earth-centered mechanisms for solar and lunar eclipses (although he also thought that dark bodies sometimes came between us and the Moon to produced a lunar eclipse). He also thought that the world was formed by a vortex in space (anticipating the Greek atomist Leucippus and the later Frenchman Descartes). In this process an inner region of air, which through consolidation produced the Earth, was separated from an outer region of a rarefied substance called "ether", whose whirling action tore stones from the Earth up into the heavens through centrifugal force to produce the heavenly bodies. The Earth thus was seen as a flat body that was supported by the surrounding air, and the Sun, Moon, and stars were stones that were on fire and were carried around by the revolving ether. Anaxagoras thought that by this same process there were other worlds that were formed in the universe and that these were inhabited by beings similar to us.

3.1.3   Pythagoras and his followers

The perception of the Earth as a flat or cylindrical body with a relatively flat surface changed with Pythagoras, who was born around 572 BC on the island of Samos, just off the coast of Ionia. He traveled to Egypt and Babylon, where he is said to have learned mathematics and science. He settled in southern Italy around 535 BC, where he founded his famous school and died around 500 BC. During his lifetime, Pythagoras made important contributions to astronomy. For example, he has been credited as being the first person to view the Earth as a sphere (Figure 3.1). In a similar manner, he thought that the universe too was spherical in shape and that the finite heavens revolved around a stationary and central Earth. Beyond the heavens was a limitless and empty void. He also wrote that the planets had motions that were independent from the stars and that the bright "morning" star and "evening" star were the same body (i.e., the planet Venus), ideas he probably learned from the Egyptians or Babylonians. Finally, Pythagoras thought that there was harmony in the universe, both in terms of the sounds that the Sun, Moon, and planets made as they moved along their orbits, and in terms of the ratios of their distances from each other, which were similar to the ratios of the notes on a musical scale.

Pythagoras influenced a number of people. In fact, it is difficult to know for sure which ideas were developed by him and which were developed by one or another of his students. One such student was Parmenides of Elea, who was active around 500 BC. Like Pythagoras, he saw the Earth as being spherical and recognized the morning/evening nature of Venus. But, unlike Pythagoras, Parmenides did not believe in the existence of an infinite void, and he thought that the movement of the heavenly sphere was an illusion. He saw the Earth as being formed from condensed air and the stars from compressed fire.

In the 5th Century BC, successors to Pythagoras developed the notion that the Earth was not the center of the universe but was a revolving planet like the others. But this was not a heliocentric system. Instead, the center of the universe was occupied by a central fire (the "Watchtower of Zeus"), near which revolved a "counter-Earth" and other bodies (postulated to account for the increased frequency of lunar over solar eclipses). These bodies could not be seen, since they were always positioned below the horizon. The Earth itself was seen as rotating on its axis in the same time as it took for it to revolve around the central fire. From the center outwards beyond the Earth revolved the Moon, the Sun, the five known planets, and the sphere of fixed stars. Outside of this finite spherical universe was the infinite void. Thus, a number of prescient ideas were introduced by these later Pythagoreans.

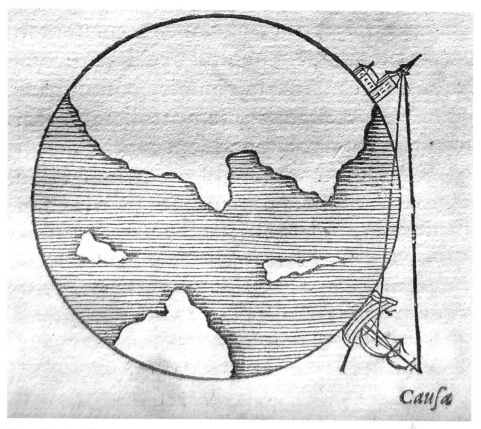

Cauſa

Figure 3.1. An illustration of the sphericity of the Earth, from the 1647 Leiden edition of Sacrobosco's *De Sphaera*. 15.2 × 9.7 cm (page size). Note that it shows how a light in a tower can be seen sooner by an observer at the top of the mast of a ship than someone on the deck, thus proving that the Earth is spherical in shape.

3.1.4   Plato

The famous Greek philosopher, Plato, was born around 427 BC and died around 347 BC in Athens. Although coming from a prominent political family, he became disgusted by the corruption of Athenian politics, particularly after the execution of his teacher Socrates in 399 BC. He saw philosophy as the source of a cure for society, and early in the 4th Century BC he founded his famous Academy, which was devoted to philosophical research and teaching. He traveled extensively and wrote over 20 dialogues and letters. Among other topics, his writings contained a number of ideas about astronomy. His most complete presentation is found in the *Timaeus*, particularly the dialogue between Socrates and Plato's surrogate, the astronomer Timaeus. Heath states that Plato was influenced more by Pythagoras than by his non-geocentric students, and his writings certainly present this view.

The Platonic system describes a universe made by a single Creator (sometimes called the "demiurge") who, wishing that all things should be good and perfect and finding around him disorder and disharmony, created a blueprint for an orderly and harmonious universe imbued with intelligence: a sort of cosmic soul or mind. In contrast to this perfect world of ideas, our corporeal version is but a reflection of this and is more finite and changing. It consists of four elements: earth, water, air, and fire. The heavenly sphere is pictured as revolving from east to west around a large, spherical, central and immobile Earth. The Sun, Moon, and planets are carried around with the heavens, but in addition each moves in its own circular orbit from west to east. Like many Pythagoreans, Plato viewed the Moon as being closest to us, followed by the Sun, Venus, Mercury, Mars, Jupiter, Saturn, and the sphere of fixed stars. The area of the sky in which the Sun, Moon, and planets move (i.e., the ecliptic) was called the circle of the Different. This area was obliquely inclined to the area of the sky that represented the equator of the sphere of the fixed stars; this was called the circle of the Same.

This basic cosmology was very influential, in two ways. First, Plato's geocentric description of the universe was picked up by later philosophers, such as Aristotle and Ptolemy, whose modified versions became the models used throughout Islamic and Byzantine areas in the Middle Ages and Europe during the early Renaissance. Second, Plato's view of a single Creator of the universe was very popular with the later Christian clergy, who could point to him as an example of a classical Greek philosopher whose views of creation were similar to those espoused by the Christian fathers. Since most of the *Timaeus* had been translated into Latin, Plato's cosmology remained popular in Europe throughout the Middle Ages.

3.1.5   Eudoxus

Eudoxus was born in Cnidus around 400 BC and died around 347 BC. He had attended lectures by Plato and was the first person we know to summarize the Greek system of constellations, which we will discuss in the next chapter. But, in addition, he made a major contribution to Greek astronomy. Although Plato's system accounted for the movement of the planets along the ecliptic, it did not account for the fact that the planets sometimes became stationary with reference to the stars, then made a retrograde motion (i.e., went from east to west), became stationary again, and then continued with their normal west-to-east motion in the sky. To solve this problem, Eudoxus articulated a geometric model of four concentric spheres with the Earth at the center to describe each planet's movement. All of the spheres rotated with uniform speed. The rotation of the outermost sphere accounted for the movement of the heavens from east to west. The next sphere toward the center rotated around an axis perpendicular to the plane of the ecliptic and accounted for the planet's basic west-to-east movement through the zodiac. The third and fourth spheres were oriented in different planes depending on the planet, which was located on the innermost fourth sphere. The combined motion of the third and fourth spheres made the planet move in a figure-of-eight curve through the sky called a "hippopede"—thought by most historians to be named for the shape of the shackles that were placed on the legs of horses to restrain them, although the great astronomer John Louis Emil [J.L.E.] Dreyer (1953) stated that the term was based on the figure produced by riding school horses while cantering. Combined with the second sphere, this reproduced the stationary and retrograde movements of the planet. Eudoxus also described the motions of the Sun and Moon but used a simpler system of three concentric spheres.

Eudoxus' work illustrates two characteristics that were to become typical of Greek astronomy: the creation of speculative geometric models for explaining the movement of heavenly bodies, and the testing of these models in terms of accurately explaining celestial phenomena through the use of observation and principles of spherical geometry.

3.1.6 Aristotle

Aristotle was born in 384 BC in Macedonia, where his father was the physician to the king. At the age of 17, he went to Athens and studied for two decades at Plato's Academy. After Plato's death, Aristotle went to Asia Minor, where he founded an academy and made biological observations on the mainland and on the nearby island of Lesbos. In 342 BC he was summoned to Macedonia by King Philip to be the tutor for his son, Alexander. When Philip conquered Greece in 338 BC, Aristotle moved back to Athens, where he founded his Lyceum and continued to receive patronage from his pupil, Alexander the Great. Throughout his life, Aristotle wrote essays on various aspects of philosophy, including natural philosophy (what we would now call science). He was a careful observer and critical thinker, as well as a respected teacher. He also articulated a cosmology that was to be very influential over the centuries. He died in 322 BC.

Reflecting the ideas of his teacher Plato, Aristotle hypothesized a geocentric cosmos with a large, spherical, and immobile Earth surrounded by a spherical universe. But, unlike Plato, who believed that reality existed not in the world of the senses but in the world of ideas, Aristotle advocated a more physical universe that could be understood and described through logic and observation. He minimized the role of a Creator since he tried to understand nature in purely natural terms, but he believed that a "Prime Mover" was responsible for keeping the heavenly spheres in motion.

He viewed that part of the universe below the sphere of the Moon as being changeable and corrupt and composed of the elements of earth, water, air, and fire, which could intermix and transform into one another. These elements had tendencies to move in a straight line, with earth moving strongly downward (and thus being concentrated in the center of the universe), water weakly downward, air weakly upward, and fire strongly upward. Comets and meteors were produced in the hot and dry fiery realm. Beyond this region were the unchangeable heavenly bodies. Each of these bodies was imbedded in a sphere, and the body and sphere were made of ether, a crystalline-like substance that was smooth, pure, changeless, and divine. Objects made of ether had a tendency to move in a circle, which the Greeks conceived of as the most perfect of all forms since it is continuous and without beginning or end. This universe contained everything; beyond the sphere of the stars was nothing.

Thus, like Eudoxus, Aristotle tried to account for the movement of the heavenly bodies through a system of revolving homocentric spheres. But, unlike his predecessors, who devised a model solely based on mathematics, Aristotle transformed it into a mechanical model, where material spherical shells physically acted upon one another. Fifty-five shells were needed, some moving the heavenly body forward, some allowing it to retrograde, and some neutralizing or decoupling the effects of one planetary sphere to allow another to move independently.

3.1.7 The "Pre-Copernicans"

Some Greek astronomers developed ideas that anticipated Copernicus' notion that the planets revolved around the Sun. An early example was Heraclides of Pontus (ca. 388 BC–ca. 315 BC). After immigrating to Athens, he became a Platonist and likely was a pupil of Plato himself. He was also influenced by Aristotle and by Pythagorean philosophers. Heraclides declared that the apparent rotation of the heavenly bodies was not produced by the actual movement of the heavenly sphere but instead was due to the daily rotation of the Earth around its own axis in a west-to-east direction. He also postulated that Mercury and Venus revolved around the Sun, although the Sun (and the other planets) still circled the Earth. This model was advocated by later scholars, such as David Origanus (1558–1628).

Following Heraclides, Aristarchus of Samos (ca. 310 BC–ca. 230 BC) also said that the Earth rotated on it axis, but he proposed a full heliocentric model where all the planets (including the Earth) revolved around the Sun. Aristarchus also wrote a treatise where he mathematically calculated the relative distances and sizes of the Sun and Moon. For example, given his estimate of the angle between the Sun and the Moon when the latter was in its half-phase, he applied trigonometric principles to the resulting triangle and calculated that the distance to the Sun was 18–20 times the distance to the Moon. From the alignment of these three bodies during a solar eclipse, he calculated that the Sun's radius was some 19 times larger than the Moon's. Both of these values were too small due to underestimates in some of the basic parameters he used, but the mathematical approaches were sound.

Why wasn't the heliocentric model that included a moving, rotating Earth more widely accepted? Perhaps because it contradicted other philosophers such as Plato and Aristotle, who argued for an immovable Earth at the center of the universe. In addition, the Earth was viewed as a large, heavy body that seemed to attract surface objects toward its center, and it was hard to conceive of it as being in motion (unlike the Sun, Moon, and planets, which were made of the lighter ether and clearly appeared to move). Also, if the Earth moved, why didn't we feel the air blowing by us, or why didn't we fall off of its surface?

Finally, as Dreyer (1953) points out, other ideas based on the geocentric model had been developed in Alexandria and elsewhere that "saved the phenomena" (sometimes said to "save the appearances"); that is, they accounted for the observed locations of the heavenly bodies in the sky regardless of representing reality. According to Dreyer, this focus on practical astronomy rather than on attempting to find the true system of the world become a characteristic of Greek astronomy for some 400 years after Aristarchus and was epitomized by the work of Apollonius, Hipparchus, and Ptolemy. We will discuss this development after a brief visit with Eratosthenes in Alexandria, Egypt.

3.1.8 Eratosthenes and the Alexandria Library

Eratosthenes was born around 276 BC in Cyrene. He went to Alexandria and Athens to pursue his studies, and he acquired a name for himself as a scholar. Around 235 BC, he was summoned to become the head librarian of the famous library in Alexandria. Founded by Alexander the Great, Alexandria had become one of the world's major centers for literary scholarship and scientific studies, and its library was the hub of these activities until its burning in 272 AD by Aurelian. As a scientist, mathematician, geographer, historian, poet, and philosopher, Eratosthenes' range of interests made him an excellent choice to be in charge of the library. He wrote treatises in many of these areas. In fact, this polymath was quite adept at a number of scholarly activities, never the best but usually second best at all of them. For this reason, his colleagues nicknamed him "beta", the second letter of the Greek alphabet. He maintained his post as head librarian until his death around 195 BC.

One important activity in Alexandria concerned time keeping and calendar development, which depended on accurately knowing and predicting the location of heavenly bodies in the sky, especially the Sun, Moon, and stars. As a result, a school of astronomy emerged that focused on observing the heavens using graduated instruments and calculating positions using principles of spherical geometry. Eratosthenes fit into this mold by devising an ingenious method of finding the circumference of the Earth. He measured the angle of the shadow made by the Sun at Alexandria at the same time that it was directly overhead at noon on a midsummer's day in Syene (modern Aswan) and cast no shadow. He found that the angle was approximately $\frac{1}{50}$th of a circle (about 7 degrees), and using geometric principles he knew that this angle also corresponded to the distance between the two cities (5,000 stades). So, he calculated the total circumference of the Earth to be $50 \times 5,000$, or 250,000 stades. Depending on which mileage equivalent is used for the stade, his results differed from our modern value by a few hundred miles or so. Eratosthenes also wrote a compilation of myths explaining the origin of the constellations called the *Catasterismi*, which will be described more fully in Section 4.1.2.

3.1.9 The eccentric model

We have seen that the model of concentric spheres became quite complicated, requiring reacting spheres to "decouple" the actions of one planet from another. In addition, the varying brightnesses of the planets, especially Venus and Mars, made it difficult to explain these heavenly bodies as always being the same distance from the Earth, which the homocentric model required. Something else was called for.

Using spherical geometry, the Greeks discovered that movements in the sky could be described in two other ways that still allowed for the presence of circular orbits that moved with uniform speed. The first was to place the center away from the Earth, making the orbit eccentric. This simple model worked well to describe the motion of the Sun. For example, it was well known to the Greeks that the time from the autumnal equinox to the vernal equinox (a little over 178 days) was shorter than the time from the vernal equinox to the autumnal equinox (187 days). How could one account for this discrepancy and still maintain a theory that depended on a perfectly circular orbit? The solution was to theorize that the Sun went around the Earth in an off-center eccentric orbit, an idea credited to Hipparchus (see Section 3.1.11). This is illustrated in Figure 3.2. Note that the line labeled *Aequinoctialis Seu Colurus Aequi-noctiorium* runs left to right through the center of the Earth, and as can be seen there is less of the Sun's orbit below than above this line, accounting for a shorter inter-equinox transit time. This hypothetical construction saved the phenomena and preserved the notion of orbital circularity.

3.1.10   Apollonius and the epicycle model

Although reasonably accurate in accounting for the locations of the heavenly bodies in the sky, the eccentric system was not perfect, and a second model was called for which has been attributed to the great Greek mathematician Apollonius. Born in Perga around 240 BC, he spent much of his life in Alexandria, where he probably died around 190 BC. Although his original works have not survived, he is associated with the theory of conic sections. He is also credited as being the first person to use deferents and epicycles in explaining irregularities in planetary movements that could not be accounted for by the eccentric theory alone. In this model a heavenly body revolves around a small circle called an epicycle, the center of which itself moves around the Earth in a circular orbit called a deferent. By adjusting the size, rotational speed, and rotational direction of the epicycle and its deferent, a model for each heavenly body could be made that accounted for its location in the sky.

This model also explained another phenomenon: the apparent retrograde motion of a planet as it moved along its path. An example of this is shown in Figure 3.3, where the path of Mars in the sky is plotted for the most of 1877. Note that it makes a big looping pattern from August to October, where it retrogrades backwards toward the west. The explanation for this using epicycles and deferents is a simple matter of geometry, as shown in Figure 3.4. Here, a planet is shown moving around the Earth "a" counter-clockwise along its epicycle "f–b–d–e–c", whose center is in turn moving counter-clockwise along its deferent "c–b". When the planet goes from "c" through "f" and on to "b", it appears to move toward the left (toward the east) as seen from the Earth. It then appears to slow down and becomes stationary at "d". As it goes from "d" to "e", it actually appears to be going retrograde (toward the west) before becoming stationary again at "e". As it moves beyond this point, it starts to go away from the observer and retraces its eastward approach.

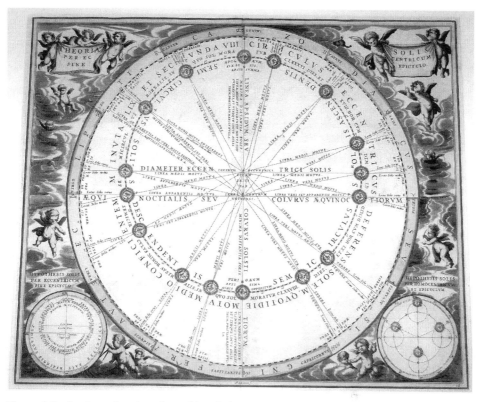

Figure 3.2. A plate showing the orbit of the Sun around the central Earth according to Hipparchus and adapted by Ptolemy, from Cellarius' *Harmonia Macrocosmica*, ca. 1661. 42.1 × 50.4 cm, 38.5 cm dia. hemisphere. Note that the eccentric orbit accounts for the unequal period of time between the equinoxes, with the lower part (autumn to spring) being shorter than the upper part (spring to autumn). *See also* Color Plate Gallery.

In many cases, the eccentric model and the epicycle model could produce equivalent results, simply by adjusting the relative parameters of the epicycle/deferent combinations in terms of their speed and direction of rotation or the relative size of the epicycle and deferent. This is shown in the diagram from the lower right corner of Figure 3.2, where if you connect the four images of the Sun as oriented in its epicycle, you define a circle that is eccentric to the central Earth. Different astronomers picked different epicycle/deferent combinations. Some even combined the models by using eccentric deferents.

We now know that the Sun is in the center of our solar system and the planets move in elliptical orbits around the Sun. This realization would not have been possible with the ancient Greeks (the "pre-Copernicans" notwithstanding), who were committed to two principles clearly articulated by Aristotle: (1) that the Earth was the center of the universe because of the apparent movement of the heavenly vault in the sky and because it was the natural tendency of solid matter to move and congeal

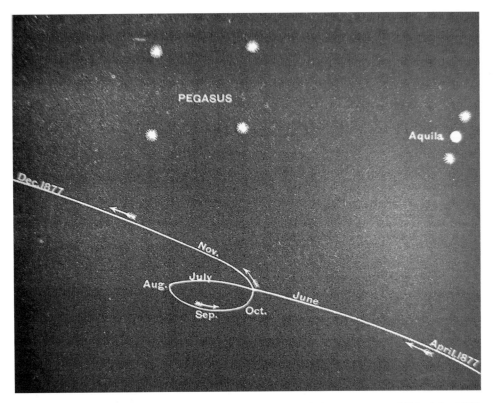

Figure 3.3. A figure from Sir Robert Ball's *The Story of the Heavens*, published in 1897. 23 × 15.3 cm (page size). Note the appearance of Mars in the heavens in the latter part of 1877, when it made its retrograde loop.

downward into the Earth; and (2) that heavenly bodies, being made from ether, had a natural tendency to move around us in perfectly circular orbits with uniform speed. With these two premises as givens, it was necessary to hypothesize eccentrics, deferents, and epicycles to save the phenomena of objects in the heavens.

Although not depicting reality, the Greek system worked mathematically for the planets because it reasonably accounted for their actual movement in the sky as seen from the perspective of an observer on the Earth. For each of the two inner planets (Mercury and Venus), the movement around the Greek deferent reflected the movement of the Earth around the Sun, and the movement around the epicycle accounted for the planet's revolution around the Sun. For each of the three outer planets (Mars, Jupiter, and Saturn), the movement around the deferent represented its revolution around the Sun, and the movement around the epicycle accounted for the Earth's revolution around the Sun. Thus, heliocentric reality was mathematically accounted for, even though it was not clearly shown in the Greek geocentric models.

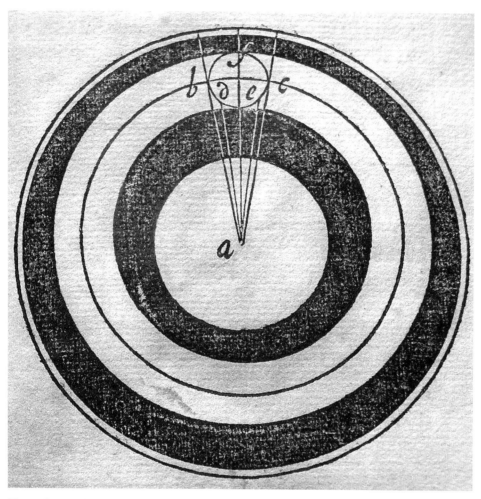

Figure 3.4. An illustration (influenced by Peurbach's *Theoricae Planetarum Novae*) explaining the retrograde motion of an outer planet in the sky, from the 1647 Leiden edition of Sacrobosco's *De Sphaera*. 15.2 × 9.7 cm (page size). See text for details.

3.1.11   Hipparchus

Hipparchus was perhaps the greatest of the pre-Christian-Era astronomers. He was born in Nicaea in Bithynia around 190 BC, but according to astronomy historian Michael Hoskin (1997, 2003) he was most active during his time in Rhodes between 141 BC and 127 BC. He was a true observer of the heavens, making improvements to astronomical instruments and compiling one of the first true star catalogs, as we shall see in Section 4.1.3. He was greatly influenced by the Babylonians, compiling a list of lunar eclipses observed at Babylon over the centuries and adopting their sexagesimal numbering system. He died around 120 BC.

Hipparchus made a number of discoveries during his lifetime. Perhaps the most important was the precession of the equinoxes, which is the slight slippage of the constellations in an eastward direction as observed at each spring equinox, which is in fact due to the top-like rotation of the Earth's axis in a cycle of nearly 26,000 years (Figure 3.5). This phenomenon also accounts for the shifting of the Pole Star over time that was discussed in the last chapter.

Hipparchus was an excellent mathematical astronomer. One precessional rate he calculated was close to one degree of arc per century, compared with the true rate of one degree per 72 years. He calculated the duration of the mean lunar month to within a second of the time we use now and the duration of a tropical year to within $6\frac{1}{2}$ minutes of modern calculations. He was aware of the inequality of the four seasons and provided accurate values for their durations (e.g., 94.5 days for spring, 92.5 days for summer), and he popularized the use of an eccentric orbit to describe the movement of the Sun around the Earth, thus accounting for the difference in time from autumnal to vernal equinox and vernal to autumnal equinox, as was mentioned earlier. He improved on Aristarchus' estimates of the sizes and distances of the Sun and Moon. Hipparchus was also one of the first people to systematically use trigonometry in his work, and he compiled a table of chords in a circle.

Because he accepted the geocentric model of the universe and the need for the orbits of the heavenly bodies to be perfect circles, he was committed to the eccentric and epicycle hypotheses. For example, he used a single eccentric orbit to describe the movement of the Sun around the Earth, and he used an epicycle/deferent combination centered on the Earth to describe the movement of the Moon.

3.1.12 Claudius Ptolemy

Classical astronomy in Europe reached its zenith in the second century after Christ in the writings of Claudius Ptolemy, a Hellenic astronomer, mathematician, and geographer, who lived in Alexandria (not to be confused with the members of the Ptolemy dynasty that ruled Egypt after the time of Alexander the Great). He was born around 100 AD and died around 178 AD. During his life, he wrote a number of books that covered a variety of topics, such as music theory, sundials, optics, stereographic projection, and astrology (*Tetrabiblos*, see Section 2.5). But two additional books were to become extremely influential: the *Geographia*, which was the first major textbook of geography and had great influence during the late Middle Ages and early Renaissance; and the *Almagest*, which summarized the state of classical astronomical knowledge at the time it was written, sometime around 150 AD.

In the *Almagest* (originally called the *Mathematical Syntaxis* but later translated to its better known name by the Arabs), Ptolemy wove together his own ideas with strands of knowledge from many of his predecessors. The result was a compendium of cosmology that was not only descriptive but was also empirically derived and mathematically precise. The book was consistent with Aristotle's geocentric and spherical model of the universe and Hipparchus' notions on epicycles and eccentrics. In the introductory first book of the *Almagest*, many of Ptolemy's basic principles were presented and defended. He pictured a spherical Earth that was immobile and

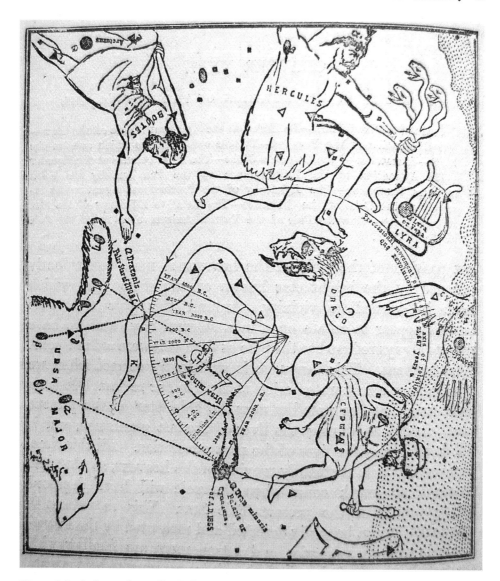

Figure 3.5. A figure from Sir Robert Ball's *The Story of the Heavens*, published in 1897. 23 × 15.3 cm (page size). Note the circle representing the precessional movement of the celestial pole. In the 3rd millennium BC, *Alpha Draconis* was the pole star. Today it is *Alpha Ursa Minoris*.

located in the center of the universe. Although massive enough so that all objects fell towards its center, the large size of the universe was such that the Earth was "like a point" in relation to the sphere of the fixed stars. There were two primary motions in the heavens. One of them accounted for the daily movement of the stars and planets from east to west. The other accounted for the motions of the spheres of the planets in

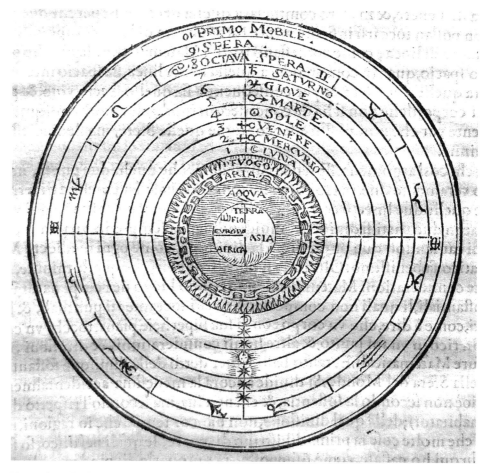

Figure 3.6. An illustration of the Aristotelian/Ptolemaic cosmological system that was used in the Middle Ages, from the 1579 edition of Piccolomini's *De la Sfera del Mondo*. 20.2 × 14.4 cm (page size), 8.9 cm dia. hemisphere. Note the map of the Earth in the center surrounded by concentric circles representing the other elements (water, air, fire) and the Moon, Sun, planets, and heavens, including the "Prime Mover" in sphere 10.

smaller west-to-east directions with reference to the fixed stars. For Ptolemy, the order of the heavenly bodies was somewhat different from that of the Pythagoreans and Plato. Like later stoic philosophers, Ptolemy advocated the following order: Moon, Mercury, Venus, Sun, Mars, Jupiter, and Saturn. This order was taken up by medieval scholars and appeared in texts showing the Aristotelian/Ptolemaic cosmological system (although Aristotle himself did not clearly describe a preferred order of the heavenly bodies in his writings—see Figure 3.6).

The remainder of the *Almagest* described the movements of the heavenly bodies through formulae and complicated diagrams that took into account the eccentricities of the orbital deferents and the direction and speed of their epicycles according to calculations based on spherical geometry and trigonomety. The details are beyond the scope of this book but are available in modern-day translations (e.g., Toomer's 1998 *Ptolemy's Almagest*). In addition to these mathematical and theoretical ideas, Ptolemy presented a catalog of stars that was to be the most influential catalog until the time of the Renaisssance. More will be said on this in Section 4.1.5.

There was one innovation in the *Almagest* that needs to be described. In order to improve on the position of some of the planets in the sky, Ptolemy introduced the concept of the equant. As shown in Figure 3.7, a planet could be seen as moving along a deferent that was in fact an eccentric circle whose center "b" was offset from the Earth "a". The equant defined a point "c" that was the same distance away from the center of the deferent but in the opposite direction from the Earth. The speed of a planet around the deferent was set to move so that from the perspective of the equant point, it was moving uniformly. This modification saved the phenomena of the planet's location in the sky. We now know from Kepler (Section 3.10.1) that the planets move in elliptical orbits with the Sun at one of the foci, and the reason the equant point worked was that its position was essentially analogous to being placed at the other focus of Kepler's ellipse. However, Ptolemy's equant model presented a problem: as seen from the point of the central Earth, the planet appeared to move at a variable speed. Since this conflicted with Aristotelian ideas about uniform speed around a circle, many critics took umbrage with its use, particularly the Arabs who "rediscovered" Ptolemy and began to make modifications of their own, which will be described in Section 3.3.

In a later work called the *Planetary Hypotheses*, Ptolemy described a universe made up of spherical shells of varying thicknesses, one contiguous with its neighbors without empty space in between. This notion was later integrated by Arab scholars with the eccentrics and deferent/epicycle combinations described for each heavenly body in the *Almagest*. This resulted in a more complete visualization of the heavens that was sometimes pictured in books printed during the Renaissance (see Figure 3.9).

3.2 EUROPEAN ASTRONOMY DURING THE EARLY MIDDLE AGES

3.2.1 Impact of the fall of Rome

Greek astronomy was taken over by the Romans, who used it and wrote comment-aries but made few new contributions. In fact, writing in Roman times, Ptolemy and his *Almagest* represented the final word in this area, possibly accounting for the dearth of astronomical work that has survived from the two centuries before him. However, the works of Ptolemy disappeared in Europe over subsequent centuries. How did this happen? To best understand this, it should be recalled that political infighting, social decay, and external invasion pressures from Germanic tribes led to a gradual decline in the western part of the Roman Empire, centered on Rome. In

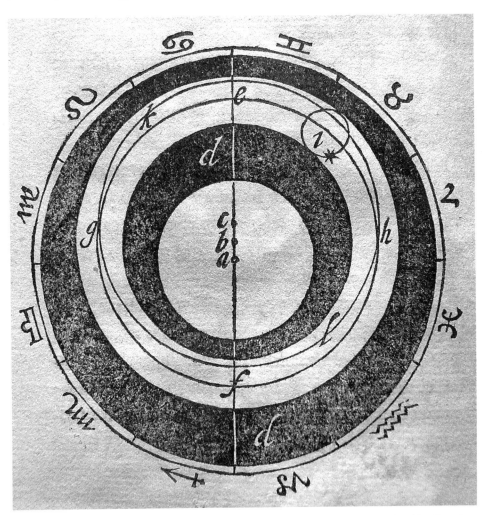

Figure 3.7. An outer planet's orbit according to Ptolemy (influenced by Peurbach's *Theoricae Planetarum Novae*), from the 1647 Leiden edition of Sacrobosco's *De Sphaera*. 15.2 × 9.7 cm (page size). Note the central Earth at "a", the center "b" of the eccentric circular deferent "e–g–f–h", and the equant point "c". The planet and its epicycle at "i" are set to move at a constant speed with reference to the equant point, which puts them at a variable speed with reference to an observer on the Earth.

330 AD, Emperor Constantine the Great moved the capital to the eastern city of Byzantium located on the Bosphorus in what is now Turkey, and he renamed it Constantinople. More on the East in Section 3.4.

As the West continued to collapse until its final fall in the latter part of the 5th Century, a number of factors contributed to the loss of classical knowledge in astronomy. According to historian Stephen C. McCluskey (1998), the educational system during the Roman period became increasingly oriented towards making civic leaders out of the sons of aristocrats who comprised the student body. In astronomy, the cosmology of Aristotle and the mathematics of Ptolemy were not emphasized as much as the poetry of Aratus or the philosophy of Plato. Mathematical astronomy did not meet the needs of people who were concerned with war and survival, and scholars began to lose the technical skills needed to comprehend Greek theory. In addition, Latin translations of earlier Greek works were imperfect, contributing to the loss of this knowledge. Finally, many of the leaders of the early Christian church advocated a strict interpretation of the scriptures in matters related to the heavens, which left little room for the writings of the barbarian Greeks.

However, some pockets of Greek mathematical astronomy survived into the 5th Century. For example, in Alexandria the Neoplatonist philosopher Hypatia (ca. 355–415) was an influential teacher of mathematics and astronomy. She wrote several important commentaries, including one with her philosopher father Theon on Ptolemy's *Almagest*.

3.2.2   Astronomy in the Latin West

Many of the poetic and philosophical aspects of Greek astronomy continued to be taught. One such advocate was the North African Martianus Capella (ca. 365–ca. 440). He wrote a popular textbook in Latin called *De Nuptiis Philologiae et Mercurii (The Nuptials of Philology and Mercury)* that used allegory and poetry to describe the seven liberal arts. In the section on astronomy he presented a model of the solar system stemming from earlier Greek sources that described Mercury and Venus as orbiting the Sun, and the Moon, Sun, and other planets as orbiting the Earth. This hybrid model followed from Heraclides and anticipated Tycho Brahe, who had a similar cosmology but had the outer planets orbiting the Sun, not the Earth (see Section 3.9.2). Interestingly, Nicholas Copernicus later cited Capella when he developed his famous heliocentric model.

In addition, there were some advocates of Greek astronomy (especially Plato and his Creator/demiurge—see Section 3.1.4) among the early leaders of the Church. One supporter was Augustine (354–430), who had studied Plato as a youth and tried to work with classical concepts that were not clearly contradicted by the scriptures. Another was Isidore (ca. 560–636), who was Bishop of Seville and presided over important Church councils. In his writings Isidore quotes classical philosophers in discussing such issues as the revolution of the spherical heavens around a central spherical Earth and the order of the heavenly bodies (using the Ptolemaic rather than the Platonic order). Similar ideas were discussed by the Venerable Bede (ca. 672–735), an English monk who had access to a number of books brought to his monastery from Rome.

In the 8th and 9th Centuries, attempts were made in the court of Charlemagne (768–814) in Aachen to systematize astronomical learning according to uniform authoritative religious standards. Schools were established for the clergy and for the children at court. Ancient texts were collected, copied, and disseminated, and anthologies were written that included solar and weather phenomena, computational tables, the structure of the heavens, and constellation descriptions. Ideas involving the geocentric universe and the spherical Earth persisted. However, unlike what was found in the work of Ptolemy, mathematical principles of spherical geometry were lacking, tables of the heavenly bodies only described mean motions and not variations from the average, and stars were inaccurately placed in the constellations without points of reference. In fact, the work of Aristotle, Hipparchus, Ptolemy, and other Greek scientists and astronomers was largely forgotten in Western Europe.

3.2.3   Time and the calendar

However, issues involving time were important during the early Middle Ages in the Latin West, primarily due to the influence of the Christian Church and its dependence on calendrical events. For example, it was necessary to establish the times of the solstices and equinoxes because these events became associated with the dates of the conception and birth of Jesus Christ and John the Baptist. In newly converted lands, pagan festivals held mid-way between these solar events became associated with Christian holidays in accord with the Julian calendar, which became the basis for the Christian ritual calendar since many religious holidays now could be given fixed dates that were independent of celestial events.

An important exception was the yearly determination of the date of Easter Sunday. Historically, Easter was defined as occurring during the first Sunday after the first full Moon after the vernal equinox. However, several systems evolved because the events related to Easter as described in the Bible were not based on a specific calendar date but occurred with reference to the time of the Jewish Passover and its associated full Moon. Political issues contributed as well to the dating confusion. Since the time of Constantine the Great, the Christian religion held dominance throughout the lands that had comprised the Roman Empire. Due to a dispute over the authority of the Pope and other religious matters, a schism occurred in the Church in 1054 that led to a western (Roman Catholic) and eastern (Eastern Orthodox) division that persists to this day. As a result of these theological, calendrical, and political issues, different systems were used to calculate the date of Easter in a given year, with the time varying by as much as several weeks. In some cases, tables simplified the calculations. Repeating sequences of letters of the alphabet (from A to G) could be used to indicate successive days of the week in counting toward Easter, with the letter representing Sunday for a given year called the "dominical" letter (Figure 8.8). Despite attempts to reconcile these different systems, discrepancies continue to the present time. For example, Easter Sunday for Roman Catholic and Eastern Orthodox Christians usually is held on different dates in a given year since the former follow the Gregorian calendar and the latter the Julian calendar in ecclesiastic matters.

Another concern of medieval astronomy had to do with determining the time for monastic prayers. In addition, monks were involved with a number of feasts and ceremonial activities throughout the month, and it was important for them to keep track of these dates. Until about the 10th Century, when water clocks began to be used again, observations of the stars were the principal means of nocturnal time keeping at most monasteries. Many monks had hand-written books of hours listing important days of the month (Figure 3.8). These manuscripts varied from relatively modest "pocket" miniatures to elaborately illustrated versions that were especially designed for members of the nobility.

3.3   ISLAMIC ASTRONOMY

Despite its absence in Western Europe, classical Greek mathematical astronomy was preserved in two other areas: the Islamic countries and Byzantium. Recall that the Islamic religion was founded by the prophet Muhammad (570–632), who was persecuted and driven out of his native Mecca but fled to Medina with his followers in 622. His teachings took hold, and through faith and warfare they rapidly spread throughout the region. The Abbasids took over the caliphate in 750, and in 762 they founded a new capital, Baghdad, as the empire continued to expand into Christian and northern Indian lands. In fact, many Indian astronomers were invited to the capital. In particular was an emissary who came to the court of Caliph al-Mansur around 773 and presented a number of ideas involving Greek mathematical astronomy and planetary theory. According to Dreyer (1953), this was followed by the direct exposure to Greek and Alexandrian astronomy through the efforts of physicians from a Nestorian Christian medical school at Khusistan. In the 9th Century the Abbasid caliphs at Baghdad commissioned the translation from Greek into Arabic of a number of scholarly manuscripts, including the works of Aristotle, Apollonius, Ptolemy, and others, and these enlightened leaders became great patrons of the arts and sciences. The distinguished Harvard historian of astronomy, Owen Gingerich (1992), singles out Caliph al-Ma'mun, who founded an academy called the House of Wisdom, which through its tolerance and acceptance of scholars from other cultures made Baghdad an important site for learning.

Besides a general desire for astronomical knowledge encouraged by enlightened caliphs, Islamic peoples were motivated by astrological, time-keeping, and religious reasons (e.g., it was important to locate the direction of Mecca for daily prayers and to determine the times of Ramadan and other religious festivals, and a knowledge of the heavens assisted with these activities). Advances were made in Ptolemaic theory and in empirical astronomy at Islamic observatories, such as at Damascus and Baghdad itself. Astronomical tables were developed that gave the mean motions and true positions of the heavenly bodies in the sky and the length of the day throughout the year based on Greek, Indian, and Islamic observations. Especially influential was the table or *zij* developed by the Baghdad astronomer and mathematician Muhammed Ibn Musa al-Khwarizmi (active late 8th Century to mid-9th Century), around 840. Islamic astronomers also refined the astrolabe (see Section 8.3.2), a calculating device originated by the Greeks that allowed for a physical representation of the geometry of the heavens to be projected onto a metal plate by means of which predictive calculations of the position of the heavenly bodies and the telling of time could be made. In mathematics the Muslims made advances over Ptolemy and his chords by introducing many of the basic trigonometric

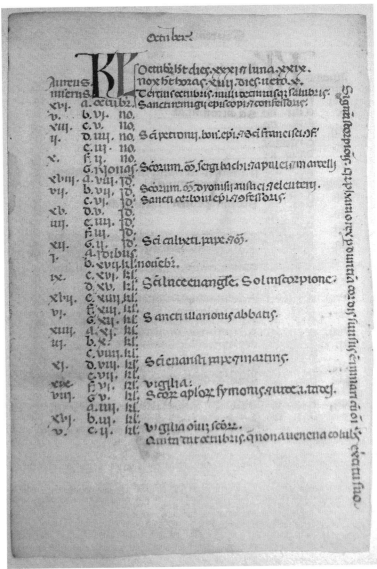

Figure 3.8. Anonymous calendar leaf for October from a Book of Hours, ca. 1350. This manuscript was written on vellum and is probably a page from a Psalter from Italy. The margins have been cut and are uneven but measure about 12.7 × 8.6 cm (page size). Note from left to right that the columns depict: Roman numerals indicating the year in the 19-year solar/lunar calendar where the full Moon falls on the given date; letters representing the days of the week, with the red "a's" probably being the dominical letter for Sunday; the Roman calendrical system, including "idib" for the "ides" of the month in the middle; and descriptions of various saints' and feast days on the right. Note also at the top the large "KL" for "Kalendarium", and the indication that October had XXXI days and XXIX lunar days that year. *See also* Color Plate Gallery.

functions (e.g., cosine, tangent) that we use today. The writings of prominent Islamic astronomers were subsequently translated into Latin in the 12th Century as the West began to re-learn classical mathematical astronomy, and the authors were also given Latinized names. Examples include Ahmed ben Muhammed al-Farghani (active 9th Century), later called Alfraganus, and Muhammed al-Battani (858–929), later called Albategnius.

In the 10th Century, the power and patronage of the Baghdad caliphs began to decline, but Islamic astronomy continued to make advances in other parts of Islam, such as in Persia, Egypt, and especially Spain. For example, in Cordova, Ibrahim Abu Ishak, or al-Zarkali (ca. 1029–ca. 1087), edited some tables of planetary locations called the Toledo Tables. He was later named Arzachel in Christian Europe. In the 11th Century, Ibn al-Haytham (Alhazen) in Cairo, and in the 12th Century, Abu Ishaq al-Betruji (Alpetragius) in Spain, both put forth critiques of Ptolemy's planetary theory, especially his use of the equant. In the 13th Century the Christian King Alfonso X of Castille (active 1252–1284) called astronomers to his court, and his patronage led to the renowned Alfonsine Tables, which were considered the most accurate planetary tables during the next 300 years.

Islamic astronomers required that a planetary system truly represent reality rather than simply being a mathematical construct to save the phenomena. For this reason, some writers like the great Muhammad Ibn Rushd (1126–1198), later called Averroes in the West, objected to the eccentrics and epicycles of Ptolemy and favored a more simple concentric spherical model based on Aristotle. Some astronomers combined the two systems into one grand concept of the universe. In this integrated model, space was provided within each planetary sphere to contain the rotating epicycle, and the sphere itself was often eccentric with respect to the central Earth (Figure 3.9). Furthermore, the upper surface of one sphere was contiguous with the lower surface of the next sphere, and there were no empty spaces in between. Although some of these ideas were introduced earlier, they received their most complete exposition in the 13th Century through the writings of al-Kazwini, Abu al-Faraj, and others. This combined system was passed onto the West and influenced many writers of the late Middle Ages and early Renaissance.

Also in the 13th Century, Nasir al-Din al-Tusi (1201–1274), the influential Persian astronomer and mathematician, critiqued Ptolemy's system and developed new geometric planetary models of his own. He also founded the great Maragha Observatory, whose foundations still survive some fifty miles south of Tabriz in what is now northwestern Iran. One of al-Tusi's more influential accomplishments in the area of planetary orbital theory was to devise a geometrical substitute for the equant by using two small epicycles. He observed that if a circle rolls inside the circumference of another circle with radius twice as large, then any point on the inner circle could be made to describe a straight line depending on its speed (Figure 3.11). This "Tusi couple" theorem could be proven geometrically (in the spirit of Ptolemy) and could be illustrated visually to create models of planetary positions. Al-Tusi's work encouraged Ibn al-Shatir (active mid-14th Century) in Damascus to devise a concentric planetary scheme of nested spheres that was free from the equants and eccentrics of Ptolemy, but according to Gingerich (1992) his work was not well known in Medieval

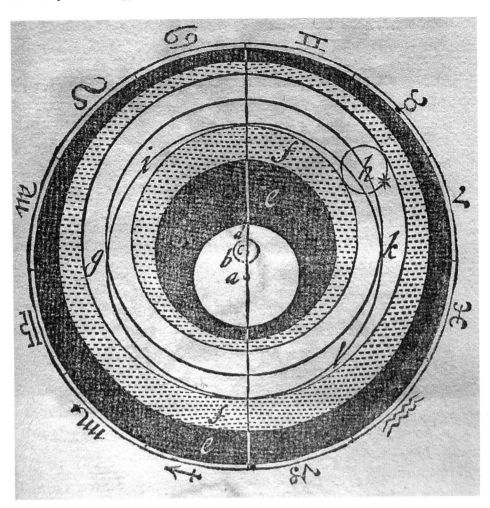

Figure 3.9. The planetary model of Ptolemy for Mercury (influenced by Peurbach's *Theoricae Planetarum Novae*), combined with the spherical model of Aristotle, from the 1647 Leiden edition of Sacrobosco's *De Sphaera*. 15.2 × 9.7 cm (page size). Note that the eccentric sphere "i–g–l–k"is wide enough to contain the epicycle of the planet "h" and that it abuts against another sphere at "f" without allowing any empty space. This integrated model was popular in Islamic astronomy.

Europe. As we shall in Section 3.4, models incorporating versions of the Tusi couple appeared in later Byzantine manuscripts, and Copernicus probably made use of its principles in his writings concerning variations in precession and the celestial latitude of heavenly bodies.

Mention should also be made of Ulugh Beg (1394–1449), the grandson of Tamerlane. Around 1420, he built a great observatory that drew a number of astronomers and other scholars to its location in Samarkand. As a result of its advanced

instrumentation and the diligence of its observers, new planetary tables and a new
star catalog were produced that were quite accurate and were used by later European
astronomers.

Interest in astronomy was continued by the Islamic Ottoman Turks after they
took Constantinople in 1453. By then, Western astronomy had regained its old
prestige, and many European ideas were subsequently adopted by the Turks. For
example, the first printed Ottoman world atlas, Katip Celebi's *Cihannuma* from 1732,
contained a number of diagrams illustrating the cosmologies of prominent Western
astronomers, such as Copernicus and Tycho Brahe (Figure 3.10).

3.4 BYZANTINE ASTRONOMY

In contrast to Islamic influences, less has been written about the impact of the
other great repository of classical learning, which was centered in Greek-speaking
Constantinople. Founded by Greek emigrants in the 7th Century BC under its
original name of Byzantium, this city became the capital of the entire Roman Empire
under Constantine the Great and remained the capital of the eastern part of the
empire when Rome fell. As the principal city of the later-called Byzantine Empire,
Constantinople became an important strategic, trade, and cultural center, a position
it holds to this day (but now renamed Istanbul, Turkey). There, a number of classical
works were preserved and discussed in their native Greek language, and many of
these were purchased by Islamic leaders and were translated into Arabic in the 8th
and 9th Centuries. In addition, there is evidence suggesting that Byzantine scholars
were not only well versed in the mathematical astronomy of Ptolemy and Islamic
writers, but that they also incorporated new elements into this work that conceptually
advanced the classical theories.

For example, physicist E.A. Paschos and philologist P. Sotiroudis (1998) have
recently translated and analyzed a late 13th-Century Byzantine manuscript uncov-
ered from the Vatican Library in Rome called *The Schemata of the Stars*. They
attribute this document to Gregory Chioniades (ca. 1240–ca. 1320), a professor of
medicine and astronomy in Constantinople who spent time studying in Persia and
later became Bishop of Tabriz, and who according to classicist Alexander Jones
brought back Arabic *zijes* from his travels, which he translated into Greek. Through
text and diagrams, the *Schemata* gives a listing of the constellations and their stars;
the mechanisms of lunar and solar eclipses; and the use of epicycles, deferents, and
eccentric orbits to describe the spherical motion of heavenly bodies around the Earth.
There is evidence of knowledge of spherical geometry and trigonometry and of
influences not only from Ptolemy but also from al-Tusi and other Arabic and Persian
scholars (Figure 3.11). But the document also contains a number of variations and
improvements to these earlier works, including novel ideas concerning the orbits of
the Sun and the Moon, a new model for the revolution of the superior planets that
uses an eccentric orbit, and improvements in the trajectory of the epicycle for
Mercury. What emerges from this work is a picture of an active and sophisticated
astronomy in Byzantium. Paschos and Sotiroudis further point out that the *Schemata*

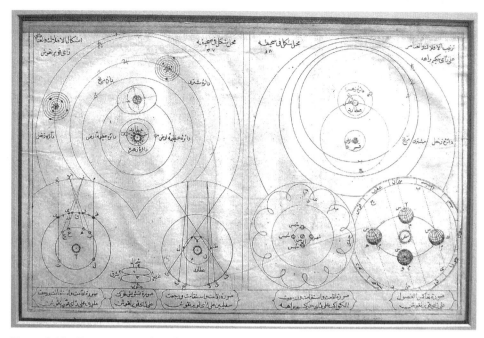

Figure 3.10. Copper engraving from the first printed Ottoman Turkish world atlas, the *Cihannuma*, produced by Katip Celebi in 1732. 16.5×26.1 cm. Note the Western influences: in the upper part are illustrations of the cosmologies of Copernicus (*left*) and Tycho Brahe (*right*), and in the lower part are diagrams showing the mechanism for the retrograde motion of a superior and inferior planet, the looped appearance of a superior planet's orbit in the sky, and the 3rd motion of the Earth according to Copernicus. *See also* Color Plate Gallery.

made its way to Italy, possibly in the 15th Century. It may there have influenced the ideas of Copernicus, who spoke Greek and studied church law, medicine, and astronomy in several Italian cities.

3.5   CLASSICAL GREEK ASTRONOMY COMES BACK TO EUROPE

3.5.1   Entry from the West: Islam

From the 11th to the 13th Centuries, much of Spain was re-taken from the Moors, who were Islamic invaders originally from Africa. The victors were Christians from independent kingdoms to the north, such as Castille. In the process, Greek and Islamic astronomical knowledge was brought back into Western Europe. For example, in the 11th Century European scholars now possessed astrolabes and were teaching their use to others. In the 12th Century a number of classical works were translated into Latin from Arabic, including works from Aristotle, Euclid's *Elements of Geometry*, and the *zij* developed by Al-Khwarizmi. Also in this century, a Latin translation of Ptolemy's *Almagest* was made from the Arabic by Gerard of Cremona (ca. 1114–1187), but it was very literal and hard to follow. Although the Latin translations of Arabic translations of Greek works were sometimes incomplete and unclear, they served to reintroduce Europeans to classical thinking.

The stage was now set for Latin astronomy to move from the poetic and philosophical to the observational and mathematical. However, this process was slow and incomplete. In part, this was due to resistance from the Catholic Church, which although tolerant of neoplatonic ideas was not so accepting of the natural philosophy of Aristotle and his followers. However, by the 13th Century, this prohibition began to pass. Dreyer (1953) attributes this in part to scholars like Thomas Aquinas (ca. 1225–1274), who wrote a commentary on the works of Aristotle that had been translated from the original Greek, and Roger Bacon (1214–1294), who tried to go beyond Aristotle in outlining principles that later were to become the elements of the scientific method of experimentation.

Although Aquinas and Bacon were familiar with the epicyclic ideas of Ptolemy, most learned men of the time such as Dante (1265–1321) were influenced more by Aristotle and his notion of concentric crystalline spheres. Dante, like Isadore, followed the Ptolemaic tradition in terms of the order of the heavenly bodies. He visualized ten spheres around the central Earth: Moon, Mercury, Venus, Sun, Mars, Jupiter, Saturn, *Stellatum* (the starry heaven composed of the fixed stars), *Primum Mobile* (the starless crystalline sphere of Aristotle's Prime Mover, which set the daily motion of all of the inner spheres from east to west), and *Empyrean* (the heaven of the blessed, the angels, and God, which extended out infinitely into space). The motions of each of the lower eight spheres were independently regulated by groups of angels in a west-to-east direction (accounting for the varying movement of each planet with respect to the stars, and, in the case of the *Stellatum*, for the precession of the equinoxes). Little is said about epicycles, and there is no indication of Greek mathematical approaches to save the phenomena.

3.5.2 Johannes de Sacrobosco

In the 13th Century the works of Aristotle began to be studied in the newly established universities, which had gained prominence with the growth of towns and the increasing secularization of cathedral schools. By the 1300s, the typical university curriculum was organized around the subjects of the classical seven liberal arts: the lower division trivium (grammar, rhetoric, logic) and the upper division quadrivium (arithmetic, geometry, music, astronomy). According to historian Lynn Thorndike (1949), in order to receive an undergraduate degree from the Universities of Paris, Oxford, Vienna, or Bologna, the student would have studied a slim book of astronomy originally named *Tractatus de Sphaera* or *Sphaera Mundi* but now generally known as *De Sphaera*. This book was written by Johannes de Sacrobosco around 1220, and it remained the most widely used textbook on astronomy from the 13th to the 17th Centuries.

Sacrobosco was probably born in Holywood, Scotland around 1195. He was educated at Oxford and may have been a canon at the Holywood Abbey in Nithesdale. Since the Latin name for the abbey was *Sacro Bosco*, his name was

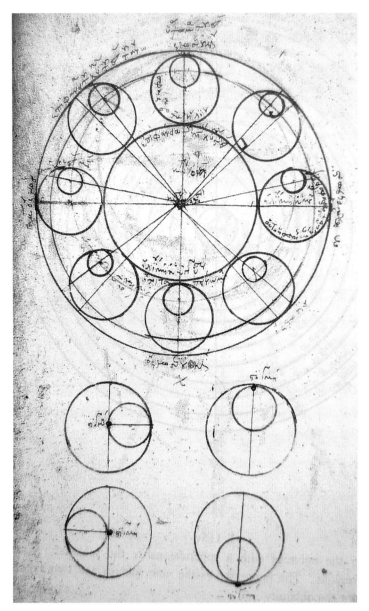

Figure 3.11. A page from a 13th-Century Byzantine manuscript. 21.6 × 15 cm (page size). Note that the bottom figure shows how the motions of the circles comprising a Tusi couple can result in a straight line. The top figure incorporates a version of the Tusi couple moving around a deferent to describe the motion of the Sun around the Earth. The resulting trajectory of the Sun, located on the smaller epicycle, is equivalent to the eccentric trajectory postulated by Ptolemy. Copyright, Biblioteca Apostolica Vaticana (Vatic. gr. 211, f. 116). Photograph and diagram from *The Schemata of the Stars* (E.A. Paschos and P. Sotiroudis, 1988). Courtesy of Professor Emmanuel Paschos and the World Scientific Publishing Company.

changed from John of Holywood to Johannes de Sacrobosco. Around 1220, he went to study in Paris. He joined the faculty at the University of Paris on June 5, 1221, subsequently becoming Professor of Mathematics. About this time, he wrote *De Sphaera*. He wrote other books as well on topics dealing with mathematics, the use of the quadrant, and time. He was also interested in the calendar, pointing out errors in the Julian system and suggesting ways that it could be reformed. He died in Paris around 1256 and likely was buried in the cloisters of the church of St. Mathurin, which was associated with the University of Paris.

De Sphaera was a short, 9,000-word manuscript written in Latin. Although generally Aristotelian in its approach, the book also contained some ideas from Ptolemy and the Arabs. In a book that went through so many editions and was in existence for over 400 years, it is important to realize that there were changes to the text and figures from edition to edition, making this book contemporary for each new generation of students. Some of the illustrations shown earlier in this chapter were from the 1647 Leiden edition of this book. Note that the "sphere" being discussed refers not only to features on the Earth itself, but also to the components of the celestial vault, such as the spheres carrying the Sun, Moon, planets, and stars.

Because of its prevalence, it is instructive to take a closer look at the contents of this important book. In the original manuscript, following a brief preface, there were four chapters in *De Sphaera*. The first dealt with the definition of a sphere, the central place of the Earth in the cosmos, the four Aristotelian elements (earth, water, air, and fire) and the spheres of the surrounding heavenly bodies (Figure 3.6), different proofs that the Earth is spherical (Figure 3.1), the small size of the Earth with reference to the cosmos, and calculations of the Earth's circumference and diameter.

The second chapter described the orientation of the heavens, the projections of the great circles on the Earth into the heavens and the movement of the heavenly bodies through them, and the important climatic zones on the Earth. There were sections on the celestial equator (i.e., the equinoctial), the north and south celestial poles, the ecliptic and the zodiac, constellations, the colures, the meridian, the zenith, the horizon, the projections in the sky of the Tropic of Cancer and Capricorn and the Arctic and Antarctic circles, and the habitability of the Earth's climatic zones.

The first part of the third chapter, the longest in the book, dealt with the rising and setting of the stars and constellations, with particular reference to the zodiac. In fact, much of this part had astrological ramifications. There were also discussions of the duration of the hours of day and night throughout the year. The middle part of the third chapter discussed the relative orientation of the Sun, zodiac, and celestial circles from the perspective of observers viewing at different latitudes on the Earth. The final part of the third chapter discussed seven climatic zones on the Earth that are inhabited by humans, including the maximum duration of their days, their breadth in miles, and the elevation in their skies of the North Pole. In later editions of *De Sphaera*, the number of these zones was increased, as explorers went northward and found people living in these regions. For example, the 1647 edition contained a supplementary table with the original seven zones plus an additional 16 that reflected settlements in Norway and Sweden.

The final chapter described the motion of the Sun in the sky and the deferent/epicycle characterization of the orbits of the Moon and planets (Figures 3.7 and 3.9). There was also an explanation of the direct, stationary, and retrograde motions of heavenly bodies according to the classical Greek geocentric model (Figure 3.4). After a description of the mechanisms of lunar and solar eclipses, the chapter concluded with a brief comment that the eclipse of the Sun that took place when the Moon was full during the Passion was due to a miracle rather than to natural events.

According to Dreyer, Sacrobosco was the first European writer in the Middle Ages to give even a short sketch of Ptolemy's ideas involving deferents and epicycles. *De Sphaera* generated a plethora of commentaries due to its popularity and use as a standard university textbook for generations. It was one of the first scientific books to be printed (in 1472) shortly after the development of movable type in Europe. It went through over 200 editions and was published in a variety of places until well into the second half of the 1600s. Copernicus used it as a student, and John Flamsteed, the first Astronomer Royal of England, was influenced as a youth by this book to pursue studies in astronomy. This longevity is especially remarkable when one realizes that this book described a geocentric universe, whereas Copernicus' great *De Revolutionibus*, which advocated a heliocentric perspective, came out in 1543, over 120 years *before* the last edition of *De Sphaera*.

3.5.3   Entry from the East: Byzantium

During the Middle Ages, the Greek language was not widely known. Except for some Irish monks and the occasional individual (such as Roger Bacon), few scholars could read Greek well enough to work with the few Greek texts that were available. Interest began to grow in the 14th Century, however. Especially in Italy, native Greek speakers were brought in as teachers, such as Manuel Chrysoloras who lectured in Florence from 1397 to 1400. This process was accelerated by a major event that has been well described by Renaissance historian Lisa Jardine (1998). On February 8, 1438, Byzantine Emperor John VIII, Eastern Orthodox Patriarch Joseph II, and an entourage of some 700 bishops, monks, and learned laymen arrived in Florence, where the court of Pope Eugenius IV was then located. The purpose of the meeting was to discuss reconciliation between the Roman Catholic and Eastern Orthodox churches. The Byzantines brought a number of books and texts in the original Greek, including the works of Plato, Aristotle, Euclid, and Ptolemy. While the leaders haggled over church doctrine and tried to negotiate (unsuccessfully) the merger of the two churches, the intellectual experts on both the Byzantine and Latin sides collaborated and shared ideas on philosophical and mathematical topics. These books and lectures given by Greek scholars during this meeting contributed to the vogue for Greek learning in Italy and led to the development of the Platonic Academy, which was founded by the wealthy Florentine patron of the arts, Cosimo de Medici.

Around the time of the fall of Constantinople to the Ottoman Turks in 1453, a number of Byzantine scholars moved to Italy and brought with them their personal libraries of rare Greek books. Venice contained so many such émigrés that the Greek scholar and immigrant Cardinal Bessarion likened the city to another Byzantium, and in 1468 he donated his magnificent collection of over 600 books and manuscripts (which included mathematical works by Archimedes, Apollonius, and Ptolemy) to St. Mark's Cathedral. In time, important classical manuscripts in the Greek language were collected in libraries at the Vatican, Florence, and Venice, and Renaissance scholars were available who could read and translate these texts into Latin and other languages.

Now, important classical documents could be read and translated into more accurate versions from the original Greek. After printing was established in Europe (see Section 3.7), these accurate editions could be disseminated more widely. In the case of Ptolemy's *Almagest*, a Greek-to-Latin version was made by George of Trebizond in 1528. Its accuracy could be tested by the printing in 1538 of a Greek original from a codex once possessed by Regiomontanus. Now, anyone knowing Greek or Latin could read this important work.

3.6 ASTROLOGY IN THE MIDDLE AGES

As discussed in Section 2.5, astrology in the West became more fragmented and unsophisticated following the break-up of the Roman world, partly due to the loss of Greek scientific astronomy and partly due to condemnations by the Church. But, in the East it was transmitted to the Arabic world and assimilated Indian, Persian, and Islamic influences. In the process, it became even more complex and mystical. In fact, later representations of the zodiac depict the syncretism of Mesopotamian/Greek notions with concepts unique to specific cultures (see Figure 2.10). By the late 10th Century, this complex astrology was brought to Spain, and it was this version that was disseminated in the West. It acquired great impetus in the 12th Century with the rediscovery and translation of Greek astronomical texts from Arabic into Latin.

Astrology generally was linked with astronomy, a characteristic that was to continue throughout the Middle Ages and into the Renaissance. In a sense, astrology covered the practical applications and astronomy the theoretical aspects of the same science of the stars. In fact, zodiacal constellations were often pictured in medieval monastic texts, books of hours, and other manuscripts of the time. After the 12th Century, the requisite astronomical and mathematical skills to construct horoscopes and interpret celestial configurations were introduced into the universities, and astrological skills were a part of the curriculum for students of astronomy and medicine. According to research fellow Sophie Page (2002), in the 15th Century Bologna even had its own Professor of Astrology who taught a four-year course in this field.

The main tool of the astrologer was the horoscope, which was a symbolic representation of the heavens at a particular time and place (Azzolini, 2011). Typically, there would be a central square surrounded by 12 contiguous triangles, which represented the "houses" of the zodiac oriented according to the zodiacal constellations that were seen in the sky at the time indicated on the horoscope (Figure 3.12).

These houses were counted sequentially in a counter-clockwise direction, starting with the triangle to the left of the center. In the central square, the astrologer would write relevant information, such as the name of the client, the date, and the issue prompting the horoscope. The positions of the planets, Sun, and Moon as they appeared in the sky were placed in their relevant house, and the degree of each zodiac sign was written at the boundary of each triangle. Each heavenly body was associated with a number of inherent properties of significance, such as gender, helpful/harmful, or quality (e.g., hot, cold, dry, moist). The zodiacal signs also represented astrological properties, such as season and element (e.g., earth, water, air, fire). Each heavenly body ruled over two of the signs of the zodiac (i.e., their houses), except for the Sun and Moon, which ruled over one. By observing the resulting horoscope, important characteristics could be found. For example, if a planet was located in one of the houses that represented a sign that it ruled, its influence in the overall picture was strengthened. Other significances were made based on the positions and relationships of the heavenly bodies with one another. Each sign in the horoscope was further subdivided into "faces" of 10 degrees each, harkening back to the decans of the Egyptian heavens (see Section 2.3.2). Based upon these and other characteristics of the horoscope, an interpretation could be made to predict the consequences of a future event, describe an individual's personality and life course, or define the best time to initiate an action.

Page has identified two main branches of medieval astrology: mundane (or natural) and judicial. Mundane astrology was concerned with celestial influences on natural phenomena (e.g., the weather) and the prediction of general events affecting everybody. Horoscopes in this branch were constructed for such things as the Sun's entry into Aries, eclipses, and conjunctions of the planets in the sky (where they all were lined up along the same longitude). Such horoscopes were often presented to the king to help him govern his kingdom with knowledge of significant future events. Mundane astrology was generally accepted, since it was easy to point out obvious influences of heavenly events upon the Earth, such as the Sun in determining the seasons and the Moon in influencing the tides.

In contrast, judicial astrology focused on the individual and provided specific information on future events affecting his or her life and the correct time to engage in various activities. Especially popular was natal astrology, where horoscopes were constructed based on the positions of heavenly bodies at the time of the person's birth. From this information, predictions could be made about one's personality, course of life, and manner of death. Initially, nativity horoscopes were made for aristocrats and the wealthy, but they became more broadly used by less prominent people as time went on. Other kinds of judicial horoscopes were horary charts, which were constructed to answer a specific question and were drawn up at the time that the individual asked the question, and electional horoscopes, which were studied to help establish the best time for beginning an activity (e.g., to begin a business venture or to leave town to go to war). Judicial astrology was more controversial than mundane astrology, since it clashed with church doctrines involving free will and personal responsibility for sins. A compromise solution was generally made that viewed personal horoscopes as predicting general tendencies and events but still gave the

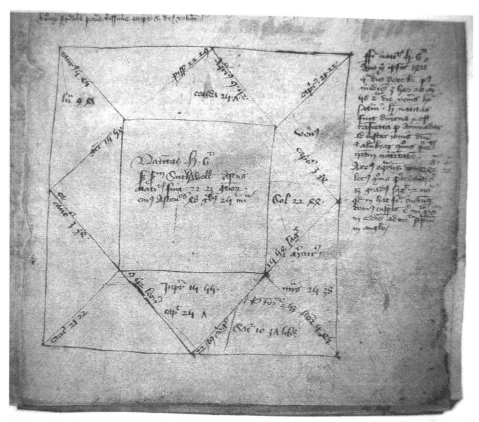

Figure 3.12. A nativity horoscope produced in 1421 for King Henry VI of England. 24.6 × 17.4 cm (page size). Note the text in the right margin and central square giving the birth date and other information on the King, and the surrounding triangles that represent the 12 zodiacal houses. Photograph taken from *Astrology in Medieval Manuscripts* (S. Page, 2002). Courtesy of the University of Toronto Press and the British Library (BL Egerton MS 889, f.5).

individual the latitude to act in a manner that preserved his ability to make choices in his life.

To the medieval mind, people and the rest of the universe were connected. Humans were viewed as a microcosm that reflected the larger macrocosm around them. For example, the four humors in the body (blood, phlegm, yellow bile, and black bile) were linked to the four temperaments (sanguine, phlegmatic, choleric, and melancholic), the four elements (air, water, fire, and earth), the four qualities of matter (hot, cold, wet, and dry), the four stages of a person's life (child, youth, adult, elderly), and the four seasons of nature. In a similar manner, the organs and areas of the body were related to a variety of sympathetic influences (e.g., medicinal plants, metals, heavenly bodies) that were depicted in diagrams that Page calls the "micro-cosmic man", which appeared in almanacs and encyclopedias of the time (Figure 3.13). In terms of the heavens, one type of relationship was between the parts of the body and the planets that governed them. University of North Carolina Professor S.K. Heninger, Jr. (2004) cites the following correspondences: Sun (heart), Moon (head), Mercury (lungs), Venus (ears), Mars (gall bladder), Jupiter (liver), and Saturn (spleen). There was a similar relationship between the houses of the zodiac and the body (here called the "zodiac man"). According to Heninger, a typical correspondence would be: Aries (head), Taurus (neck), Gemini (arms), Cancer (chest, lungs, and stomach), Leo (heart and liver), Virgo (intestines and rectum), Libra (kidneys and buttocks), Scorpio (versus Scorpius, the constellation name—genitals), Sagittarius (pelvis), Capricorn (versus Capricornus, the constellation name—knees), Aquarius (lower legs), and Pisces (feet). Such relationships often guided medical decisions. For example, in medieval medical lore, it was considered unsafe to bleed a patient when the Moon was in that part of the zodiac that ruled the body part that was diseased. According to Daniele Bini (1977), Curator at the Biblioteca Estense in Modena, Italy, in 1427 Charles VIII of France commanded that the barbers in his kingdom (who were the surgeons of the time) possess and utilize a chart of the zodiac man when deciding when to perform a surgical procedure.

As Europe progressed through the Renaissance, the influence of astrology began to decline. Astronomy began to be seen as a separate science, in contrast to astrology, which began to take on a more subjective and superstitious quality. This process was accelerated as Aristotelian natural philosophy lost its influence and as the universe moved from a geocentric to a heliocentric orientation. Today, astrology (especially its natal version) continues as a popular pastime, but most educated people rarely confuse it with the science of astronomy.

3.7   PRINTING AND THE RENAISSANCE

There were a number of factors in the early Renaissance that influenced the re-birth and advancement of classical astronomy, especially its more mathematical and scientific aspects. First, adequate translations now were freely available in Latin, in many cases made directly from the original Greek by native Greek-speakers from Constantinople. Second, these translations included commentaries and additions from Islamic and Byzantine scholars that enhanced the originals. Third, the spirit of the Renaissance encouraged the advancement of knowledge for its own sake rather than solely for the sake of spiritual and religious needs. Fourth, secular universities were well established and now were more open to transmitting new information to their students.

But how could these ideas reach people who were not involved with universities? The answer is the fifth factor that explained how classical astronomical learning prospered and advanced during this time: the explosion of printing in Europe in the late 1400s. A central figure in this development was Johann Gutenberg.

3.7.1   Johann Gutenberg

Gutenberg was born in Mainz around 1398, the son of a wealthy merchant from an old patrician family who was named Friele Gensfleisch zur Laden. Johann later took

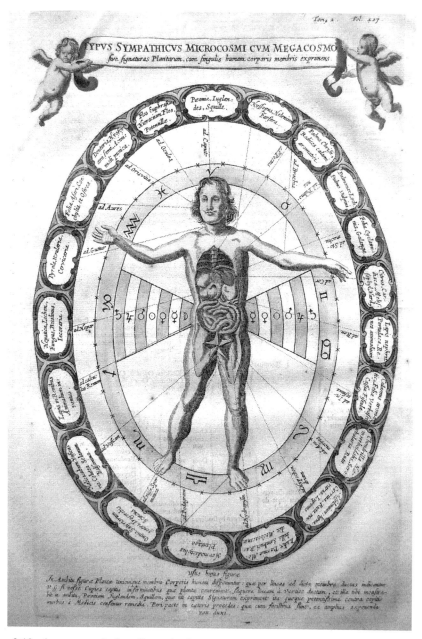

Figure 3.13. An engraved plate from the first Dutch edition (1682) of *Mundus Subterraneus*, first published in 1665 by Athanasius Kircher. 35.6 × 20.8 cm (plate size). This shows the correspondence between the microcosm (the "microcosmic" or "zodiac" man in the center) and the macrocosm (symbols of planets and zodiac in the periphery). Note the dotted lines linking the sympathies of the parts of the human body and various medicinal plants (identified in the Baroque-styled labels in the outer circle). *See also* Color Plate Gallery.

the name of Gutenberg after the neighborhood in which the family lived. As a child, he was exposed to a number of hand-written manuscripts and books printed from fixed woodblocks. He moved to Strasburg as a young man and was trained as a goldsmith. However, he became interested in printing and began experimenting with ways to make it more efficient and inexpensive. He subsequently returned to Mainz around 1444, secured some investment money, and formed a printing company.

His focus was on making types with single letters, first out of wood, then out of metal, which was more durable and easier to produce using a mould. Unlike block printing, where the words of the whole page were carved out of one piece of wood, the use of individual letters inserted into a wood frame to produce the text was fast and flexible, since the types could be quickly reused to form words for other pages. It should be noted that the Chinese and Koreans had been printing books from woodblocks for centuries, and the Koreans had developed an iron printing press in the 13th Century. Movable clay type was used in China since the 11th Century, and movable metal type was used in Korea since the end of the 14th Century. However, the use of movable type in the Orient was restricted owing to the complexity of the writing system, particularly in China.

Gutenberg gets the credit for revolutionizing printing in Europe through his use of accurately cast metal types, oil-based inks, and a sturdy printing press that was based on the wine press. He began applying his technique around 1448, when he published a poem and an astronomical calendar. But the *coup de gras* occurred in 1455, when he began mass-producing a beautiful Latin Bible. It was in two volumes, each with some 300 pages of 42 lines. The letters were sharp and clear and had even spaces between them. Some 180 Gutenberg Bibles were produced on rag cotton linen paper and vellum. Gutenberg became famous as a result, and his technique of printing using metal movable type quickly spread. Unfortunately, he quarreled with his business partner, Johann Fust, who took over control of the business as a result of a court action. Gutenberg was subsequently impoverished and had to be subsidized by the Archbishop of Mainz. He died in relative poverty in 1468.

3.7.2 The spread of printing

Printing quickly spread throughout Germany and from there to other countries. Within 15 years of the appearance of Gutenberg's Bible, printing presses were operating in many major urban areas, including Strasburg, Cologne, Basel, Augsburg, Paris, Rome, and Venice. Grendler (1984) estimates that the number of European towns with printing presses was 17 by the end of 1470, 120 by the end of 1480, and 255 by 1500. He goes on to say that Venice became a major center, publishing nearly 15% of all books in Europe and nearly twice as many as second-place Paris by 1501. It was also the home of two great publishers who produced books that included the first printed images of constellations: Erhard Radholt and Aldus Manutius, whom we shall meet in Section 5.2.

Although printed books were relatively expensive initially, they found readership among people involved in commercial interests (such as shipbuilding and navigation) and members of aristocratic families. In fact, among Renaissance men of power, having a great library became an important status symbol. These new collectors competed for rare books, which stimulated the book trade and led to further translations and printings of old masterpieces. Gradually, new markets opened up for more affordable textbooks in schools and universities, where the less privileged were exposed to the works and bought less expensive editions. Some of these books included volvelles, which consisted of movable attachments on the page that permitted calculations to be made and were a more affordable alternative to astrolabes made out of metal (see Section 8.2).

Thus, by the time of the Renaissance, classical Greek astronomy was retrieved from Islamic and Byzantine sources, sometimes in improved versions. Through the influences of the secular universities, the availability of printed books, and the humanism of the times, the stage was set for new developments in the arts and sciences. Much of this took place in Italy, especially in areas such as painting and sculpture. However, in the area of astronomy, a number of important books were written and published by scholars living in Central Europe north of the Alps.

3.8 ASTRONOMY AND CENTRAL EUROPE

During the second half of the 14th Century, a number of new universities sprung up in Germanic areas, leading to a new humanism in the region. There developed a great interest in the ideas of ancient Greece and Rome, especially in the natural sciences. Later, this *weltanschauung* would pave the way for a series of remarkable Central European astronomers, such as Copernicus and Kepler (see Sections 3.9.1 and 3.10.1, respectively). But, even in the 1400s, scholars began to speculate again about the universe, sometimes very perceptively. For example, Nicolaus de Cusa (1401–1464), who studied astronomy and mathematics at Heidelberg and in Italy and who later became the Bishop of Brixen in the Tyrol, wrote about a universe that was infinite in size and could therefore not have a center, and he advocated that the Earth and the heavenly bodies all were composed of the same elements. But others brought back the ideas of Aristotle and Ptolemy, both in their original forms as well as in versions modified by Islamic sources. This led to the printing of a number of very influential cosmological books by scholars from this region in the 15th Century.

3.8.1 Georg Peurbach

Georg Peurbach was born near Linz, Austria, on May 30, 1423. Although born Georg Aunpekh, he later took the name of his home town as his surname. He attended the University of Vienna, where he became interested in astronomy. After graduating, he began traveling throughout Europe in 1448, where he met with important astronomers such as the Italian Giovanni Bianchini and where he acquired an international reputation based on his astronomical lectures. Upon returning to Vienna in 1453, he was awarded a Master's Degree and subsequently began teaching at the University. He also enjoyed royal patronage, being appointed court astrologer to King Ladislas V of Hungary from 1454 to 1457, then to the Holy Roman Emperor, Frederick III.

During his lifetime, Peurbach reported on a number of astronomical events, such as planetary positions, comets (including Halley's comet in 1456), and eclipses. Based on discrepancies he found between the actual time of the lunar eclipse of September 3, 1457 and that predicted from the Alphonsine Tables, he subsequently produced his own tables of eclipse calculations. He also recognized the advantage of using sines instead of chords in calculating planetary positions (an idea he learned from Islamic sources), and he consequently computed his own table of sines. He also constructed celestial globes and made improvements to astronomical instruments.

In 1454, Peurbach completed a textbook entitled *Theoricae Planetarum Novae (New Theories of the Planets)*. Influenced by Islamic sources, this manuscript presented an integrated planetary theory that combined a basic exposition of the eccentrics and epicycles of Ptolemy with the concentric crystalline spheres of Aristotle. In its later printed form beginning around 1474, this influential book went through over 50 editions, and it was one of the most widely used textbooks on astronomy until well into the mid-1600s.

In 1460, Peurbach met with Cardinal Bessarion (ca. 1395–1472), a native of Trebizond in Byzantium who came to Italy and had a mission to promote classical Greek works in Europe (and who himself had a marvelous library of classical manuscripts). Bessarion came to Vienna to (unsuccessfully) generate support for a crusade to recapture Constantinople from the Turks, which they had taken in 1453. However, he also wanted to produce a shortened translation of the *Almagest* from the original Greek. Together with his student Regiomontanus, Peurbach began this translation, but he died in Vienna on April 8, 1461 before it could be completed.

3.8.2   Regiomontanus

Regiomontanus was born near Königsberg, Lower Franconia (now in Bavaria, Germany) on June 6, 1436. He was the son of a miller and was initially named Johann Mueller, but he later took the Latinized version of Königsberg for his name. He was a mathematical and astronomical prodigy, studying dialectics at the University of Leipzig from ages 11 to 14 and at the University of Vienna from ages 14 to 16, from where he received a baccalaureate degree. In 1457, when he reached the required minimum age of 21, he was awarded a Master's Degree and appointed to the faculty of the University of Vienna. Both as a student and as a faculty member, he collaborated with Georg Peurbach on a number of projects, including observing the heavens, creating more accurate celestial tables, and constructing astronomical instruments.

This collaboration also included the writing of a shortened translation of Ptolemy's great astronomical work at the behest of Cardinal Bessarion. Regiomontanus completed the manuscript of this *Epitome of the Almagest* after Peurbach died, and its printed version appeared in 1496. But this book was not just a simple translation of Ptolemy's work; it also included revised calculations and critical comments on the original. It attracted wide attention in Europe, including from a young student then at the University of Bologna named Copernicus.

From 1461 to 1465, Regiomontanus lived mainly in Rome at the household of Cardinal Bessarion, who became his patron. He continued to study Greek manuscripts, and he lectured widely in Italy and elsewhere. By 1467, he had been given an appointment as custodian of the Royal Library in Buda, Hungary, which contained manuscripts from Constantinople. While there, he wrote a major book on trigonometry and constructed two tables of sines.

In 1471 he moved to Nuremberg, where he built an observatory to continue his studies of the heavens and a workshop to make astronomical instruments, such as quadrants, astrolabes, and armillary spheres. Typical of astronomers of his day, he also was interested in astrology and made nativities and predictions for patrons. He set up his own printing press and began publishing a number of books on mathematics, geography, and astronomy. These included the first printed edition in 1472 of Peurbach's *Theoricae Planetarum Novae*; his own *Ephemerides* for the years 1474–1506, which included predictions of the location of the Moon in the heavens and had implications for determining longitude at sea; and in 1474 his *Calendarium*, which also included tables that were useful for navigation. In fact, his *Ephemerides* were later used by Christopher Columbus and Amerigo Vespucci on their explorations to the New World.

In 1475, Regiomontanus was summoned by the Pope to Rome to work on calendar reform. It was there that he died on July 6, 1476, probably from the plague but possibly from poison administered by relatives of a scholar he had insulted.

3.8.3   Hartmann Schedel

One of the major printed books of the late 1400s was written by Hartmann Schedel. Born in Nuremberg on February 13, 1440, he studied in Leipzig and Padua, where he received degrees in the humanities, law, and medicine. He re-settled back at Nuremberg around 1481, where he remained and practiced medicine until his death on November 28, 1514. He maintained a number of connections with scholars and artists throughout his life, as well as an extensive library that included books on astronomy, mathematics, medicine, religion, philosophy, and rhetoric.

Schedel's own scholarship culminated in his history of the world, entitled *Liber Chronicarum*, popularly referred to as the *Nuremberg Chronicle*. It was first published in Latin in Nuremberg in July 1493, and a German translation followed in December of the same year. The work was divided into six ages, going back to the biblical creation of the world. It was a progressive chronology, with subsequent paragraphs dealing with later periods of time. The content followed a medieval point of view that included legends, traditions, the Bible, and the occasional scientific fact. The book was quite popular, and a number of copies still survive.

The text was illustrated by over 1,800 wonderful woodcut images, many of which were repeated. Some of them probably were created by Dürer, who was an apprentice to Michael Wolgemut in one of the workshops that produced them. Notably, many

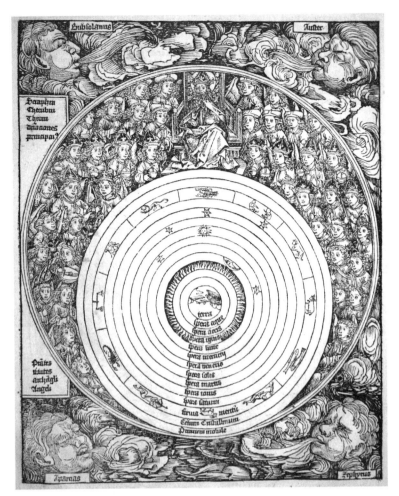

Figure 3.14. Woodcut illustration depicting the 7th day of Creation, from a page of the 1493 Latin edition of Schedel's *Nuremberg Chronicle*. Note the Aristotelian cosmological system that was used in the Middle Ages, below, with God and His retinue of angels looking down on His creation from above. Approximately 27.4 × 22.2 cm. Courtesy of the Collection of Owen Gingerich.

of the woodcuts were integrated in with the text rather than being set off in their own space. They depicted a number of things, including biblical scenes, images of popes and kings, and vast cityscapes, most of which were not truly accurate representations but more idealized indications related to the importance of the subject. Astronomically, there were cometary representations that accompanied the text describing the appearance of a notable comet and its usually disastrous influence on the people and animals of the time. In addition, there was a series of seven woodcuts that illustrated the seven days of Creation, which concluded with a spectacular image that showed the Aristotelian celestial spheres surrounded by the image of God and His heavenly host (Figure 3.14).

3.8.4   Peter Apian

Peter Apian (also known as Petrus Apianus or Peter Bennewitz) was born on April 16, 1495, in Leisnig, Saxony. He matriculated into the University of Leipzig and studied mathematics, astronomy, and cosmography. After graduating, he moved to Vienna. Around 1522, he published a world map (based on a 1520 work by Waldsemuller), followed by a geographical commentary on his map. In 1524 he published his famous *Cosmographia* (see below), which led to his being appointed Professor of Mathematics at the University of Ingolstadt in 1527. The same year he published a small practical book on arithmetic, which is notable because the title page contained the first printed appearance of Pascal's triangle. This was followed by other publications, including a mathematics book in 1534 that contained sine tables calculated for every minute of arc, and the first large-scale map of Europe, which no longer exists. Apian also was a talented instrument maker and made improvements to the designs of surveying quadrants and armillary spheres. In the 1530s, Apian came to the attention of Holy Roman Emperor Charles V, perhaps as a tutor, and he began to receive special privileges that included being knighted, being appointed court mathematician, and being given the power to grant higher academic degrees. Apian's fame and wealth continued to grow until his death in Ingolstadt on April 21, 1552. His son Phillip was subsequently appointed to fill his now-vacant position at the University.

His famous 1524 book *Cosmographia* was based on the work of Ptolemy. It dealt with a variety of topics, such as astronomy, geography, cartography, navigation, weather, the shape of the Earth, and mathematical instruments. It was lavishly illustrated and contained a number of maps of the continents, including America. Warner (1979) describes a figure in the book that contained two 5.3 cm diameter side-by-side images of the stars and constellations around the north equatorial pole. Using an external orientation, the left image showed Ursa Minor as a bear and Ursa Major as a four-starred wagon being pulled by three starred horses, and the right image showed both constellations as bears and the first European-labeled representation of the star Alcor in the tail of Ursa Major. A similar diagram was included in Apian's 1533 *Instrument Buch*.

An edited edition of the *Cosmographia* was published in 1529 by Gemma Frisius (1508–1555), a Dutch physician serving on the medical faculty at Louvain who was also active in cosmography and the production of maps, globes, and mathematical instruments. His patron was Emperor Charles V, and he tutored such notables as Gerard Mercator and John Dee. In 1533, Gemma published another edition of the *Cosmographia* that was enlarged and included two sections that he wrote (Figure 3.15). With this edition, the popularity of the book soared. It became widely distributed throughout Europe; was translated from Latin into Dutch, French, and Spanish; and generated over 40 subsequent editions printed in 7 different cities. Gemma's contributions to the *Cosmographia* over the years included a description

Figure 3.15. Page from the 1533 edition of Apian's *Cosmographia*. 12.4 cm dia. Note the volvelle (i.e., a printed calculating device—see Section 8.2), which here is similar in appearance to an astrolabe, with zodiac and month circles on the periphery, unequal hour lines in the center left, and a shadow square in the center right. There is a string attached to the center that serves as a pointer.

of the principle of triangulation in surveying and new designs for observational instruments, such as astronomical rings and cross staffs. His editions were in quarto size (21 × 15.5 cm) and added more paper volvelles (see Section 8.2) to the three that appeared in the first edition. He also published an abridged version in 1531 entitled *Cosmographiae Introductio* that was in octavo size (15.5 × 10.5 cm) and did not contain any volvelles.

In 1533, Apian published two books, *Horoscopion Apiani Generale* and *Instrument Buch*, that according to Warner contained a 19 cm diameter planisphere that showed the sky from the north equatorial pole down to the Tropic of Capricorn using a polar equidistant projection with geocentric orientation. Although 19 constellations were shown in traditional Ptolemaic form, three others were depicted according to Bedouin tradition (e.g., Draco was four camels rather than a serpent). Warner states that although Bedouin constellations were described in both the Islamic and European literature, this was the only known illustration of constellations based on this tradition (but see Figure 4.4).

In 1540, Apian published one of the most beautiful books ever printed, the *Astronomicum Caesareum*. It was dedicated to Emperor Charles V, who received it with much pleasure and gratitude. The book presented much of the same material found in the *Cosmographia* but in a more lavish and elegant manner. It was larger, in color, and contained some of the most elegant and complicated volvelles ever produced (see Figure 8.7). It also included some new material not found in the *Cosmographia*, such as the use of solar eclipses to determine longitude, and a description of five comets (including Halley's comet), with the observation that the cometary tail always points away from the Sun.

According to Warner, one unlabeled figure in the *Astronomicum* that was made into a volvelle had been previously published as an independent map in 1536 and was labeled *Imagines Syderum Coelestium ut Sun ...* It was a 29 cm diameter planisphere that extended from the north ecliptic pole down to about 60 degrees S lat. in polar equidistant projection with external orientation. Many of the 48 constellations in this map resembled those of Dürer, and the star and constellation names were set in stereotype. In addition, seven of the stars were labeled using Islamic names. This reflects the merging of Islamic and Ptolemaic traditions in many of the celestial maps of the time.

3.9 PARADIGM SHIFT: HELIOCENTRISM WITH CIRCULAR ORBITS

At the start of the 1500s, Aristotle and Ptolemy continued to dominate astronomy. For over 1,300 years, the Earth was seen at the center of the universe, with the Moon, Sun, and planets embedded in spheres that rotated so that heavenly bodies defined circular orbits around the Earth. These spheres were either real and made of some exotic crystalline or ethereal substance (Aristotle), or they were hypothetical and could mathematically be viewed as circular epicycles revolving in a geocentric or eccentric deferent (Ptolemy), or they were real and made up of concentric spheres in which space was provided for embedded epicycles and deferents (Islamic integrated model, as popularized in Europe by Peurbach and Regiomontanus—see Section 3.8). Although these various models were reasonably accurate in predicting celestial positions and saving the phenomena, they were not perfect, and anomalies resulted that required counter-spheres, additional epicycles, or novel approaches such as the equant. The problem was the need to strictly adhere to two principles that we now know are wrong: (1) the Earth is at the center of the universe, and (2) the heavenly bodies revolve in perfectly circular orbits. Each of these would require a paradigm shift to change, the first by Copernicus and the second by Kepler.

3.9.1 Nicholas Copernicus

Originally named Niclas Kopernik before his name was later Latinized, Copernicus was born on February 19, 1473 in the town of Torun, which at the time was part of Royal Prussia, an autonomous province of the Kingdom of Poland. His father was a merchant who had emigrated from Cracow, and he was the youngest of four children. His father died when Nicholas was 10 years old, and he and his siblings were taken under the wing of his maternal uncle, who was a church canon. In 1491, Copernicus entered the University of Cracow, and in 1497 he continued his studies at the University of Bologna. There he pursued canon law, Greek and the classics, mathematics, and astronomy under the famous astronomer Domenico Maria Novara. To provide financial support, his uncle (now Bishop of Warmia) obtained for Copernicus a position as a canon in the Cathedral at Frauenburg in 1497, a position he kept for the rest of his life even though his travels frequently took him away from the area. From 1501 to 1503, he took a leave of absence from his duties in Frauenburg to study medicine and jurisprudence at universities in Padua and Ferrara, obtaining a degree of Doctor of Canon Law from the latter. He subsequently practiced medicine at Heilsberg from 1506 to 1512. After his uncle died in 1512, Copernicus returned to Frauenburg, and for the next 11 years he held a series of administrative posts there and in Allenstein. On the strength of some financial work he had done and a treatise he wrote on monetary reforms, he was appointed as a deputy counselor involved with financial regulations in Royal Prussia from 1522 to 1529.

These various administrative and clerical positions served to support his growing fascination with astronomy. Throughout the early 1500s, he made his own observations of the Sun, Moon, and planets, and he began formulating his ideas about a heliocentric universe. By 1514, he had produced a short commentary, *Commentariolus*, which summarized his ideas. Although unpublished, he disseminated this manuscript to his friends and close colleagues. They urged him to publish a more complete and mathematical treatise, but being a perfectionist, and fearing ridicule, he delayed.

However, in 1539 a young mathematician named Georg Rheticus came from the University of Wittenberg to Frauenburg to study under Copernicus. Recognizing the genius of his master, Rheticus began spreading his ideas to others, and at their urging for more information Copernicus yielded and began preparing his various manuscripts for publication. Rheticus was given the task of editing the resulting book, and he took the completed version to Nuremberg for printing. This was proofread by the Lutheran theologian Andreas Osiander, who unbeknownst to Rheticus and Copernicus inserted a prologue stating that the heliocentric perspective was a hypothesis and not necessarily reality, which was not Copernicus' intent. The reason for this was probably to avoid censure by religious groups (mainly Protestant) who could accept scientific hypotheses but not stated realities that went against the geocentric Scriptures. Printed copies of this monumental work,

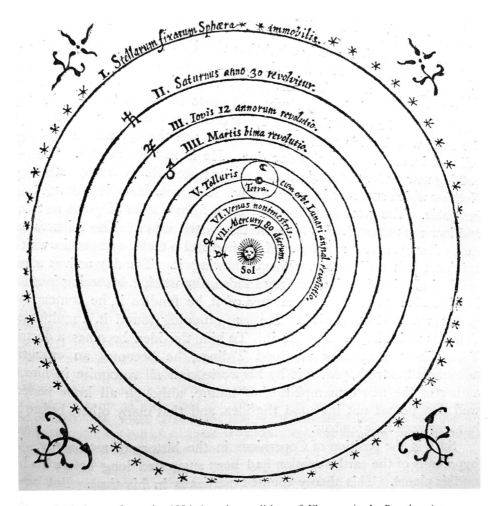

Figure 3.16. Image from the 1894 American edition of Flammarion's *Popular Astronomy*. 23.2 × 15.5 cm (page size). This image is a facsimile of the Copernican solar system taken from the 1617 third edition of his monumental book, *De Revolutionibis*.

entitled *De Revolutionibus Orbium Coelestium (On the Revolutions of the Heavenly Spheres)*, were presented to Copernicus in Frauenburg on his deathbed on May 24, 1543.

In *De Revolutionibus*, Copernicus presented a solar system in which the planets, including the Earth, revolved around the fixed Sun (Figure 3.16). In addition, the spherical Earth rotated on its axis once a day from west to east, accounting for the observed east-to-west movements of the Sun during daytime and the celestial vault during nighttime. He stated the correct order of the planets. He explained the precession of the equinoxes as being due to a slow change in the position of the Earth's rotational axis. He also explained the seasons as resulting from the tilt of the

Earth's axis with respect to the plane of its orbit (which also accounts for the angle of the ecliptic in the night sky). He defined three motions of the Earth: its rotation around its axis, its revolution around the Sun, and its "motion in declination" to explain the tendency of its axis to point in the same direction *vis-à-vis* the heavens during its yearly trip around the Sun (which in fact is not a motion but is a characteristic of a rotating sphere with gyroscopic action). In the mathematical sections, Copernicus replaced Ptolemy's equant with more epicycles. In fact, his system continued the Greek ideas of heavenly bodies moving in circular orbits and of using epicycles and deferents to account for their positions in the sky. Although the Sun was at the center of the planetary spheres, it was not at the exact center of the universe in his model.

Despite the intentions of Osiander, opposition was raised against the Copernican solar system by Protestant theologians, who thought it contradicted the Scriptures. Catholic opposition followed, and in 1616 *De Revolutionibus* was placed on the Index of Prohibited Books, and only "corrected" versions (with offensive passages deleted) were tolerated. Between this religious opposition and the opposition from conservative scholars favoring Aristotle and Ptolemy, it wasn't until well into the 17th Century that the ideas of Copernicus began to achieve wide acceptance. Even in the 1700s, the Copernican model was being contrasted with other models, including a hybrid proposed by Tycho Brahe that was more acceptable to religious groups. It is to this model that we now turn.

3.9.2   Tycho Brahe

Tyge (later Latinized as Tycho) Brahe was born on December 14, 1546 in Knudstrup, Skane (then part of Denmark and now Sweden). He was the eldest son of parents who both came from Danish noble families, although he was brought up by his childless paternal uncle and his wife. He studied Latin as a child and at age 12 began taking classes at the University of Copenhagen, which included law following the wishes of his uncle. On August 21, 1560, he observed an eclipse, which kindled his interest in astronomy. Even at this early age, he began to appreciate the value of making careful and systematic observations of the heavens, even before the invention of the telescope. He purchased or made other astronomical instruments (e.g., astrolabes, quadrants, sextants), books, and ephemerides to assist him with his observations. He continued his studies in law and other subjects at a number of universities, including Leipzig, Wittenberg, and Rostock. While at Rostock in 1566, he lost the bridge of his nose in a duel and wore a metal covering over the defect for the rest of his life.

He returned to Denmark at the end of 1570 to be with his ailing father, who died the next year. In 1572 he fell in love with a commoner and lived with her and their eight children in a common law marriage until his death. Also in 1572, he observed the nova in Cassiopeia and wrote a report about it in 1573, where he concluded that it must be far away since he could detect no change in its parallax with respect to the background stars from night to night. The next year, he gave a series of lectures in astronomy at the University of Copenhagen. He continued to observe and made a "grand tour" to Germany to visit prominent astronomers.

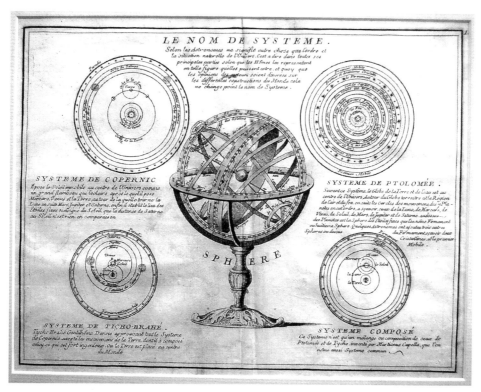

Figure 3.17. Print entitled "Le Nom de Systeme", from Le Rouge's *Atlas Nouveau Portatif à L'Usage des Militaires et du Voyageur*, ca. 1761. 19.4 × 26.1 cm. Note the central armillary sphere surrounded by cosmographic models taken from Copernicus, Ptolemy, Tycho Brahe, and Martianus Capella, which is labeled "Compose". Note that the models of Tycho Brahe and Capella are both hybrids of Copernicus and Ptolemy, allowing the Earth to remain in the center of the universe but having all of the other planets (Tycho Brahe) or just Mercury and Venus (Capella) revolve around the Sun. *See also* Color Plate Gallery.

On May 23, 1576, Danish King Frederick II made a decree to fund and support an observatory on the island of Hven, headed by Tycho (people in Scandinavia at the time were commonly referred to by their Christian names). Called Uraniborg, this became the finest observatory in Europe, and it was visited by many scholars and attracted many students. It included a main castle building with a library, printing press, alchemy furnace, observing rooms, and a satellite observatory (itself named Stjerneborg). The fame of this complex was due to Tycho's commitment to careful observational astronomy and the wonderful instruments that were there, which he designed, built, and calibrated himself. In fact, Tycho wrote a book entitled *Astronomiae Instauratae Mechanica* in 1598 that summarized his career and described in some detail the instruments at his observatory. More will be said about these instruments and this book in Section 8.3.5.1. Like Regiomontanus and other astronomers of the time, Tycho developed an interest in astrology, and he furnished his patrons (such as Frederick II) with nativities and other predictions. He was also interested in alchemy and performed metal experiments at his observatory.

In 1577, Tycho observed and reported upon a prominent comet. The absence of detectable parallax made him conclude that it was located in the heavens beyond the Moon, thus challenging Aristotle's notion that comets were sub-lunar phenomena and not part of the "perfect" and unchanging heavens (an idea also challenged by the nova of 1572). In addition, the movement of the comet through the heavens literally and figuratively shattered the notion that the planets were carried on real crystalline spheres in the heavens that were fitted adjacent to each other.

Tycho was aware of Copernicus and his heliocentric theory. However, he rejected this notion for his Aristotelian belief that the large heavy Earth must be immobile and at the center of the universe since it naturally attracted objects towards its center, and for his failure to observe parallax in nearby stars, which would be the case if the Earth orbited the Sun. (Such parallax actually occurs, but even Tycho's instruments were not sensitive enough to detect it.) However, he appreciated some advantages of having planets revolve around the Sun, such as providing a simpler explanation of retrograde planetary motion. He therefore proposed a model that was a hybrid between Ptolemy and Copernicus: the Earth remained in the center of the cosmos and was orbited by the Moon and Sun, but all of the other planets revolved around the Sun (Figure 3.17). The sphere containing the fixed stars was centered around the central Earth. This hybrid system became quite popular, since its geocentric focus satisfied religious groups and scholars who could not give up Aristotelian physics, yet who could appreciate some of the advantages of Copernicus' ideas. However, there was a flaw in the system, in that Mars was sometimes observed to penetrate through the Sun's orbital sphere (*Orbita Solis* in Figure 3.18). This observation, along with the observation of comets that seemed to go through the spheres of several planets as they moved toward and away from the Sun, demonstrated that these spheres were not solid, as Aristotle had surmised.

Due to progressive lack of support from the new King Christian IV, Tycho packed up his instruments in 1597 and left Denmark, finally settling in Prague in 1599, where he became the chief mathematical advisor to Emperor Rudolph II. He also set up an observatory nearby and continued his work. In 1600 he hired a new assistant named Johannes Kepler to calculate planetary orbits based on his observations, which not only were accurate and extensive, but also made corrections for atmospheric refraction. As we shall see, Kepler made good use of these observations (see Section 3.10.1).

Tycho died in Prague on October 24, 1601, probably due to uremia from the rupture of an extended bladder. A book in two volumes entitled *Astronomiae Instauratae Progymnasmata* and edited by Kepler was published posthumously the next year. It presented some of Tycho's observations, it included a catalog of 777 stars, and it summarized some of his astronomical ideas. It also contained two woodcut charts showing an external orientation of the constellation Cassiopeia and the nova of 1572 (according to Warner, one chart was a trapezoidal projection 11 cm high, 13 cm wide at the top, and 17.5 cm wide at the bottom; the other was rectangular, 18 cm high, and 12.2 cm wide).

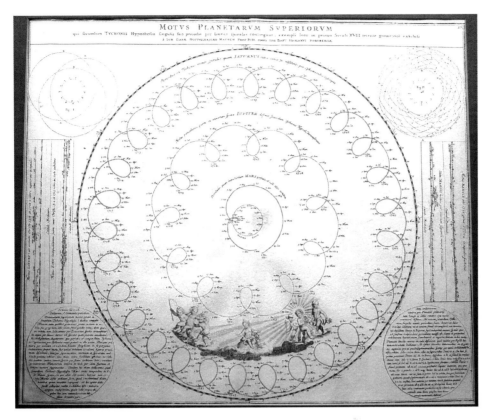

Figure 3.18. Print entitled "Motus Planetarum Superiorum Qui Secundum Tychonis Hypothesin ...", from the 1742 edition of Doppelmayr's *Atlas Coelestis*. 48 × 57.2 cm, 43.8 cm dia. hemisphere. Note the movement of the outer planets in the early 1700s according to the view of Tycho Brahe. The loops represent the retrograde actions of the planets as seen from the Earth. Note also that Mars sometimes passes through the sphere of the Sun's orbit (here labeled *Orbita Solis*) demonstrating that it is not made of solid crystalline material, which was proposed by Aristotle. *See also* Color Plate Gallery.

3.9.3 Galileo

Although it took Copernicanism over 100 years to be widely accepted, there were some prominent scientists and philosophers who embraced it earlier. One such person was the famous astronomer Galileo Galilei. He was born in Pisa, Italy on February 15, 1564, the eldest child of a musician. Although coming from nobility, the family was not wealthy. In the early 1570s, they moved to Florence. In 1581, Galileo enrolled at the University of Pisa to study medicine, but his interests soon turned to mathematics and the physical sciences and to explaining phenomena through experimentation. For example, he became fascinated with the pendular movement of a lamp in the Cathedral in Pisa, and this led to his discovery of the law of the isochronism of pendular motion; that is, that the period of large and small swings is the same. Although he never obtained a formal degree from Pisa, he was given a position in the Department of Mathematics in 1589. It was during this period that he allegedly dropped objects of different weights from the leaning tower of Pisa to see if they fell at the same rate. He later found that balls of different weights rolled down an inclined plane at the same speed. These experiments disproved the Aristotelian idea that heavy objects fall faster than lighter ones.

In 1592, Galileo accepted the prestigious position of Chairman of the Department of Mathematics at the University of Padua. During his many visits to nearby Venice, he became interested in nautical technology, and over the next two years he solved a problem involving the correct placement of oars in galleys and patented a model for a pump to raise water from aquifers. He also continued his experiments in the physical sciences. Like many other learned men of his time, he was able to read horoscopes and even constructed his own natal horoscope. Although he never married, he lived with his mistress, with whom he had three children.

Galileo had a strong interest in astronomy. He accepted the heliocentric ideas of Copernicus. He observed and lectured on the supernova that had appeared in the sky in 1604. His skill in working with mechanical devices led him to make his first telescope in 1609, which was modeled after a telescope produced in Holland but which was superior in terms of its magnification. In 1609 and 1610 he made a number of astonishing discoveries through his telescopes that argued against the Aristotelian/ Ptolemaic geocentric universe and supported Copernicus, such as the occurrence of hills and valleys on the Moon (disproving the Aristotelian notion that the heavenly bodies were smooth and made of unchanging ether); the presence of four moons revolving around Jupiter (suggesting that there were other bodies in the heavens than those seen by the ancients and providing an example that not all heavenly bodies revolved around the Earth); and the appearance of more distant stars with increasing aperture (supporting Copernicus' notion that the universe was larger than previously thought). With these observations, his fame escalated, and in 1610 he moved to Florence to work as chief mathematician and philosopher to the Grand Duke of Tuscany, Cosimo de Medici II. Also in 1610, he published his findings in a slim but revolutionary book entitled *Sidereus Nuncius (The Starry Messenger)*, which included drawings made at the telescope of the Moon, Jupiter's satellites (which he named "the Medicean stars" after his patron), and four star fields showing geocentric views of Orion, Praesepe, and the Pleiades. In subsequent years he made more discoveries that argued against Aristotle and supported Copernicus, such as the appearance of movable spots on the surface of the Sun and the pattern of phases observed in Venus, which could only occur if this planet orbited the Sun.

However, the heliocentric model of Copernicus was declared heretical by the Catholic Church in 1616, and Galileo was warned not to advocate it as truth since this would violate the Scriptures. However, he was under the impression that he could discuss this model as a mathematical theory, which he did in dialogue form in his 1632 book entitled *Dialogue Concerning the Two Chief World Systems*. But this book was not well received by the Vatican, partly because of its content and partly because it seemed to parody the Pope. Galileo was called to Rome in 1633 and tried for heresy by the Inquisition. He was allowed to return home to Florence, but he was placed under house arrest for the rest of his life. He died on January 8, 1642.

3.9.4 René Descartes

Another advocate of Copernicus was René Descartes, who was born at La Haye (now Descartes), France, near Tours, on March 31, 1596. His mother died of tuberculosis when he was one year old, and he was raised by his father, a judge in the High Court. Between 1606 and 1614, he studied at the Jesuit College at La Flèche. He then enrolled at the University of Poitiers, graduating with a degree in law in 1616. In 1618 he joined the army of Prince Maurice of Nassau, leader of the United Provinces of The Netherlands, in order to travel and gain worldly experiences. In the process, he met the Dutch mathematician Isaac Beeckman, who sparked his interest in mathematics and physics. On November 10, 1619, while serving with the army of Maximillian, Duke of Bavaria, in Germany, he had a dream whose interpretation indicated that he would found a new science based on mathematics, and this experience had a major influence on his life course.

After his military adventures, he returned to France in 1622, spending time in Paris but also traveling to Italy and other parts of Europe. Through his writings and debates, he began to gain a reputation for his philosophical ideas. From 1628 to 1649, he relocated to Holland due to its liberal intellectual climate, living in a number of locations within the country. By selling his property in France and making wise investments, he was financially secure and able to pursue his scientific and philo-sophical interests. His ideas ranged widely, from mathematics, science, and optics to philosophy, theology, and psychology, and the details of his accomplishments go beyond the scope of this book. Some of his more notable accomplishments include the development of the Cartesian grid, the use of equations to classify mathematical curves, his discussions concerning the mind–body dichotomy, the development of rationalistic methods in philosophy, and the phrase "I think, therefore I am." However, many of his freethinking ideas were considered radical by some religious and educational groups. For example, in 1643 Cartesian philosophy was condemned at the University of Utrecht, and in 1667 after his death some of his works were placed on the Index of Prohibited Books by the Pope. But his work was respected in other circles, and in 1647 he was awarded a pension by the King of France. In 1649 he was invited to Stockholm to tutor Queen Christina of Sweden. He subsequently contracted pneumonia and died there on February 11, 1650.

In astronomy, Descartes was a confirmed Copernican. From 1629 to 1633, he worked on a scientific treatise entitled *Le Monde (The World)*, which in part outlined his support for a heliocentric system. However, in 1633 he learned that Galileo was condemned by the Catholic Church for advocating the ideas of Copernicus, so Descartes decided to abandon his plans to publish this book. In 1644 he published a book entitled *Principia Philosophiae (The Principles of Philosophy)*, which was the most complete statement of his philosophy and views of the universe. In this book he outlines a theory of vortices, whereby the universe is entirely filled with matter in various states whirling around central stars. Bodies in motion continue to move in a

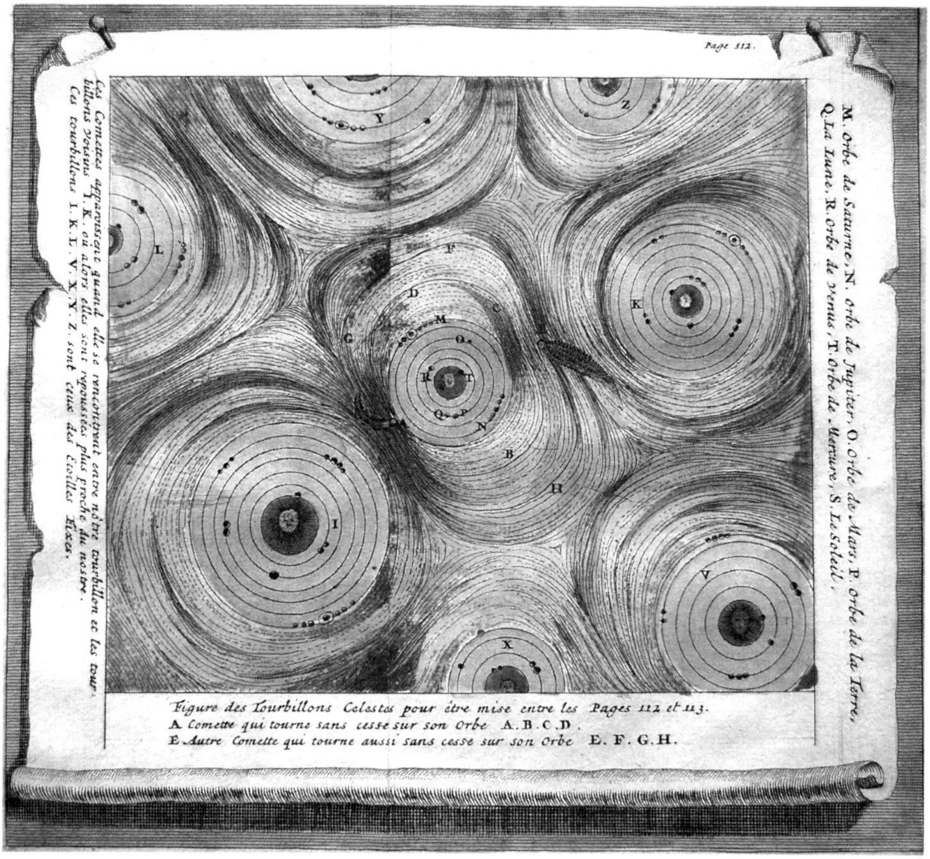

Figure 3.19. Copper engraving from Bion's *L'Usage des Globes Celestes et Terrestres*..., which was published in Amsterdam and bound into the 1700 edition of Nicolas Sanson's *Description de tout l'Univers en Plusieurs Cartes.* 17.7 × 19.8 cm. Note the depiction of how comets enter into orbits around a star according to the cosmological system of Descartes, which postulates swirling vortices surround the stars. *See also* Color Plate Gallery.

straight line unless they are interfered with by another body or unless they are caught up in a vortex. Thus, objects on Earth fall straight down unless deflected, whereas planets move around stars because they are caught up in the whirlpool of subtle matter surrounding the star. Comets move through space in a straight line at the border of two vortices, but when they enter the influence of one, they begin to curve with the whirlpool (Figure 3.19). This system was created by God but now continues to operate by itself for eternity. Although difficult to prove empirically, the prestige of Descartes ensured that these ideas would be popular in France and elsewhere for some 100 years, even after Newton disproved the vortex theory of planetary motion in his *Principia*.

3.10 PARADIGM SHIFT: HELIOCENTRISM WITH ELLIPTICAL ORBITS

3.10.1 Johannes Kepler

Johannes Kepler was born on December 27, 1571 in Weil-der-Stadt, which is now in southwestern Germany. His childhood was not very happy: his reckless father was often away as a soldier of fortune, his mother was erratic and later was tried for witchcraft, and he had smallpox at age three, leaving him with impaired eyesight and crippled hands. He lived with his grandparents for two years until 1576, when his parents moved to Leonberg and enrolled him in the Latin school the following year. He entered the Protestant Seminary at Adelberg in 1584. In 1589 he enrolled at the Protestant University of Tubingen, where he studied theology and graduated with a Master's Degree in 1591. He remained to do graduate work and was thereby exposed to the heliocentric system by an early advocate of Copernicus, Michael Maestlin, who was his teacher in mathematics. In 1594, Kepler was appointed as a Professor of Mathematics at the Protestant Seminary in Graz. To make extra money, he prepared an almanac with prognostications, and he made astrological nativities for patrons.

In 1596, the year before he was married for the first time, he published a book entitled *Mysterium Cosmographicum (The Cosmographic Mystery)*. In this important work, he argued that the distances of the six known planets from each other in Copernicus' system were in the same ratio as the distances resulting from imbedding the five regular solids within one another in the following order (from outside to inside): cube, tetrahedron, dodecahedron, icosahedron, and octahedron. For example, the sphere of Saturn was circumscribed on a cube, in which the sphere of Jupiter was inscribed; the sphere of Jupiter was circumscribed on a tetrahedron, in which the sphere of Mars was inscribed; and so on to Mercury. He hit upon this relationship a year earlier in a flash of insight while giving a lecture, and he thought that this association between geometry and cosmology was part of God's plan for a unified universe, which suited Kepler's predisposition toward mystical speculation. In fact, this relationship was simply a coincidence that broke down later as additional planets were discovered. Nevertheless, Kepler's reputation as a mathematician and astronomer grew as a result of this book, putting him in correspondence with two other prominent astronomers, Tycho Brahe and Galileo Galilei.

Tycho invited Kepler to Prague to help him calculate new orbits for the planets, and he served as Tycho's assistant from 1600 until the latter died in 1601. Kepler subsequently became Imperial Mathematician (and astrologer), a position he held until 1612 when Emperor Rudolph was deposed. During this time he wrote treatises on optics and the nova of 1604 in Serpentarius (which came out in 1606 and was illustrated with a map of the constellations in geocentric orientation measuring 17 cm high and 23.5 cm wide). In 1609 his *Astronomia Nova* was published, in which he presented his first two laws: that the planets revolve in elliptical orbits with the Sun as one of the foci, and that a planet sweeps out equal areas in equal times. He discovered these laws in the process of accounting for the observational data he inherited from Tycho, especially in trying to explain the position of Mars based on its having an assumed circular orbit, as the Greeks and Copernicus had maintained. But the data would not conform until he hit upon the idea of the ellipse as the best figure for the orbit of a planet (like Mars). Furthermore, he did not need to postulate eccentrics, epicycles, or equants, so long as the Sun was placed in one of the foci of the ellipse. Suddenly, the solar system became much simpler and more accurately reflected reality. This paradigm shift revolutionized astronomy. He also introduced a dynamic component, whereby he tried to explain planetary motion as being due to magnetic influences sent out by the rotating Sun, producing a circular movement, which then became elliptical through the Sun's push/pull interaction with each planet's own magnetic axis.

Kepler had been in written communication with Galileo, and when he read about the latter's discoveries, he borrowed a telescope and made his own observations of Jupiter's satellites and the heavens. Kepler wrote two treatises that provided support for the Italian astronomer's work. In 1611 he wrote a book entitled *Dioptrice* that discussed some of the theoretical aspects of how a telescope worked.

In 1612, the same year his wife died, Kepler accepted the position of district mathematician in Linz, a position he occupied until 1626. He remarried in 1613 after reviewing the qualifications of a number of candidates for his hand. While in Linz, he published a number of treatises and books on a variety of subjects, including comets and their condensation in space from the ether, the date of Jesus' birth, an epitome of the astronomy of Copernicus, a treatise on logarithms, and the Pythagorean-like relationship between musical harmonics and planetary distances and periods (again suggesting that he had made a discovery relating to God's master plan). In this latter work, entitled *Harmonices Mundi (Harmony of the World)* and published in 1619, he presented his third law: the squares of the periods of the planets are proportional to the cubes of their mean distances from the Sun. Kepler also constructed new planetary tables based on Tycho's careful observations and calculations made according to his new elliptical astronomy. These were published in 1627 as the *Rudolphine Tables*, named for his former patron, Emperor Rudolph. Included was a star catalog giving the ecliptic coordinates and magnitudes of over 1,440 stars, which Warner (1979) breaks down as including Tycho's group (but expanded to 1,005), 136 stars from 12 new southern constellations of Pietr Keyser (see Section 4.3.2.3), and additional Ptolemaic stars and stars in the Ptolemaic constellations added to by Dutch navigators.

Persecution from the Counter Reformation and the Thirty Years' War forced Kepler and his family to leave Linz in 1626. He tried unsuccessfully to obtain court appointments and to pry loose back salary that was owed him as Imperial Mathematician. In 1628 he entered into the services of the Duchy of Sagan, and he published some ephemerides that included predictions of future transits of Mercury and Venus. Following a long horseback ride, he became ill and died from a fever in Regensburg on November 15, 1630.

3.10.2 Follow-up to Kepler

The stage was now set for Newton (1642–1727) to explain how the planets moved in accordance with Kepler's laws. Although Kepler's own magnetic theory was rejected by most scientists of the time, Descartes' concept of vortices was more acceptable, and Newton initially considered it as an explanation of how the planets moved through space. However, he gradually developed his notion of universal gravitation, noting that the orbital motions of celestial objects (such as the planets and comets) and the kinds of free-fall forces on Earth that were studied by Galileo seemed to follow the same principles. He outlined his ideas in his monumental *Philosophiae Naturalis Principia Mathematica (Mathematical Principles of Natural Philosophy)*, which was published in 1687. In this book he not only disproved the existence of Descartes' vortices, but he showed how Kepler's three laws could be derived from his ideas concerning the gravitational attraction of physical objects (e.g., planets) at a distance.

The specifics of Newtonian physics and its development are beyond the scope of this book, since the formulae and arguments in the *Principia* and other writings by Newton did not generate images that became widely used in celestial prints. In addition, Newton's solar system generally looked like Kepler's, although the dynamical explanation behind it was quite different. In contrast, the invention of the telescope in the first decade of the 1600s begged for illustrations that showed the wonders of the universe and the new ability to position stars in the sky more accurately. It is to this mapping of the stars and the development of the constellations that we now turn.

3.11 BIBLIOGRAPHY

Azzolini, M. (2011) Refining the astrologer's art: Astrological diagrams in Bodleian MS. Canon. Misc. 24 and Cardano's *Libelli Quinque* (1547). *Journal for the History of Astronomy*, **42**, 1–25.

Bini, D. (ed.) (1977) *Astrologia: Art and Culture in the Renaissance*. Modena, Italy: Biblioteca Estense Universitaria di Modena.

Brown, B. (1932) *Astronomical Atlases, Maps and Charts*. London: Search Publishing.

Campion, N. (2000) Babylonian astrology: Its origin and legacy in Europe. In: H. Selin (ed.), *Astronomy across Cultures: The History of Non-Western Astronomy*. Dordrecht, The Netherlands: Kluwer Academic.

Cornish, A. (1993) Dante's moral cosmology. In: N.S. Hetherington (ed.), *Cosmology: Historical, Literary, Philosophical, Religious, and Scientific Perspectives*. New York: Garland Publishing.

Crowe, M.J. (2001) *The Theories of the World from Antiquity to the Copernican Revolution*, 2nd revised edn. Mineola, NY: Dover Publications.

Deakin, M.A.B. (2007) *Hypatia of Alexandria: Mathematician and Martyr*. New York: Prometheus Books.

Donahue, W.H. (1993) Kepler. In: N.S. Hetherington (ed.), *Cosmology: Historical, Literary, Philosophical, Religious, and Scientific Perspectives*. New York: Garland Publishing.

Drake, Stillman (1993) Galileo. In: N.S. Hetherington (ed.), *Cosmology: Historical, Literary, Philosophical, Religious, and Scientific Perspectives*. New York: Garland Publishing.

Dreyer, J.L.E. (1953) *A History of Astronomy from Thales to Kepler*, 2nd edn. New York: Dover Publications.

Dzielska, M. (1995) *Hypatia of Alexandria*. Cambridge, MA: Harvard University Press.

Edison, E. and Savage-Smith, E. (2004) *Medieval Views of the Cosmos*. Oxford, U.K.: Oxford University Press.

Evans, J. (1993) Ptolemy. In: N.S. Hetherington (ed.), *Cosmology: Historical, Literary, Philosophical, Religious, and Scientific Perspectives*. New York: Garland Publishing.

Evans, J. (1998) *The History and Practice of Ancient Astronomy*. Oxford, U.K.: Oxford University Press.

Ferguson, K. (2002) *Tycho and Kepler: The Unlikely Partnership that Forever Changed Our Understanding of the Heavens*. New York: Walker & Co.

Flammarion, C. (1894) *Popular Astronomy* (1st American edition, translated by J.E. Gore). New York: D. Appleton & Co.

Gingerich, O. (1992) *The Great Copernicus Chase and Other Adventures in Astronomical History*. Cambridge, MA: Sky Publishing.

Gingerich, O. (1993) *The Eye of Heaven: Ptolemy, Copernicus, Kepler*. New York: American Institute of Physics.

Gingerich, O. (2004) *The Book Nobody Read*. New York: Walker & Co.

Goddu, A. (2006) Reflections on the origin of Copernicus's cosmology. *Journal for the History of Astronomy*, **37**, 37–53.

Grant, Edward. (1993) Medieval cosmology. In: N.S. Hetherington (ed.), *Cosmology: Historical, Literary, Philosophical, Religious, and Scientific Perspectives*. New York: Garland Publishing.

Grendler, P.F. (1984) *Aldus Manutius: Humanist, Teacher, and Printer*. Providence, RI: The John Carter Brown Library.

Heath, T. (1981) *Aristarchus of Samos: The Ancient Copernicus*. New York: Dover Publications.

Heath, T.L. (1991) *Greek Astronomy*. New York: Dover Publications.

Heninger, S.K. Jr. (2004) *The Cosmographical Glass: Renaissance Diagrams of the Universe*. San Marino, CA: Huntington Library Press.

Hetherington, N.S. (1993) The presocratics. In: N.S. Hetherington (ed.), *Cosmology: Historical, Literary, Philosophical, Religious, and Scientific Perspectives*. New York: Garland Publishing.

Hetherington, N.S. (1993) Plato's cosmology. In: N.S. Hetherington (ed.), *Cosmology: Historical, Literary, Philosophical, Religious, and Scientific Perspectives*. New York: Garland Publishing.

Hetherington, N.S. (1993) Aristotle's cosmology. In: N.S. Hetherington (ed.), *Cosmology: Historical, Literary, Philosophical, Religious, and Scientific Perspectives*. New York: Garland Publishing.

Hoskin, M. (1997) *Cambridge Illustrated History of Astronomy*. Cambridge, U.K.: Cambridge University Press.

Hoskin, M. (2003) *The History of Astronomy: A Very Short Introduction*. New York: Oxford University Press.

Jardine, L. (1998) *Wordly Goods: A New History of the Renaissance*. New York: W.W. Norton & Co.

Jones, A. (1996) Later Greek and Byzantine astronomy. In: C. Walker (ed.), *Astronomy before the Telescope*. New York: St. Martin's Press.

Kanas, N. (2002) Mapping the solar system: Depictions from antiquarian star atlases. *Mercator's World*, **7**, 40–46.

Kanas, N. (2003) From Ptolemy to the Renaissance: How classical astronomy survived the Dark Ages. *Sky and Telescope*, January, 50–58.

Kanas, N. (2005) Are celestial maps really maps? *Journal of the International Map Collectors' Society*, **101**, 19–29.

Kanas, N. (2007) Sacrobosco's *De Sphaera*: Required reading for astronomy in early European universities. *Griffith Observer*, **71**(3), 2–14.

King, D.A. (1996) Islamic astronomy. In: C. Walker (ed.), *Astronomy before the Telescope*. New York: St. Martin's Press.

King, D.A. (2000) Mathematical astronomy in Islamic civilization. In: H. Selin (ed.), *Astronomy across Cultures: The History of Non-Western Astronomy*. Dordrecht, The Netherlands: Kluwer Academic.

Lee, D. (transl. and ed.) (1977) *Plato: Timaeus and Critias*. London: Penguin Classics.

Lerner, M-P. and Verdet, J-P. (1993) Copernicus. In: N.S. Hetherington (ed.), *Cosmology: Historical, Literary, Philosophical, Religious, and Scientific Perspectives*. New York: Garland Publishing.

Mair, A.W. and Mair, G.R. (transl.) (1989, [1955]) *Callimachus (Hymns and Epigrams), Lycophron, Aratus*. Cambridge, MA: Harvard University Press [Loeb Classical Library].

McCluskey, S.C. (1998) *Astronomies and Cultures in Early Medieval Europe*. Cambridge, U.K.: Cambridge University Press.

Neugebauer, O. (1969) *The Exact Sciences in Antiquity*, 2nd edn. New York: Dover Publications.

North, J. (1995) *The Norton History of Astronomy and Cosmology*. New York: W.W. Norton & Co.

Omodeo, P.D. (2011) David Origanus's planetary system (1599 and 1609). *Journal for the History of Astronomy*, **42**, 439–454.

Page, S. (2002) *Astrology in Medieval Manuscripts*. Toronto: University of Toronto Press.

Pannekoek, A. (1989) *A History of Astronomy*. New York: Dover Publications.

Paschos, E.A and Sotiroudis, P. (1998) *The Schemata of the Stars: Byzantine Astronomy from A.D. 1300*. Singapore: World Scientific Publishing.

Robbins, F.E. (transl. and ed.) (1994 [1940]) *Ptolemy: Tetrabiblos*. Cambridge, MA: Harvard University Press [Loeb Classical Library].

Saliba, G. (1994) *A History of Arabic Astronomy: Planetary Theories during the Golden Age of Islam*. New York: New York University Press.

Stolzenberg, D. (2001) *The Great Art of Knowing: The Baroque Encyclopedia of Athanasius Kircher*. Stanford, CA: Stanford University Libraries.

Swerdlow, N.M. (1996) Astronomy in the Renaissance. In: C. Walker (ed.), *Astronomy before the Telescope*. New York: St. Martin's Press.

Swerdlow, N.M. (2004) Galileo's horoscopes. *Journal for the History of Astronomy*, **35**, 135–141.

Swerdlow, N.M. (2004) The empirical foundations of Ptolemy's planetary theory. *Journal for the History of Astronomy*, **35**, 249–271.

Thorndike, L. (1949) *The "Sphere" of Sacrobosco and Its Commentators*. Chicago: University of Chicago Press.

Thurston, H. (1996) *Early Astronomy*. New York: Springer-Verlag.

Toomer, G.J. (1996) Ptolemy and his Greek predecessors. In: C. Walker (ed.), *Astronomy before the Telescope*. New York: St. Martin's Press.

Toomer, G.J. (transl. and ed.) (1998) *Ptolemy's Almagest*. Princeton, NJ: Princeton University Press.

Van Helden, A. (transl.) (1989) *Sidereus Nuncius or the Sidereal Messenger: Galileo Galilei*. Chicago: University of Chicago Press.

Warner, Deborah J. (1979) *The Sky Explored: Celestial Cartography 1500–1800.* Amsterdam: Theatrum Orbis Terrarum.

Westfall, R.S. (1993) Newtonian cosmology. In: N.S. Hetherington (ed.), *Cosmology: Historical, Literary, Philosophical, Religious, and Scientific Perspectives.* New York: Garland Publishing.

4

European constellation development

Previously, we considered the indigenous constellation systems that developed in China, Mesopotamia, Egypt, and India. Here, we will trace the development of European constellations, which form the basis of the star maps that will be discussed in subsequent chapters. Like many other things astronomical, we will begin with the ancient Greeks, whose classical 48 constellations greatly influenced star map development and form the core of the system we use today.

4.1 CLASSICAL GREEKS

The Greek constellation system was strongly influenced by the system that developed in Mesopotamia and then came to Greece, probably through Egypt. British astronomy writer Ian Ridpath (1988) has provided an argument that some of the Mesopotamian constellations may have been transported in the 2nd Millennium BC by the seafaring Minoans of Crete and accompanying islands, such as Thera (now Santorini). The argument is based on the fact that the nautical lore included in some constellation mythology, as well as the latitude of the sky represented, referred to the skies over Crete and the expansion of the Minoan empire into the eastern Aegean at that time.

By the 8th Century BC, Homer and Hesiod were mentioning star groups in their works that corresponded to our Ursa Major, Orion, and the Pleiades and Hyades star clusters, along with the stars Arcturus and Sirius. However, a complete set of Greek constellations was not described until the 4th Century BC by the astronomer Eudoxus in two works called the *Enoptron* and *Phaenomena*. He may have been introduced to these constellations during his travels in Egypt. In addition to his writings, it is also possible that he constructed a celestial globe that showed these star patterns and identified the Mesopotamian zodiac.

Based on his review of the appearances and locations of the constellations in the sky taken from the *Mul Apin* clay tablets and other sources, astronomy historian Bradley Schaefer (2004, 2006) has concluded that the bulk of the Mesopotamian constellations were probably developed between 1300 BC and 1100 BC by Assyrian observers in the northern part of the region. Thus, the complete set may not have arrived in Greece until sometime between the writings of Homer and Hesiod (who describe only a few constellations) in the 8th Century BC and the descriptions made by Eudoxus in the 4th Century BC. A likely time would be around the 5th Century BC, when it is thought that the zodiac also arrived in Greece. According to Schaefer, 20 of the 48 classical Greek constellations appear to be direct copies from the Mesopotamians, such as the Assyrian Goat-Fish (Capricornus) and Great Twins (Gemini). Another 10 have the same stars but different names, such as the Assyrian Hired Man (Aries) and Swallow (Pisces). The remaining 18 Greek constellations were home-grown. These include Hercules and the creatures he defeated (Leo and Draco), Ophiuchus (representing Asclepius, the God of Medicine) and the Serpent he carries, Delphinus the dolphin, and the constellations involved with Perseus' rescue of Andromeda (including Cepheus, Cassiopeia, and Cetus). Some of the constellations were altered by the Greeks to suit their interests. Schaefer gives the example of Triangulum, which in the *Mul Apin* is designated as a plow but which was converted to a triangle by the geometry-loving Greeks.

4.1.1   Aratus' *Phaenomena*

Although the works of Eudoxus are lost to us, his constellations have lived on in poetic form through the activities of Aratus, who was born around 315 BC at Soli in Cilicia (an area of southern Turkey). After studying in Athens, Aratus joined the court of King Antigonus of Macedonia, who requested that he produce a poetic version of Eudoxus' *Phaenomena* around 275 BC. In his version, also called the *Phaenomena*, he identified and described 47 constellations, which also included the Pleiades star cluster, the Water (which is now part of our Aquarius), and the star Procyon in Canis Minor. He also named 5 other stars: Arcturus, Aix (now called Capella), Sirius, Stachus (now called Spica), and Protrygeter (now called Vindemiatrix), the last of which was an important calendar star whose rising was associated with the start of the grape harvest. Aratus died around 245 BC.

According to classicist Theony Condos (1997), the *Phaenomena* consisted of some 1,150 verses whose main purpose was to describe the appearance and the organization of the constellations in the sky with reference to each other. Occasionally, the locations of specific stars in a constellation and the mythology behind its origins were presented. However, this was not the main focus of this poetic work.

The *Phaenomena* was very popular and was subsequently translated by the Romans into Latin, including versions by Cicero and Avienus. One of the more influential translations was by Germanicus Caesar (15 BC–19 AD), whose adaptation included more information about some of the constellations than even the original. This version formed the basis of the Leiden *Aratea*, which will be described in Section 5.1. Another Roman author influenced by Aratus was Marcus Manilius, who around

15 AD wrote a book called *Astronomica* that contained astrological material as well as information on constellation lore. Latin versions of Aratus' work continued to appear in manuscripts of the Middle Ages and early Renaissance, and some of these were accompanied by constellation illustrations that were the predecessors of later star maps. This work also stimulated long commentaries in both Greek and Latin.

4.1.2 Eratosthenes' *Catasterismi*

A second major Greek work on constellations that achieved popularity through the centuries is the *Catasterismi* (i.e., "constellations"), a compilation of myths explaining their origin that is attributed to Eratosthenes, the "beta" who was discussed in Section 3.1.8. The original work does not survive. What we do have is a summary of the original, and because the author of this work is not known, it is attributed to a "pseudo-Eratosthenes", who probably wrote it in the 1st or 2nd Century AD. Unlike Aratus' great work, the major focus of the *Catasterismi* was on constellation mythology, not simply poetic description. In addition, the location of the principal stars in each constellation is mentioned. In all, there are 42 stories dealing with the constellations, and two additional stories dealing with the Milky Way and the five planets known to the Greeks (Mercury, Venus, Mars, Jupiter, and Saturn). Like Aratus' *Phaenomena*, the *Catasterismi* was later translated into Latin and influenced a number of subsequent writers, such as Hyginus. More will be said about the *Catasterismi* and its constellations below in Section 4.1.5 on Ptolemy.

4.1.3 Hipparchus' star catalog

Besides his contributions to mathematical astronomy and the epicycle and eccentric hypotheses, Hipparchus (see Section 3.1.11) also compiled a catalog of at least 850 stars, which included a listing of stellar positions in the sky in terms of a coordinate system based on celestial latitude and longitude and centered along the ecliptic. Heath (1991) states that he was the first person to use such a system, at least in the West. Hipparchus also was thought to have created a celestial globe that showed a number of constellations, but this no longer exists.

Some writers such as Allen (1963) have suggested that Ptolemy later took over Hipparchus' system for his own celestial catalog (as well as for much of his cosmological system), although a careful comparison of the two catalogs made by astronomy historian Owen Gingerich (1992) casts doubt on this assertion. Nevertheless, Ptolemy was clearly influenced by the work of Hipparchus, as is indicated by the numerous references to the earlier astronomer that appear in the *Almagest* (for example, see Figure 3.2).

4.1.4 Geminos' *Introduction to the Phenomena*

Recently, a translation in English has appeared of a 1st-Century-BC textbook in astronomy that was written by a Greek scholar named Geminos. Entitled *Introduction to the Phenomena*, it provided a descriptive, non-mathematical view of Greek astronomy in the period of time between Hipparchus and Ptolemy. Included in the text were subjects related to the zodiac, the celestial sphere, lunar cycles, and eclipses. Also included was a listing of prominent stars and asterisms (without further information on their location or magnitude) and a list of the 12 constellations of the zodiac, 22 northern constellations, and 18 southern constellations, for a total of 52. According to James Evans and J. Lennart Berggren (2006), the translators, the four constellations mentioned by Geminos but not found in Ptolemy (see Section 4.1.5) were: the hair Lock of Berenike (our Coma Berenices); Thyrsus-lance (an ivy-wreathed branch with a pine cone at the top, which is part of Centaurus); Water (which is the water being poured out in Aquarius); and the Caduceus (the double-serpent wand carried by Hermes, whose location Geminos does not specify).

4.1.5   Ptolemy's star catalog

What we think of today as the classical Greek constellations were described by Ptolemy. His *Almagest* included a catalog of 1,022 stars that were arranged into 48 constellations, with a listing of each star's descriptive location in the constellation figure, its longitude in degrees, its latitude in degrees and direction north or south of the ecliptic, and its brightness or magnitude on a 1 (brightest) to 6 scale. Some of the more prominent stars also had separate individual names that were popularized by earlier writers.

The notion of describing the location of stars in terms of their positions in a constellation was a tradition followed by earlier Greek astronomers. It suggested that the constellation images were true pictorial representations in the skies that everyone knew about. So when a star was said to be "on the end of the tail" or "above the right knee", people were supposed to visualize the constellation image first, and then imagine where the star was in the heavens. This cumbersome method became problematic later on, when some writers pictured a constellation facing us, whereas others pictured it from behind. Thus, the right knee was on the left side of the constellation image in the one case as we looked up at the sky, whereas in the other case it was on the right side. It wasn't until later that astronomers almost exclusively identified stars according to locations on accurate coordinate systems in the sky and according to Greek or Latin letters and numbers based on stellar brightness.

Ptolemy's classic Greek constellations are listed in Table 4.1. Constellations 1–27 are in the northern celestial hemisphere and are listed in Book VII. Constellations 28–48 are in the southern celestial hemisphere and appear in Book VIII. The 12 zodiacal constellations are numbered 22–33. Most of these constellations are shown in the two celestial hemispheres in Figures 4.1 and 4.2 that are centered on the north and south equatorial poles, respectively. These are reproduced from Schaubach's 1795 book entitled *Eratosthenis' Catasterismi* (by "pseudo-Eratosthenes") and reflect the constellations described in this source. These hemispheres have been widely reproduced—for example, in Appendix 3 in Theony Condos' (1997) book and on the cover of Sir Thomas Heath's (1991) book.

Although the constellations listed in Table 4.1 and shown in Figures 4.1 and 4.2 are by and large the same, there are some differences. First, neither Equuleus nor

Table 4.1. Classical 48 Greek constellations from Ptolemy's catalog. The order and constellation names reflect modern usages and are taken from Toomer's (1998) translation of Ptolemy's *Almagest*. Constellations still in use today are shown in **bold**. The attributions for the constellations are taken from Ridpath (1988) and from Bakich (1995).

1. **Ursa Minor**: little bear or bear cub
2. **Ursa Major**: great bear
3. **Draco**: dragon
4. **Cepheus**: King of Ethiopia
5. **Bootes**: herdsman or bear driver
6. **Corona Borealis**: northern crown
7. **Hercules**: hero or kneeling man
8. **Lyra**: lyre or harp
9. **Cygnus**: swan
10. **Cassiopeia**: Queen of Ethiopia
11. **Perseus**: hero holding head of Medusa
12. **Auriga**: charioteer
13. **Ophiuchus**: serpent bearer
14. **Serpens**: serpent
15. **Sagitta**: arrow
16. **Aquila**: eagle (included the now-obsolete Antinous)
17. **Delphinus**: dolphin or porpoise
18. **Equuleus**: foal or small horse
19. **Pegasus**: winged horse
20. **Andromeda**: Princess of Ethiopia
21. **Triangulum**: triangle or letter *delta*
22. **Aries**: ram
23. **Taurus**: bull
24. **Gemini**: twins
25. **Cancer**: crab
26. **Leo**: lion
27. **Virgo**: virgin
28. **Libra**: scales or claws of Scorpius
29. **Scorpius**: scorpion
30. **Sagittarius**: archer
31. **Capricornus**: sea goat
32. **Aquarius**: water bearer
33. **Pisces**: fishes
34. **Cetus**: sea monster or whale
35. **Orion**: hunter
36. **Eridanus**: river
37. **Lepus**: hare
38. **Canis Major**: great dog
39. **Canis Minor** (Procyon): little dog
40. Argo: Jason's Argonaut ship—now split in three (see Section 4.3.2.5)
41. **Hydra**: water snake
42. **Crater**: cup
43. **Corvus**: crow
44. **Centaurus**: centaur
45. **Lupus** (Bestia): wolf or beast
46. **Ara**: altar
47. **Corona Australis**: southern crown
48. **Piscis Austrinus**: southern fish

Libra is shown in the figures. Eratosthenes did not mention the small horse, nor did he conceive of Pegasus as being winged, and so the figures depict a non-winged Pegasus that is labeled Equus. The stars that were considered to constitute Libra were originally viewed as the claws of Scorpius, but by the time of Ptolemy Libra had been liberated as its own constellation (the only one in the zodiac that was not an animate entity). Second, as with Geminos above, the hair of Berenike is included in Figure 4.1 as the constellation Coma Berenices, but it is not in Ptolemy's list as a separate constellation (although he mentioned some of its stars as a sub-division of Leo). Third, there are several synonymous name differences. For example, in Figure 4.1 Ptolemy's Triangulum is called Deltoton (reflecting its shape as the Greek letter "delta"); Bootes is labeled along with his alternate name Arctophylax (translated as "bear driver", reflecting his position behind the tail of the Great Bear); Canis Minor is also labeled with its alternate name Procyon (the name of its brightest star, which Ptolemy actually preferred); and Hercules is called Engonasin (translated as the

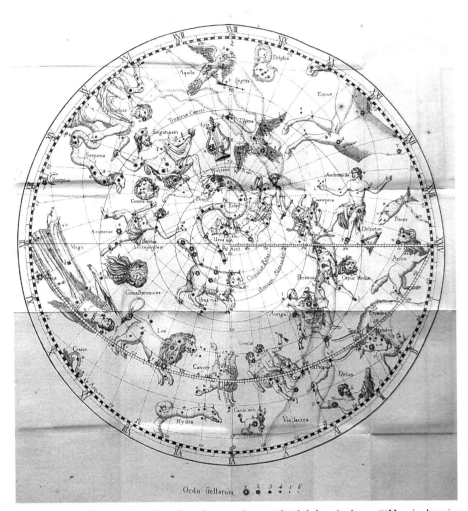

Figure 4.1. A pull-out plate showing the northern celestial hemisphere ("Hemisphaerium Boreale") centered on the north equatorial pole, from Schaubach's *Eratosthenis' Catasterismi*, 1795. 31.8 × 29.1 cm image, 24.6 cm dia. hemisphere. Note the classical Greek constellations according to this source. See text for details.

"kneeling one", as he was conceived before he became the hero Hercules). In Figure 4.2, Lupus appears as Bestia, the beast, which was the name Ptolemy preferred. Note the area around the south pole that is devoid of constellations. This area represents the part of the sky that was always below the horizon in the Greek world (see, for example, the circle representing the horizon of Alexandria). It would not be until the Age of Exploration that the southern sky began to be filled in with constellations (see Section 4.3.2). Finally, Ptolemy included with Aquila the constellation Antinous (the drowned boy lover of Hadrian), which appeared in later star atlases but is now obsolete (see Figure 4.8).

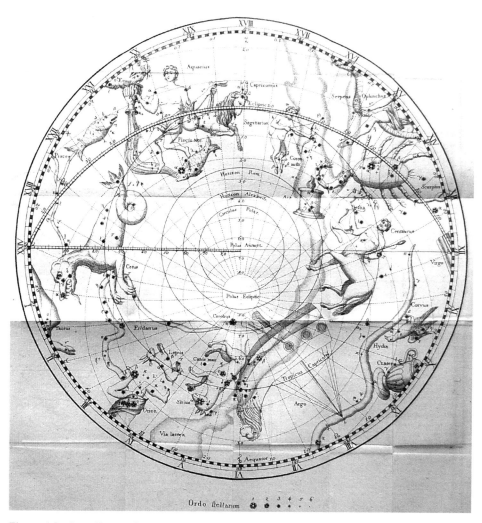

Figure 4.2. A pull-out plate showing the southern celestial hemisphere ("Hemisphaerium Australe") centered on the south equatorial pole, from Schaubach's *Eratosthenis' Catasterismi*, 1795. 31 × 28.9 cm image, 24.6 cm dia. hemisphere. Note the classical Greek constellations according to this source and the blank area around the pole that represents the sky below the horizons of Rome and Alexandria. See text for details.

The differences between the constellations described in the *Catasterismi* and the *Almagest* illustrate the point that some constellations had different mythologies, and some writers preferred one to another. In addition, since the constellations were subjectively determined, new ones were added or old ones were omitted or altered at the whim of the writer. We shall see later how in subsequent centuries some astronomers added a constellation to a star chart to please a patron or to celebrate an important event. Some of the constellations were mythologically linked together.

For example, Andromeda, the daughter of King Cepheus and Queen Cassiopeia of Ethiopia, was rescued from Cetus, the sea monster, by Perseus. To celebrate this myth, all of the principles found their place in the sky. The various mythologies of the constellations go beyond the scope of this book but are reviewed in Allen (1963) and Ridpath (1988).

4.1.6   Hyginus' *Poeticon Astronomicon*

Many of the Greek contributions to star mapping were later translated into Latin during the time of the Romans. One such writer was Hyginus, who assembled a collection of Greek myths about the stars entitled *Poeticon Astronomicon*. It used to be thought that this was the historian C. Julius Hyginus, who lived in the 1st Century AD. However, it is now thought that a later Hyginus wrote this work, possibly in the 2nd Century AD since the constellation order follows that in the *Almagest*.

The *Poeticon Astronomicon* was similar to the *Catasterismi* in focusing on the mythology of the constellations and the location of the principal stars within each constellation, although it contained some additional stories. It also was influenced by Aratus' *Phaenomena*. In the Middle Ages and Renaissance, there were many illustrated versions of the *Poeticon Astronomicon*, including the famous Ratdolt edition of 1482 with its splendid woodcuts, which were among the earliest printed depictions of the constellations (see Figures 4.3 and 5.3 and Section 5.2.1).

4.2   ISLAMIC CONSTELLATIONS

As we learned earlier, Ptolemy's *Syntaxis* was translated into the Arabic *Almagest* as Baghdad became the new center of astronomical learning. Although Ptolemy's conceptual star patterns and number of constellations did not change throughout the Islamic era, there were some changes in star naming and constellation design.

4.2.1   Arab star names

The Arabs introduced a number of names for stars that persist to this day. Many of these were simple translations of the Greek descriptive terms from Ptolemy's catalog. But others came from old Bedouin traditions, where stars that shone brightly over the desert skies were seen to represent animals or people. Different tribes might have different representations, but—as Allen (1963) has pointed out—they tended to be pastoral in nature, dealing with topics such as shepherds, horses, sheep, camels, birds, reptiles, tents, boats, wells, rivers, etc. Also, unlike the Greek imports, the old Arabic traditions did not group stars together into large constellations. Animate objects tended to be represented by just one star. For example, Ridpath (1988) points to the stars *Alpha* and *Beta Ophiuchi*, which were regarded a being a shepherd and his dog, respectively, herding a number of celestial sheep, represented by fainter

Figure 4.3. The constellation of Argo Navis, from Firmicus Maternus' *Scriptores Astronomici Veteres*, Aldine Press, 1499. 27.4 × 17.8 cm (page size). Note the simple but dynamic figure and the lack of relationship of the stars to their actual position in the sky. This image is nearly identical to that in Hyginus' *Poeticon Astronomicon*, except that it is left-to-right reversed.

neighboring stars. There were some exceptions, however, such as two three-star constellations that were in our Aquilla and Lyra. In contrast, inanimate objects were usually represented by more than one star, but again the number was not generally very high.

Allen has cited another class of names that were very specific to old Arabic traditions and pointed to characteristics of people. Examples include Al Saidak, Al Simak and Al Suha, which stood for the Trusted One, the Lofty One, and the Neglected One, respectively. However, such names were not used by later, more scientific Islamic scholars and were seen as interesting curiosities.

According to Allen, some examples of Arabic names for stars that persist today include Achernar (from Al Ahir al Nahr, or "End of the River") in Eridanus; Aldebaran (from Al Dabaran, "the Follower") in Taurus; Altair (from Al Nasr al Tair, or "Flying Eagle"), which was the same name that was given to their three-star version of our constellation of Aquilla; Fomalhaut (from Fum al Hut, or the "Fish's Mouth") in Piscis Australis; and Vega (Waki, which derived from Al Nasr al Waki, or the "Stone Eagle of the Desert") which was their three-star constellation of our Lyra.

4.2.2   Al-Sufi and his *Book of Fixed Stars*

Abd-al-Rahman al-Sufi (later called Azophi in the West) collected many of these names into an influential book. This famous Persian scholar and astronomer was born on December 7, 903 AD. He lived and worked at the court of the Emir Adud ad-Daula in Isfahan. He was very knowledgeable about Greek astronomical manuscripts, especially Ptolemy's *Almagest*. In fact, he made corrections to Ptolemy's catalog based on his own observations of the stars, particularly those involving stellar magnitudes and colors. He also tried to reconcile the Greek and traditional Arabic star names and constellations. He died on May 25, 986 AD.

Around 964 AD, al-Sufi published a manuscript entitled the *Book of Fixed Stars*, wherein he summarized much of his work. This important manuscript included both text and pictures of the stars and constellations. It also had a catalog of 1,018 stars that included their approximate positions, magnitudes, and colors. In the section describing the constellation of the Big Fish (our Andromeda), he referred to a small cloud located before the mouth, which was actually the first written description of the Andromeda Galaxy (M31). He also referred to the Large Magellanic Cloud, which he called Al Bakr (the White Ox), and to the Beehive Cluster (Praesepe or M44) in Cancer. He recorded other star clusters and nebulous objects as well.

One edition of the *Book of Fixed Stars* that has come down to us was produced around 1009 AD, after the death of al-Sufi (Figure 4.4). It was a beautiful manuscript that included constellations with sharp outlines and vivid colors. The figures were not dressed in classical Greek attire but in contemporary Arabic clothes and jewelry. Constellation stars were drawn in red; other stars were in black. The book contained

Figure 4.4. Two constellations from a manuscript copy of al-Sufi's *Book of Fixed Stars*, ca. 1009 AD. 29.7 × 17.8 cm (page size). Note the clothing contemporary with the times rather than classical Greece. On the left is Perseus holding the head of a male demon instead of Medusa. On the right is Andromeda with two fish in front of her, reflecting a Bedouin tradition. Courtesy of the Bodleian Library, University of Oxford (MS. Marsh 144 p. 111 & 167). *See also* Color Plate Gallery.

two illustrations of each constellation: one from the geocentric perspective and one from the external perspective.

4.3 CONSTELLATIONS AND THE AGE OF EXPLORATION

As Europe reasserted itself and regained its lands from Islamic control, the Greek constellation system of Ptolemy was retranslated from Arabic into Latin. However, many of the new Arabic star names were imported and kept their identity. But the number of constellations remained the same until navigational requirements during the Age of Exploration (1400 to 1800) made people realize that they needed more accurate charts of the heavens, particularly of the southern sky where large areas were devoid of constellations.

4.3.1 Navigating the oceans

In the 15th Century, after the land routes were blocked by the Ottoman takeover of Constantinople and Western Asia, the Portuguese began sailing south along the west coast of Africa with the intent of opening up trade routes to the Orient. Over time, they managed to round the Cape of Good Hope and proceed on to India and China. Other countries followed suit, particularly Spain, Holland, and England. In such a route, where one was hugging a north–south coastline, longitude was not as important as latitude in navigation, and since the time of the Greeks determining latitude at sea was an easy task. The method used was based on measuring the elevation above the horizon of the Sun at noon or of the Pole Star at night. This is because as one goes due south, the north polar star appears to fall proportionally (i.e., a 5-degree sail to the south leads to a 5-degree drop in elevation of the Pole Star as more of the southern sky becomes visible above the horizon). Hence, by using an instrument for measuring degrees, such as an astrolabe or sextant (see Section 8.3 for a description of instruments used in navigation), one can determine how many degrees latitude one has traveled with respect to the latitude of their home port.

But longitude was not so easy to determine. Its calculation depended on knowing the local time aboard ship (set each day at noon when the Sun was at its zenith) and comparing this with the time at the port from which the ship embarked. Since each one-hour difference accounted for 15 degrees of longitude, the observer's position on longitude could be determined. Ideally, one would have an accurate timepiece that could withstand the rigors of a sea voyage that could be set with other accurate timepieces at the time of embarkation. However, such a timepiece would not be developed until the mid-18th Century by the Englishman John Harrison (1693–1776). The story of this development has been wonderfully told by science writer Dava Sobel (1995).

But there were alternative approaches for determining longitude that depended upon the difference in time that a significant celestial event (e.g., the occultation of stars or planets by the Moon, eclipses of Jupiter's satellites) was observed at sea versus the time it was scheduled to occur at a place whose longitude was known.

Again, the longitude could be determined from this time difference. But this method depended on having celestial maps with accurate stellar positions. The most glaring problem was in the Southern Hemisphere, where the positions of the stars were not well charted and where large areas of the sky were devoid of known constellations. This situation needed to be remedied.

4.3.2   Filling gaps in the Southern Hemisphere

4.3.2.1   Amerigo Vespucci

Amerigo Vespucci (1454–1512) was an Italian navigator and successful businessman who made several voyages to the New World, mainly to the Caribbean and South America. Although other explorers like Alvise da Mosto and João Faras had made drawings of the southern sky (Dekker, 1990), Vespucci was the first European to measure the positions of many of the stars.. According to Warner (1979), after returning from his second voyage to the New World, Vespucci allegedly presented a catalog of these stars to King Manuel I of Portugal, which unfortunately has not survived.

However, Vespucci also wrote a brief description of important southern stars and sent this to his patron in Florence, Lorenzo de Medici. This has come down to us as his *Mundus Novus*, which was published in 1504. It was widely printed and circulated during the 16th Century. Warner gives four excerpts of this work from an English translation entitled "Of the Pole Antartike and the starres about the same", which appeared in 1555 in Richard Eden's *The Decades of the Newe World, or West India . . .* These excerpts are: (1) a text and diagram of four stars arranged in a square around the south equatorial pole, which probably belong to our current constellation Octans; (2) a text and diagram of three stars arranged in a right triangle, which probably belong to Chamaeleon; (3) a textual description of three stars that probably are in Apus or Triangulum Australe; and (4) a text and diagram of six stars, four in a line and two below, which probably are the Southern Cross and *Alpha* and *Beta Centauri*.

4.3.2.2   Andreas Corsali

Andreas Corsali was an Italian navigator who was active in the early 1500s. According to Warner, from 1515 to 1517 he sailed to China and the East Indies in an expedition sponsored by King Manuel I of Portugal. In 1516 he wrote a letter describing his observations of the southern sky labeled *Lettera di Andrea Corsali allo Illustrissimo Signore Duca Iuliano de Medici . . .* Included was a 7.5 × 4.5 cm map showing 18 stars and the two Magellanic Clouds in a geocentric orientation. There were no constellation figures, but there was a vivid and beautiful description of the Southern Cross in the text. The letter and map were published in Ramusio's 1550 *Navigationi et Viaggi* and in Eden's 1555 *The Decades of the Newe World or West India*, and a ca. 1530 pirated map is shown in Kanas (2008).

Although the ancient Greeks had known about the stars in the Southern Cross and included them as part of the hind legs of Centaurus, Corsali's was the first clear description of them as a separate constellation. According to astronomy writer Ian

Ridpath (1988), this constellation of Crux first appeared in its modern form and location on celestial globes by Petrus Plancius (see below) in 1598 and Jodocus Hondius (1563–1612) in 1600.

4.3.2.3 Petrus Plancius, Pieter Keyser, and Frederick De Houtman

The Dutch were also involved in charting the southern sky. One such person was Petrus Plancius (1552–1622). Originally a Flemish monk who later became a Calvinist theologian, he became interested in cartography and globe production. He consequently moved to Amsterdam in 1585, which by then had become a major center of these activities, and Plancius' reputation grew.

Plancius also was a promoter of Dutch naval expeditions and became involved with the Dutch East India Company. Before the departure of the first Dutch trading expedition to the East Indies in 1595, Plancius encouraged one of the navigators and chief pilots, Pieter Dirkszoon Keyser (1540–1596), also known as Petrus Theodorus, to chart the southern skies. He even gave him instruments with which to do this. As a result, Keyser made a catalog of 135 southern stars. Although Keyser died during the trip, his catalog was given to Plancius when the fleet returned in 1597, and from these observations 11 new constellations were included on a globe that Plancius made the following year (see Section 8.1.3), along with the correct positions for Crux and Triangulum Australe (van Gent, 2006). The constellations were also included in a globe made by Hondius in 1600, in the famous atlas produced by Bayer in 1603, and in Kepler's 1627 *Rudolphine Tables*.

During the expedition, Keyser was assisted by a crewmember, Frederick De Houtman (1540–1627). During another sailing expedition in 1598, De Houtman was imprisoned in Sumatra, during which time he made his own astronomical observations. After returning to Holland, he published his own celestial catalog in 1603, which according to Ridpath included the new constellations and stars and added another 168 stars, for a total of 303 measured southern star positions. Willem Janszoon Blaeu used this catalog for his celestial globes. Today, Keyser, Plancius, and De Houtman are given joint credit for the constellations, which are still recognized and are listed in Table 4.2.

The constellation of Musca has an interesting story. According to Ridpath (1988), it was originally called Apis, the bee, and was placed near the Southern Cross. When Plancius introduced another bee in the north (spelled Apes), the southern version became the southern fly, or Musca Australis, in contrast to the northern fly Musca Borealis (which had its own transition from Apes—see below). But, when the latter constellation was dropped, the southern fly simply became Musca, the name it retains today. However, one sometimes runs into Apis on old star charts (e.g., Bode's *Uranographia*).

The new constellations of Keyser, Plancius, and De Houtman are shown but labeled in French in the hemisphere of the southern sky in Figure 4.5, which is from the 1795 edition of Fortin's celestial atlas.

Table 4.2. Constellations from the work of Keyser, Plancius, and De Houtman. The constellation names reflect modern usages. Constellations still in use today are shown in **bold**. The attributions for the constellations are taken from Ridpath (1988) and from Bakich (1995).

1. **Apus**: bird of paradise

2. **Chamaeleon**: chameleon
3. **Dorado**: goldfish or swordfish
4. **Grus**: crane
5. **Hydrus**: little water snake
6. **Indus**: Native American indian
7. **Pavo**: peacock

8. **Musca**: fly. Originally it was called Apis (bee), then it was called Musca Australis (southern fly) when there was a Musca Borealis (northern fly). Now it is simply called Musca.
9. **Phoenix**: mythical bird reborn from ashes
10. **Triangulum Australe**: southern triangle
11. **Tucana**: toucan
12. **Volans**: flying fish

Figure 4.5. The southern celestial hemisphere ("Hemisphere Austral") centered on the south equatorial pole, from the 1795 edition of Fortin's *Atlas Céleste de Flamsteed*. 17.5×22.4 cm image, 16.9 cm dia. hemisphere. Note the additional constellations from those in Figure 4.2, but still the presence of a blank area below and to the left of the pole. See text for details. *See also* Color Plate Gallery.

Table 4.3. Constellations introduced by Petrus Plancius. The constellation names reflect modern usages. Constellations still in use today are shown in **bold**; obsolete constellations are not. The attributions for the constellations are taken from Ridpath (1988) and from Bakich (1995).

1. **Camelopardalis**: giraffe	8. Musca Borealis: northern fly. Plancius
2. **Columba**: dove	called it Apes (bee), Bartsch called it
3. Gallus: cockerel (near Puppis)	Vespa (wasp), Pardies called it Lilium
4. Jordanus: Jordan River (near Ursa Major)	(fleur-de-lis), Hevelius called it Musca
5. **Monoceros**: unicorn	(fly) (near Aries)
6. Polophylax: guardian of the south pole	9. Sagitta Australe: arrow (near Aquila)
7. Cancer Minor: small crab	10. Tigris: Tigris River (near Cygnus)

Plancius invented ten constellations of his own. Three of these (Camelopardalis, Columba, and Monoceros) still exist and are listed in bold in Table 4.3. They are also shown in Figure 4.5 (Columba and the body of Monoceros as "la Licorne") and in Figure 4.6 (the head of Monoceros and the giraffe Camelopardalis). According to Ridpath, Columba was formed in 1592 from some stars that Ptolemy had listed as being outside of Canis Major. Bakich states that Polophylax, Cancer Minor, and Sagitta Australe were formed around 1614. Plancius' other six constellations were introduced on a globe that he completed in 1613. Many of Plancius' constellations were included in the 1624 celestial maps of the German astronomer Jacob Bartsch (1600–1633)—see Appendix B.

Plancius had originally called Musca Borealis Apes, the bee, and had placed it near Aries. However, Bartsch changed the name to Vespa, the wasp, in 1624, and the French cartographer Ignace-Gaston Pardies (1636–1673) patriotically called these same stars Lilium, the fleur-de-lis, in a map that was published posthumously in 1674. To confuse matters, the great Polish astronomer, Johannes Hevelius (1611–1687), renamed it Musca, the fly. Since there was another Musca in the south, some people referred to this constellation as Musca Borealis, or the northern fly, before it was finally dropped.

4.3.2.4 Edmond Halley

Edmond (or Edmund) Halley was born near London on October 29, 1656 (according to the Julian calendar being used in England at the time). His father was a prosperous soap maker, purveyor of salt, and landowner who gave Edmond a telescope and financially supported his interests in astronomy. He attended the prestigious St. Paul's School, where he developed an interest in astronomy. Upon graduation, Halley entered Queens' College, Oxford in 1673. He began corresponding with John Flamsteed, the first English Astronomer Royal, and assisted him in observing two lunar eclipses. In 1676, Halley published a paper that extended Kepler's ideas concerning the elliptical orbits of planets.

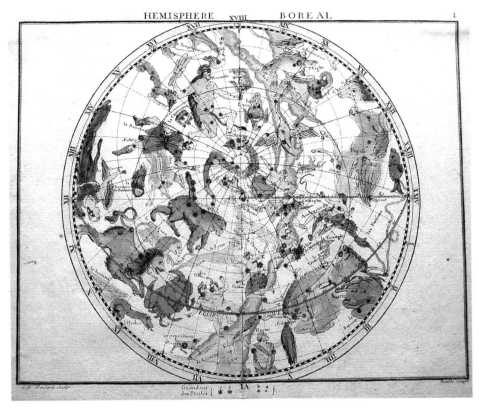

Figure 4.6. The northern celestial hemisphere ("Hemisphere Boreal") centered on the north equatorial pole, from the 1795 edition of Fortin's *Atlas Céleste de Flamsteed*. 17.5 × 22.4 cm image, 16.9 cm dia. hemisphere. Note the additional constellations to those in Figure 4.1. See text for details. *See also* Color Plate Gallery.

Also in 1676, Halley left Oxford before receiving a degree with financial support from his father to undertake a voyage to the island of St. Helena in order to chart the stars in the Southern Hemisphere. According to geography professor Norman Thrower (1981), he began his observations shortly after arriving the following year using a large sextant with a telescope attached, a quadrant, a pendulum clock, a number of freestanding telescopes, and two micrometers. While there, he recorded the longitudes and latitudes of 341 southern stars and observed several eclipses and a transit of Mercury across the Sun. His measurements were more accurate than those made by Keyser and de Houtman. After he returned home in 1678, he produced a celestial catalog that contained the first telescopically derived locations of southern stars, along with an accompanying celestial hemisphere (see below). As a result of this work, he was elected Fellow of the Royal Society in 1678 at the young age of 22. Halley was married in 1682 to Mary Tooke, the daughter of an officer at the Exchequer, and the couple subsequently had three children.

As a result of his gregarious nature, his growing fame, and his activities with the Royal Society, Halley came into contact with a number of prominent scientists during his professional life, including Johannes Hevelius, Jean Dominique Cassini, and Isaac Newton. In fact, Halley edited and financially supported the publication of Newton's *Principia* in 1687. But for personal and professional reasons, he had a falling-out with the more reserved Flamsteed, which probably contributed to his failure to be appointed as Professor of Astronomy at Oxford in 1691. Nevertheless, he continued with his work in astronomy. Citing the work of Jeremiah Horrocks, Halley published several papers of his own on using observations made during transits of Venus to measure the distance between the Earth and the Sun. In 1704 he was appointed Savilian Professor of Geometry at Oxford, although he remained active in astronomy. After Flamsteed died, Halley was named Astronomer Royal in 1720, a position he held until his own death on January 14, 1742. Among his astronomical feats, he is credited with discovering two globular clusters: Omega Centauri and the Hercules Cluster (M13). He also did very important work on comets, which is discussed in Section 8.5.4.2.

Halley pursued other interests as well, such as optics, the Earth's magnetism and compass variations, oceanic winds, tidal action, and barometric pressure. He also designed a diving bell and a diving helmet. From 1698 to 1701, he took command of HMS *Paramore* to test his geomagnetic theories and investigate tidal phenomena. As a result, he produced the first published isogonic map of the Earth's magnetic declinations and an important tidal chart of the English Channel. The story of these voyages is beyond the scope of this book but has been well-documented by Thrower (1981).

Halley produced two celestial hemispheres, which according to Warner (1979) were 46.6 cm in diameter each and were centered on the north and south ecliptic poles using a polar stereographic projection and geocentric orientation. The northern map was made around 1678 and was labeled *The Right Ascensions and Declinations of the Principal Fixed Stars ...* It was dedicated to Jonas Moore, who had arranged for government support of Halley's St. Helena venture. With the addition of the star Cor Caroli (the brightest star in Canes Venatici, named by Sir Charles Scarborough to honor the deposed Charles I), and some minor modifications, this map was later reissued by John Seller.

The southern map was produced in 1678 and was labeled *Carolo II D. G. Mag. Brit. ...* It was dedicated to Charles II and introduced the constellation Robur Carolinum, which depicted the oak tree in which the monarch had hidden from Cromwell's soldiers in 1651 (this constellation no longer exists but is shown in Figure 4.5 just below the centaur). The map was endorsed by the Royal Society and originally was published in Halley's 1679 *Catalogus Stellarum Australium sive Supplementum Catalogi Tychonici*. Warner states that it was copied and reissued by Royer in Paris, Lamb in London, and bound in Seller's *Atlas Maritimus*. It also was issued in a version that included Via Lactea. Halley's work on the southern skies was incorporated into the maps of later cartographers, such as Hevelius and Zahn (see Section 6.3).

4.3.2.5   Nicolas de Lacaille

Nicolas Louis de Lacaille was born in 1713 at Rumigny, France. As a student, he pursued mathematics and theology, ultimately acquiring the title of Abbé. He was also a pupil of Cassini at the Paris Observatory, and in 1740 he was appointed Professor of Astronomy at what is now called the Collège de France. During his lifetime, he wrote several books on astronomy and published several papers in the prestigious *Mémoires, Académie Royale des Sciences*. He died in 1762.

Backed by the *Académie Royale des Sciences* and the French East India Company, Lacaille left on a voyage to South Africa on November 21, 1750 in order to participate in a program of determining the solar and lunar parallaxes and to telescopically fix the positions of the stars in the southern sky. He arrived in Cape Town in April 1751, where he set up his observatory near Table Mountain (which he commemorated in a new constellation, Mensa). In one year's time—from 1751 to 1752—he observed and recorded the positions of some 9,800 stars and created a number of new constellations, all of which still exist today.

After returning to France, Lacaille presented his work to the *Académie Royale des Sciences* on November 15, 1754. According to Warner (1979), his talk was accompanied by a large painted celestial planisphere measuring 192 cm in diameter and showing the stars from the south equatorial pole to the Tropic of Capricorn using a polar stereographic projection and geocentric orientation. This map now hangs in the Paris Observatory.

In keeping with the spirit of the Enlightenment, Lacaille did not use mythological themes for his new constellations but instead used tools and instruments from the arts and sciences. A list of these constellations is shown in Table 4.4. All but three resulted from his voyage to Cape Town. The exceptions were Carina, Puppis, and Vela, which were invented when he broke up the large constellation Argo (a.k.a. Argo Navis, meaning the "ship Argo") into more manageable components. These three constellations were introduced in his star catalog of 1763, which was entitled *Coelum Australe Stelliferum*. This catalog also eliminated Halley's constellation Robur Carolinum.

Table 4.4. Constellations introduced by Nicolas de Lacaille. The constellation names reflect modern usages. Constellations still in use today are shown in **bold**. The attributions for the constellations are taken from Ridpath (1988) and from Bakich (1995).

1. **Antila**: air pump	10. **Octans**: octant
2. **Caelum**: chisel	11. **Pictor**: painter's easel
3. **Carina**: keel of Argo Navis	12. **Puppis**: stern of Argo Navis
4. **Circinus**: pair of compasses	13. **Pyxis**: compass
5. **Fornax**: furnace	14. **Reticulum**: net
6. **Horologium**: pendulum clock	15. **Sculptor**: sculptor
7. **Mensa**: Table Mountain near Cape Town	16. **Telescopium**: telescope
8. **Microscopium**: microscope	17. **Vela**: sails of Argo Navis
9. **Norma**: level	

Lacaille published his Cape Town constellations in a star map that appeared in the *Mémoires, Académie Royale des Sciences* in 1756. According to Warner, it was entitled *Planisphère contenant les Constellations Célestes ...*, measured 19 cm in diameter, and showed the stars and constellations from the south equatorial pole to the Tropic of Capricorn in a polar stereographic projection with geocentric orientation. It included not only his new constellations, but also those from Keyser and De Houtman. A slightly smaller 18.6 cm diameter version also appeared in Lacaille's 1763 book *Coelum Australe Stelliferum*.

Lacaille's chart was reproduced in Fortin's 1776 *Atlas Céleste de Flamsteed ...* and is shown in Figure 4.7. It is interesting to compare this plate with Fortin's own, which is shown in Figure 4.5. Note the filling in of constellations in the Lacaille plate around the pole.

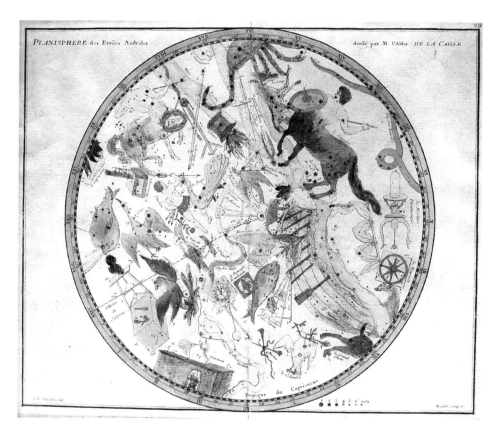

Figure 4.7. Lacaille's famous map of the south celestial polar region, from the 1776 edition of Fortin's *Atlas Céleste de Flamsteed.* 19.4 × 23.9 cm image, 18.4 cm dia. hemisphere. Note the new constellations invented by Lacaille that picture scientific instruments and tools of the period. In comparison with Figure 4.5, the whole area has now been filled in with constellations. See text for details. *See also* Color Plate Gallery.

4.3.2.6   John Herschel

John Herschel (1792–1871) was the son of William Herschel, the discoverer of Uranus (see Section 8.5.1). John achieved his own fame in observational astronomy. Using his father's 20-foot reflector, he completed a telescopic survey of the southern sky at the Cape of Good Hope. In 1847 he published his work as the *Results of Astronomical Observations Made during the Years 1834–8 at the Cape of Good Hope*, which added some 2,100 double stars and over 1,700 deep-sky objects to what were known before. In addition, he published his *Catalogue of Nebulae and Clusters of Stars* in the 1864 *Philosophical Transactions*. He also advanced the theory of the nebulae and made a map of the Orion Nebula (Ashbrook, 1984c).

4.3.3   Filling gaps in the Northern Hemisphere

4.3.3.1   Caspar Vopel

Compared with the southern celestial hemisphere, the stars of the northern hemisphere had been observed for a long time, and many constellations had been identified. However, there were still a few gaps that could be filled by enterprising individuals. One such person was Caspar Vopel (1511–1561), a German cartographer and instrument-maker who taught mathematics at the Gymnasium in Cologne. He was also a globe-maker, and in 1536 he produced a celestial globe that added two new constellations to Ptolemy's group: Antinous and Coma Berenices. Both of these constellations appeared in Mercator's 1551 globe (see Section 8.1.3) and were also mentioned in Tycho Brahe's star catalog of 1602. See also Appendix B.

Antinous was the boy lover of Roman Emperor Hadrian, whose tragic death in a drowning in 130 AD devastated the Emperor and led to his believing that the boy's soul entered a star in the heavens. Ptolemy actually commemorated Antinous as a sub-division of Aquila, but he did not devote a separate constellation to him. For a time, Antinous was represented as being carried in the claws of Aquila by later cartographers (Figure 4.8). Antinous is now obsolete. As we have seen, the other of Vopel's constellations, Coma Berenices, was included as a separate constellation by some Greek astronomers but not by Ptolemy. Nevertheless, it remained popular and still exists today (see Figure 6.6).

4.3.3.2   Johannes Hevelius

Other gaps in the Northern Hemisphere were filled in by Johannes Hevelius (1611–1687), the famous Polish astronomer whose work we will discuss in more detail in Sections 6.3 and 8.5.2. In 1690, after his death, his wife and collaborator Elisabeth published his great star catalog and atlas (which was completed in 1687), which introduced eleven new constellations. These are listed in Table 4.5. Hevelius originally conceived of Vulpecula, the fox, as having a goose, called Anser, in its jaws, and this is sometimes depicted on old star charts (Figure 1.1). Scutum was actually introduced in 1684 as Scutum Sobiescianum, or Sobieski's shield, to honor King John III Sobieski of Poland, who in 1683 had defeated the Ottoman Turks in their

Table 4.5. Constellations introduced by Johannes Hevelius. The constellation names reflect modern usages. Constellations still in use today are shown in **bold**; obsolete constellations are not. The attributions for the constellations are taken from Ridpath (1988) and from Bakich (1995).

1. **Canes Venatici**: hunting dogs
2. Cerberus: 3-headed monster (in Hercules)
3. **Lacerta**: lizard
4. **Leo Minor**: little lion
5. **Lynx**: lynx
6. Mons Maenalus: Mt. Maenalus in Greece (in Bootes)
7. Musca: fly (for history, see Musca Borealis in Table 4.3 and in accompanying text)
8. **Scutum**: shield
9. **Sextans**: sextant
10. Triangulum minor: little triangle (near Triangulum)
11. **Vulpecula**: fox, sometimes carrying Anser, the goose, in its jaws

attempt to take over Vienna (Figure 4.8). Ridpath (1988) points out that this is the only constellation introduced to honor a patron for political reasons that is still recognized today.

Figure 4.8. "Fig. Q." constellations from Hevelius' *Firmamentum Sobiescianum, sive Urano-graphia*, first published in 1687. 29.2 × 36.3 cm. Note on the left Scutum Sobiescianum (Sobieski's shield) and above it a prominent cartouche honoring King John III Sobieski of Poland for defeating the Ottoman Turks at Vienna in 1683. Note on the right the now obsolete constellation of Antinous soaring in the sky with Aquila the eagle, and near the bottom the horizontal line of the ecliptic.

4.4   OBSOLETE CONSTELLATIONS

The notion of creating constellations to please patrons, to honor national symbols, or to commemorate events started to become the norm and led to what Ridpath has called "constellation mania". Since Lacaille's splitting up of Argo Navis in 1763, none of these newly created constellations has survived to the present day, although they often appeared for years in early star maps. For this reason, it is useful to summarize some of these (see also the obsolete constellations discussed above and listed in the preceding tables). More information on most of the constellation creators and their maps will be given in future chapters and in Appendix B.

In 1627 an atlas was published posthumously by Julius Schiller (d. 1627) entitled *Coelum Stellatum Christianum* ... In this atlas, all of the constellations were given Christian names and images, which reflected Schiller's attempt to conceive of the sky in less pagan terms. Although the atlas was beautiful and accurate, the idea of Christianizing the heavens never caught on (see Section 6.2.3.1 and Figure 6.2 for more on this atlas).

In 1679 the French astronomer Augustin Royer (active mid-1600s) introduced the constellation of Sceptrum in his north celestial planisphere. The constellation was located near Lacerta and represented the French scepter and hand of justice in honor of Louis XIV. Royer's celestial maps are discussed in more detail in Section 6.2.3.4.

Three new constellations were introduced by Gottfried Kirch (1639–1710), who was the Director of the Berlin Observatory. The first was published in 1684 and was named Gladii Electorales Saxonici. It was located near Bootes and pictured the crossed swords of the Electors of Saxony to honor German Emperor Leopold I. The other two were introduced in 1688. One was located near Aquila and was called Pomum Imperiale. It represented Leopold's orb. The other was located near Eridanus and was called Sceptrum Brandenburgicum. It pictured a scepter to honor the Brandenburg province of Prussia, where Kirch lived. See Appendix B for more details on Kirch and these constellations.

The famous French astronomer Pierre-Charles Le Monnier (1715–1799) was responsible for two new constellations. In 1743 he introduced Rangifer (a.k.a. le Renne or Tarandus), the reindeer (Figure 6.13). In 1776 he introduced Solitaire at the tip of Hydra's tail to represent an extinct flightless bird from an island in the Indian Ocean. However, on his star map the image looked more like a rock thrush of the genus *Turdus*, so it was later called Turdus Solitarius. Subsequent cartographers have seen this area as representing a dodo, a mockingbird, and Noctua the owl (see Section 6.5.5.2 and Figure 9.5). Section 7.13 contains more information on Le Monnier.

In 1754 the English naturalist John Hill (ca. 1707–1775) introduced 13 new constellations in his book *Urania: Or, A Compleat View of the Heavens* ... These constellations were all animals, such as Bufo (toad), Aranea (long-legged spider), Lumbricus (earthworm), and Limax (slug). See Section 6.4.4.1 for more on Hill and his book.

In 1777, Martin Odlanicky Poczobut (1728–1810), Professor of Astronomy and Director of the Royal Astronomical Observatory at Vilna, published a book labeled

Cahiers des observations astronomiques faites à l'Observatoire Royal de Vilna ... It included a catalog of 16 stars that comprised a new constellation located near Ophiuchus, Taurus Poniatovii, or Poniatowski's bull. This constellation was in honor of Poczobut's patron, King Stanislas II Poniatowski of Poland. Images of this constellation appear in Lalande's 1779 celestial globe, in Fortin's 1795 *Atlas Céleste de Flamsteed*, and in Bode's 1801 *Uranographia*.

In 1785, Karl-Joseph König (1751–ca. 1821), astronomer of the observatory at Mannheim, Germany, introduced in his *Nova Constellatione Coelo Inlatus* the new constellation Leo Palatinus, an imperial lion near Aquarius. This was to honor his patrons, the Elector Charles Theodore and his wife Elisabeth Augusta, and was accompanied by another constellation comprised of their initials. Both are shown in Warner.

Two constellations were introduced in 1789 by the Austrian Jesuit astronomer, Maximilian Hell (1720–1792). One was called Psalterium Georgianum and honored King George III of England; it was located near Taurus. The other was Telescopium Herschelii, or Herschel's telescope, located near Auriga in the area where Herschel discovered Uranus. Originally, Hell showed two telescopes: Tubus Herschelii Major and Minor, but in his 1801 atlas *Uranographia* (see below), Bode reduced this to one located in the spot of Tubus Herschelii Major. There is more on Hell in Section 6.4.4.4.

The French astronomer Joseph-Jerôme de Lalande (1732–1807) created several constellations. These included: Custos Messium (the harvest keeper, which was sometimes called Messier for the famous comet hunter—it first appeared on his celestial globe of 1779 and is shown in Figure 6.13); Quadrans Muralis (the mural quadrant, honoring the instrument he used to record star positions, which first appeared near Bootes in the 1795 edition of Fortin's *Atlas Céleste de Flamsteed*); Felis (the cat, which was located near Hydra and first appeared in Bode's 1801 atlas); and Globus Aerostaticus (the balloon, which honored the Montgolfier brothers, who were pioneering balloonists—it first appeared in Bode's 1801 atlas and is shown in Figure 4.9). Lalande also produced maps of cometary paths, an example of which for the comet of 1762 is shown in Warner (see also Appendix B).

The champion of extraneous constellations was the famous German astronomer, Johann Bode (1747–1826), who in his gigantic 1801 atlas *Uranographia* showed over 100 constellations, many of which were new (Figures 4.9 and 6.15). In addition to the two mentioned above that were suggested (but not shown) by Lalande, these included: Harpa Georgii (George's harp, actually a renamed version of Hell's Psalterium Georgianum—see Figures 6.18 and 9.7); Honores Friderici (Frederick's glories, the Latinized name of Friedrich's Ehre, originally introduced by Bode in 1787 as a scepter topped with a Prussian eagle to commemorate the death of King Frederick the Great of Prussia—it is shown in Figures 4.9 and 6.13); Lochium Funis (the nautical log and the line used to measure distance traveled by a ship—this was located near Argo Navis); Machina Electrica (the electrical machine, an electrostatic generator—see Figure 4.9); and Officina Typographica (a printing office just east of the star Sirius honoring the 350th anniversary of the invention of movable type). Bode's life and celestial atlases are discussed in Section 6.5.

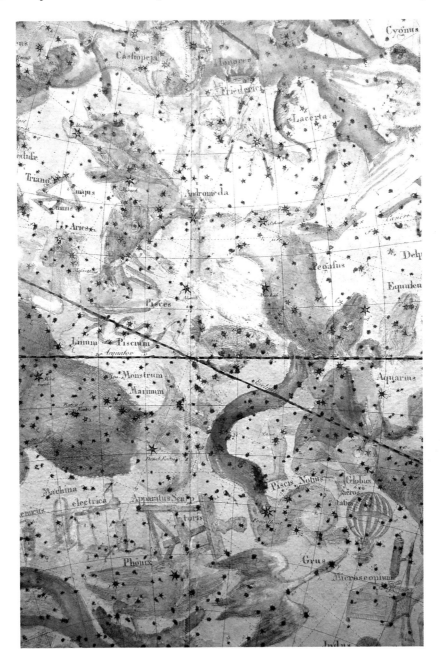

Figure 4.9. A section of the hemisphere centered around the first point of Aries (where the ecliptic and celestial equator lines cross), from Bode's *Uranographia*, published in 1801. 56.7 cm dia. hemisphere. Note the new constellations that appear here for the first time: Honores Friederici (*top right*), Machina Electrica (*bottom left*), and Globus Aerostaticus (*bottom right*). *See also* Color Plate Gallery.

During the rest of the 1800s, constellation mania subsided, although here and there a new constellation would appear. One example was Noctua the owl, referred to above. Another was the Battery of Volta, introduced near Equuleus in 1807 by British physician and physicist Thomas Young (1773–1829). There is more on Young in Appendix B. There were also two constellations introduced by William Croswell (1760–1834), who Warner credits as producing the first American star map in 1810: Marmor Sculptile, a bust of Columbus made up of the stars of Reticulum, and Sciurus Volans, the flying squirrel in the tail of Camelopardalis. See Section 9.3.3 for more on this early American.

4.5 BIBLIOGRAPHY

Allen, R.H. (1963) *Star Names: Their Lore and Meaning*. New York: Dover Publications.

Ashbrook, J. (1984a) Chapter 9. John Herschel's expedition to South Africa. *The Astronomical Scrapbook*. Cambridge, U.K.: Cambridge University Press, pp. 37–41.

Ashbrook, J. (1984b) Chapter 42. Edmond Halley at St. Helena. *The Astronomical Scrapbook*. Cambridge, U.K.: Cambridge University Press, pp. 213–218.

Ashbrook, J. (1984c) Chapter 73. The visual Orion Nebula. *The Astronomical Scrapbook*. Cambridge, U.K.: Cambridge University Press, pp. 379–388.

Bakich, M.E. (1995) *The Cambridge Guide to the Constellations*. Cambridge, U.K.: Cambridge University Press.

Condos, T. (1997) *Star Myths of the Greeks and Romans: A Sourcebook*. Grand Rapids, MI: Phanes Press.

Dekker, E. (1990) The light and the dark: A reassessment of the discovery of the Coalsack Nebula, the Magellanic Clouds, and the Southern Cross. *Annals of Science*, **47**, 529–560.

Dekker, E. (2010) Caspar Vopel's ventures in sixteenth-century celestial cartography. *Imago Mundi*, **62**, 161–190.

Evans, D.S. (1980) Nicolas de la Caille and the southern sky. *Sky & Telescope*, July, 4–7.

Evans, J. and Berggren, J.L. (transl.) (2006) *Geminos's Introduction to the Phenomena*. Princeton, NJ: Princeton University Press.

Gingerich, O. (1992) *The Great Copernicus Chase and Other Adventures in Astronomical History*. Cambridge, MA: Sky Publishing.

Heath, T.L. (1991) *Greek Astronomy*. New York: Dover Publications.

Kanas N. (2002) Mapping the solar system: Depictions from antiquarian star atlases. *Mercator's World*, **7**, 40–46.

Kanas, N. (2005) Are celestial maps really maps? *Journal of the International Map Collectors' Society*, **101**, 19–29.

Kanas, N. (2008) Celestial mapping of the southern heavens. *Journal of the International Map Collectors' Society*, **114**, 7–13.

Krupp, E.C. (2006) Heavenly reward. *Sky & Telescope*, January, 55–56.

Kunitzsch, P. and Smart, T. (1986) *Short Guide to Modern Star Names and Their Derivations*. Wiesbaden, Germany: Otto Harrassowitz.

Lovi, G. (1985) Rambling through the August skies: Edmond Halley's planisphere. *Sky & Telescope*, August.

Mair, A.W. and Mair, G.R. (transl.) (1989) *Callimachus Hymns and Epigrams, Lycophron, Aratus*. Cambridge, MA: Harvard University Press [Loeb Classical Library].

Ridpath, I. (1988) *Star Tales.* New York: Universe Books.

Schaefer, B.E. (2004) The latitude and epoch for the origin of the astronomical lore of Eudoxus. *Journal for the History of Astronomy,* **35**, 161–223.

Schaefer, B.E. (2006) The origin of the Greek constellations. *Scientific American,* November, 96–101.

Sobel, D. (1995) *Longitude.* New York: Penguin Books.

Thrower, N.J.W. (ed.) (1981) *The Three Voyages of Edmond Halley in the Paramore: 1698–1701,* two volumes. London: The Hakluyt Society, Second Series, Volume 157.

Toomer, G.J. (transl) (1998) *Ptolemy's Almagest.* Princeton, NJ: Princeton University Press.

Van Gent, R.H. (2006) *Andreas Cellarius. Harmonia Macrocosmica of 1660: The Finest Atlas of the Heavens.* Cologne, Germany: Taschen.

Varisco, D.M. (2000) Islamic folk astronomy. In: H. Selin (ed.), *Astronomy across Cultures: The History of Non-Western Astronomy.* Dordrecht, The Netherlands: Kluwer Academic.

Warner, D.J. (1979) *The Sky Explored: Celestial Cartography 1500–1800.* Amsterdam: Theatrum Orbis Terrarum.

5

Early European star maps

In previous chapters we discussed the cosmologies and constellation developments of both non-European and European countries. We have considered ancient ways of mapping the heavens using such media as paintings on the walls and ceilings of temples, markings on oracle bones or clay tablets, and representations on unusual flat surfaces, such as papyrus. We now are in a position to begin our journey of describing and illustrating how the heavens were depicted on flat surfaces that we usually associate with such activities in Europe, such as vellum or paper. In fact, it was in Europe that star maps reached their zenith in terms of accuracy and beauty, especially during the Golden Age of celestial cartography in the 17th and 18th Centuries. But prior to this time, especially in the 16th Century, there were a number of "firsts" that set the stage for the Golden Age, and this will form the subject of this chapter.

5.1 THE MANUSCRIPT ERA

Typically, early depictions of the constellations in Europe were drawn by hand on vellum or parchment and go back to the Middle Ages, although in his classic book on astronomical maps author Basil Brown (1932) has shown a manuscript planisphere by Geruvigus that dates from a 2nd-Century-AD Roman copy of Aratus' *Phaenomena* that is located in the British Library. When stars were included in old manuscripts, they usually were inaccurately represented without benefit of a coordinate system. Thus, they were not true star maps, since they did not allow for an accurate representation of where the stars were located in the heavens. In fact, the constellation images were often simply used as decorative illustrations for medieval texts, such as religious books, astrological expositions, and classical poems. It would be some time before the rebirth of mathematical astronomy in the Renaissance would create the need for accurately locating stars in the heavens and representing these locations on two-dimensional star maps.

An example of an early hand-made image of the constellation of Aries is shown in Figure 5.1. This decorated manuscript on vellum was produced in the 10th Century in Italy and was used to illustrate Cicero's translation of Aratus' *Phaenomena*, which is shown in the text at the bottom of the page. Interestingly, the text within the body of this and the other constellations is an extract from the constellation myths of Hyginus, who we shall meet again below. Thus, in addition to its aesthetic beauty, this constellation print exposes the reader to the mythological astronomical ideas of two famous classical writers.

Another manuscript using constellation images to illustrate Aratus' poem is the beautiful *Aratea*, which is housed at the University of Leiden Library. Stimulated by the "Carolingian Renaissance" from Charlemagne's court, this parchment manuscript contained images of gods, heroes, or animal representations of the constellations, the seasons, and the planets. The text is likely from Germanicus' 1st-Century-AD translation, and this manuscript is probably a 9th-Century copy of a 4th or 5th-Century work. Although the stars (accented in gold leaf) provided the framework

Figure 5.1. The constellation of Aries, from a 10th-Century manuscript written on vellum used to illustrate the text of Aratus' *Phaenomena*. 38 × 29.9 cm (page size). Note the text in the constellation, which is an excerpt from the writing of Hyginus. Photograph taken from *Celestial Charts* (C. Stott, Studio Editions, 1995). Courtesy of Ms. Stott and the British Library (Harley MS 647). *See also* Color Plate Gallery.

for 39 figures containing 43 constellation images, they were only loosely modeled on the star patterns we now use, and the text gave no precise indication of stellar positions. Some of the constellation figures were shown from the back, and some of the star patterns have an external orientation, like on a celestial globe. Again, the constellations in the *Aratea* are not true star maps, since they do not represent what we see when we look up at the night sky, nor do they give accurate information about the location of the stars in the heavens (Figure 5.2).

Figure 5.2. The constellation of Aquarius, from Hugo Grotius' *Syntagma Arateorum Opus Poeticae et Astronomiae*, 1600. Approximately 15.5 × 13.5 cm. The image is based on that found in the famous Leiden *Aratea*. Note that the stars are inaccurately placed, and there is no coordinate system indicating their positions in the sky. Courtesy of the Collection of Owen Gingerich.

5.2 EARLY PRINTED CONSTELLATION IMAGES

5.2.1 Erhard Ratdolt and Hyginus' *Poeticon Astronomicon*

As described in Section 3.7.1, the activities of Gutenberg in the mid-1450s catalyzed printing in Europe, leading to a revolution in the production and use of books. One of the most important figures in the early days of printing was Erhard Ratdolt. He was born in Augsburg around 1443 in a family of artists. He took up the printer's trade, and according to astronomy historian Owen Gingerich (pers. commun.), he likely apprenticed to the pioneering astronomical printer Regiomontanus (see Section 3.8.2). In 1475, Ratdolt moved to Venice, where he founded a printing business. By then, printing had spread to Italy from Germany, and in the 1470s Venice was the center of the industry, being the home to over 100 printing companies. Ratdolt's business flourished, in part because of his publication of Euclid's *Elements of Geometry* in 1482, which went through many editions and was one of the first printed books to include mathematical figures. Ratdolt returned to Augsburg in 1486, where he continued to be successful and was to become perhaps the finest printer of scientific works of his time. He died around 1528.

Also in 1482, Ratdolt published an edition of Hyginus' *Poeticon Astronomicon* (see Section 4.1.6), which included a set of woodcuts of the constellations to illustrate the text. The edition in the British Library, which is bound with a copy of Sacrobosco's *De Sphaera*, has 47 separate plates, 39 of which are constellations. The remaining figures are allegorical representations of the Milky Way, the Sun, and the planets. The images vary in size from 3.5×9 cm to 9×10 cm.

Hyginus' book was very popular and went through many editions, well into the early 1500s. All of the Greek constellations were depicted, except for Equuleus, and all were shown from the front using a geocentric orientation. Although crude, the figures were lively and energetic (Figure 5.3). Warner (1979) has pointed out that not all of them conformed to earlier descriptions based on Ptolemy. For example, Ara had been described classically as simply an altar, whereas Hyginus followed medieval mythology in having the altar surrounded by demons. There were also some peculiarities, such as the wrapping of Draco around Ursa Major and Ursa Minor in one image. But in other cases, Ratdolt's woodcuts led to stylistic conventions that endured. One example was that of Bootes, who was shown with a sheaf of wheat at his feet, not the hair of Berenice as in some earlier manuscripts, and this convention was later used by Bayer (Figure 6.1). Although stars were shown in these woodcuts, they were primarily decorative and had no relationship to the description in the text or the actual positions of the stars in the heavens. Nevertheless, these images were among the first printed constellation figures and were widely copied, as we shall now see.

5.2.2 An early printed edition of Aratus' *Phaenomena*

Aratus' *Phaenomena* (see Section 4.1.1) continued to be important in the printing era. In 1488, Antonius de Strata de Cremona published in Venice a version of this poem

Figure 5.3. The constellation of Cetus, from Firmicus Maternus' *Scriptores Astronomici Veteres*, Aldine Press, 1499. 27.4 × 17.8 cm (page size). Note the simple but dynamic figure and the lack of relationship of the stars to their actual position in the sky. This image is nearly identical to that in Hyginus' *Poeticon Astronomicon*, except that it is left-to-right reversed.

by Avienus entitled *Arati Phaenomena* ... The edition in the British Library contains 38 woodcuts as part of the text: a northern planisphere, allegorical prints of the Sun and Moon, and 35 images of constellations. Except for the planisphere, the woodcuts in this book were nearly identical to those in Hyginus' *Poeticon Astronomicon* in size and style, although not all were present in this edition. Such plagiarism was not uncommon in the Renaissance, with publishers and authors often borrowing from each other. But, unlike in Hyginus' work, most of the constellation images were left-to-right reversed, they were not labeled, and the order was somewhat different.

There was no planisphere in Hyginus, and according to Warner the Cremona planisphere was the first printed celestial map to show multiple constellations arrayed *vis-à-vis* each other. It was not a true star map, since individual stars were not included. The planisphere was centered on the north equatorial pole down to around 40 degrees S dec. and used a polar equidistant projection in geocentric orientation (see Section 6.1 for map terms and conventions used in this book). There were concentric circles representing the Arctic Circle and the Tropics of Cancer and Capricorn projected into the sky, as well as the celestial equator. The diameter was 10.0 cm, and no stars accompanied the constellation figures.

5.2.3 Aldus Manutius and Aratus' *Phaenomena*

A version of Aratus' *Phaenomena* was also published by the Aldine Press, whose Dolphin and Anchor printer's device became recognized throughout Europe. The founder of this famous publishing house was Aldus Manutius (originally Teobaldo Mannucci). Born around 1450 in a hill town south of Rome, he studied Latin at Rome, Greek at Ferrara, and Greek literature at Mirandola. In the late 1470s he became the tutor for two young Italian princes from the court of Carpi, who supported him financially. His interest in ancient writings led him to become enamored with the idea of publishing and preserving the classics of Greek literature in their original language, and for this he moved around 1489 to the mecca of printing, Venice. There, he formed a printing company that became filled with Greek scholars, and instructions to his printing staff were in the Greek language. From 1495 to 1498, he published five volumes of Aristotle's work. Not only were the texts accurately checked by native Greek speakers, but they were also beautifully done, setting a high standard in terms of quality of paper and binding and beauty of type. Over 30 famous Greek works followed (e.g., Aristophanes, Plato, Plutarch, Sophocles, Thucydides), as well as a number of Latin and Italian classics (e.g., Cicero, Dante, Erasmus, Petrarch, Virgil). He also created and patented the italic typeface style, which was especially suitable for his cheaper and smaller, more portable books. In 1500 he founded the new Academy of Hellenists, where everyone was required to speak Greek. Since his wife Maria came from a family who was also in the printing trade, she and their sons continued the business after Aldus' death in 1515.

In 1499 the Aldine Press published a book entitled *Scriptores Astronomici Veteres . . .* , which was a collection of classic texts related to astrology and included works by Julius Firmicus Maternus, Marcus Manilius, Aratus Solensis, and Proclus. Typically for multiple books bound together, this book is sometimes referenced by the name of the author of the first entry, in this case Firmicus Maternus. In the British Library edition the *Phaenomena* contains 41 woodcuts: the same planisphere appearing in de Strata de Cremona's work mentioned above, allegorical images of the Sun and Moon, and 38 constellation images. The constellation images were nearly identical to those of Hyginus, except that they were left-to-right reversed (Figures 4.3 and 5.3). However, there was no one-to-one correspondence in terms of the constellations included or their order, and—at least in the British Library editions—the Aldine Press version had three more constellations.

5.3   ALBRECHT DÜRER AND THE FIRST PRINTED STAR MAPS

5.3.1   Life and times

The great Renaissance artist and polymath, Albrecht Dürer, was born on May 21, 1471 in Nuremberg. He learned the goldsmith trade from his father, but he soon developed an interest in painting. In 1486 he was apprenticed to the painter and printmaker, Michael Wolgemut, where he learned to work with woodcuts and copper engravings and probably contributed to the *Nuremberg Chronicle* (see Section 3.8.3). Beginning in 1490, he began to travel throughout Europe to enhance his artistic education. He also studied music, anatomy, mathematics, and astronomy. In 1494 he married Agnes Frey of Nuremberg, but the couple had no children. He began to demonstrate great artistic growth and productivity, characterized by his publication of a portfolio of woodcuts from 1496 to 1498. During a trip to Italy from 1505 to 1507, his work was influenced by the great Venetian artists Bellini and Mantegna, and he began to integrate the Gothic traditions of Germany with achievements in the perspective and color of the Italians. From 1512 on, he became noted for his self-portraits and his portraits of rich and famous people, including Emperor Maximilian I and King Christian II of Denmark. He experimented with tempura and oil glazes, and his well-known works also included religious themes and altarpieces. Altogether, this prolific artist produced over 800 drawings, 200 woodcuts, 100 line engravings, and some 80 paintings. He also wrote a series of books on proportions, which were published a few months after his death in Nuremberg on April 6, 1528.

Figure 5.4. The northern celestial hemisphere produced by Albrecht Dürer in 1515. Approximately 42.9 × 42.9 cm. Note the radial lines resulting in a primitive coordinate system, allowing the stars to be located in the heavens. This woodcut is the first printed star map. Courtesy of the Collection of Robert Gordon. *See also* Color Plate Gallery.

5.3.2 Dürer's celestial hemispheres

In 1515, Dürer collaborated in the printing of two woodcuts showing images of the Greek constellations, one for the northern celestial hemisphere (which according to map historian Peter Whitfield (1995) was influenced by a mid–15th Century hemisphere now in the Austrian National Library) and one for the southern. He was assisted by two noted mathematicians and astronomers: Johann Stabius, who drew the maps' coordinate system, and Konrad Heinfogel, who positioned the stars. The hemispheres were labeled *Imagines coeli Septentrionales* (or *Meridionales*) ... Each was centered on an ecliptic pole using polar stereographic projection with external orientation. According to Warner (1979), they were 35 cm in diameter on sheets that were 43 cm square. Most of the stars in the constellations were positioned according to Ptolemy's catalog, and over 1,000 were shown. The constellation images were typically viewed from the rear but were depicted according to their classical Greek and Roman attributes (e.g., Perseus held the head of Medusa). Radial lines extended outward from the center at 30-degree intervals. The northern map showed the 12 zodiac constellations on the periphery of the hemisphere, and all were accompanied by their astrological symbols (Figure 5.4). In the corners of this map were depicted images of four historical individuals who were involved with constellation description. The southern map lacked zodiac figures and had coats of arms in the corners on the left and dedications on the right.

These hemispheres were in fact the first printed star maps, in that they included a coordinate system from which a star's position could be read in celestial longitude. However, the value of these woodcuts as celestial maps was limited because the constellation patterns were depicted mirror-reversed from what we see looking up at the heavens, the stellar coordinate system was primitive, and there was no indication of a star's magnitude.

Nevertheless, Dürer's style influenced that of later cartographers, such as Honter (see Section 5.4), Apian (see Section 3.8.4), and Middoch and Volpaia (see Appendix B). A number of states of these two hemispheres were later printed that included minor alterations to the constellation figures or to the corner inserts, and these are reviewed by Warner. She also mentions that these hemispheres were indicated as accompanying Noviomagus' 1537 Ptolemaic star catalog, although they are not found in any known copies of this book.

5.4 JOHANNES HONTER AND HIS GEOCENTRIC HEMISPHERES

5.4.1 Life and times

Johannes Honter was born in 1498 in Brasov (or Kronstadt), Transylvania, which is now Romania. He studied at the University of Vienna from 1520 to 1525, receiving a "Magister Atrium" degree. With the threat of the Turks' impending attack, he left Vienna in 1529 for Regensburg. He then registered at the Jagiellonian University in Krakow in 1530, where he wrote a book on Latin grammar and a manual on cosmography. From 1530 to 1532 he lived in Basel, where he became proficient in

Figure 5.5. The northern celestial hemisphere produced by Johannes Honter in 1541. Approximately 27.9 × 27.9 cm. Note the geocentric orientation, the addition of lines for the Tropics and the celestial equator, and the medieval clothing on some of the figures. Courtesy of the Collection of Robert Gordon. *See also* Color Plate Gallery.

wood engraving and produced an influential map of his native Transylvania that was the first of the region and made famous in a later copy by Ortelius. He moved back to Brasov for good in 1533, where he became involved in introducing Lutheranism to the region. In 1539 he set up a printing press and began issuing a number of books, including many that he wrote. One of these was a new version in verse form of his cosmography manual entitled *Rudimenta Cosmographica*, published in 1542. According to his entry in *Wikipedia.org*, it contained 13 maps of the world and became so popular that it went through some 39 editions and was partially reprinted in other books. Honter died on January 23, 1549.

5.4.2 Honter's celestial hemispheres

While in Basel in 1532, Honter produced two important celestial hemispheres that were later bound and distributed in a book published by Heinrich Petri in 1541 of the first collected works of Ptolemy: *Omnia, quae extant opera, Geographia excepta*. The hemispheres were labeled *Imagines Constellationum Borealium* or *Australium*, and according to Warner each was 25 cm in diameter on plates that measured 26×27 cm. They were centered on an ecliptic pole using a polar stereographic projection. Honter knew about Dürer's maps, and his hemispheres show the influence, but with three important changes (Figure 5.5). First, rather than an external orientation, Honter was the first major cartographer to use a geocentric orientation in printed form for the stars in his constellations. Second, instead of using Arabic or classical clothing, Honter employed contemporary Renaissance clothing (e.g., heavy beards, bulky clothes and tunics, fancy hats) for five of his male constellation figures. Finally, like Dürer, radial lines extended outward allowing for the measurement of a star's celestial longitude. But in addition, Honter included a depiction of the Arctic Circle, the Tropic of Cancer, and the celestial equator to give an approximate location of a star's latitude in the heavens. As Warner points out, there was a problem with his maps, in that he used an older, more classical radial coordinate system that did not take into account precession, so it was some 30 degrees off in longitude for the time it was printed.

Nevertheless, his hemispheres were very influential (see, for example, Gefugius, Maggi, and Postel in Appendix B), especially since—unlike Dürer's rare single sheets—Honter's were bound in books that went through many editions and were widely copied. His geocentric orientation was taken up by Piccolomini (see Section 5.5). Warner speculates that the woodblocks may have passed on to Paris, where the maps were reissued with a new label (*Arati Solensis Phaenomena*) by several publishers (e.g., de Gourmont, in Postel's 1553 *Signorum Coelestium*; Morel's 1609 issue and his successor Bienne's 1626 issue of *Phaenomena et Prognostica*).

5.5 ALESSANDRO PICCOLOMINI AND THE FIRST PRINTED STAR ATLAS

5.5.1 Life and times

Alessandro Piccolomini was born on June 13, 1508 and was a member of an old and noble family of Siena in northern Italy. In his youth he studied and translated classical literature into Italian, and he wrote over 100 sonnets, comedies, and rhymes. In 1538 he left Siena to study and then teach philosophy at the University of Padua. In 1545 he left Padua for Rome, where he worked as a secretary for a cardinal and an archbishop and was himself ordained in 1555. He also wrote a number of commentaries on the theories of ancient and medieval philosophers (including Aristotle and Ptolemy), and he published a scientific treatise on the relative proportion of sea and land on the Earth. In 1574, Pope Gregory XIII appointed him coadjutor to the Archbishop of Siena and titular Archbishop of Patras (Greece). After obtaining a Doctorate in Theology, he participated in calendrical reform and wrote on this topic until his death on March 12, 1578.

During his lifetime, Piccolomini wrote several works on astronomy. One of these was *La Sfera del Mondo (The Sphere of the World)*, which was published in 1540 in a large octavo format. It was written in the Italian vernacular of Tuscany rather than in Latin, thus being accessible to more people in the Italian peninsula. This book was an exposition of the geocentric cosmology that was commonly used at the time, which largely reflected Aristotle and Ptolemy. Woodcuts illustrated the text, which dealt with such topics as the Aristotelian cosmological system (Figure 3.6), armillary spheres and other astronomical instruments used to determine the positions of heavenly objects, the great circles and climatic zones of the Earth, and the mechanics of eclipses. Essentially, this book was a summary of the state of astronomy that was prevalent in the Medieval and early Renaissance periods, and it went through several editions.

5.5.2　*De le Stelle Fisse*

But more important to celestial cartography was Piccolomini's *De le Stelle Fisse (On the Fixed Stars)*. This book too was published in 1540 in the Italian vernacular and used a large octavo format. It was often bound together with *La Sfera del Mondo*. *De le Stelle Fisse* was the first printed star atlas and was very popular, going through at least fourteen editions by the end of the 16th Century. There were four major sections to this book: a star catalog organized by constellation, a series of constellation woodblock plates, a set of tables that indicated stellar locations throughout the year, and a text dealing with the risings and settings of stars with reference to the constellations of the zodiac. The first two parts of the book have special relevance for celestial mapping.

The star catalog gave a mythological description for 47 Ptolemaic constellations (Equuleus was absent), followed by a list of the brighter stars using Roman letters (with "a" typically being the brightest, "b" being second, etc.). This was the first time that such a categorization occurred in a book that focused on the constellations. Each star in turn was followed by a brief description of its location in the constellation and an indication of its "grandezza" (or magnitude) from 1 to 4.

Each constellation in the catalog was keyed to one of the plates in the next section, which were numbered from I to XLVIII. But there were only 47 plates, since Equuleus was missing. The plate number for this missing constellation varied with the edition (e.g., the 1540 and 1559 editions lacked Figura XXIIII; the 1553 edition lacked Figura XVIII, and the 1579 edition lacked Figura XVI). Otherwise, the plates were in the same sequence and looked similar across different printings, despite some stylistic and spelling differences. The plates from the 1579 edition are listed in Appendix C. The plates illustrating Ursa Minor and Ursa Major are shown in Figure 5.6.

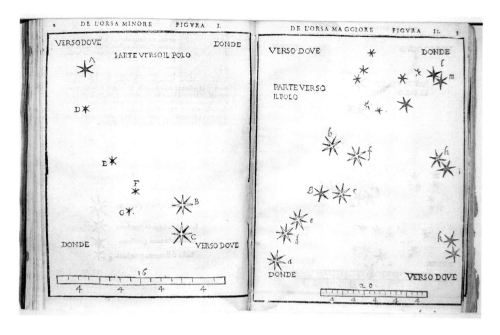

Figure 5.6. The constellations of Ursa Minor (*left*) and Ursa Major (*right*), from the 1579 edition of Piccolomini's *De le Stelle Fisse*. 17.5 × 13.9 cm (*left image*), 17.7 × 14.1 cm (*right image*). Note the geocentric orientation, the different symbols representing the stars' magnitudes, and the accuracy of the star patterns in the absence of constellation figures.

Each plate featured an accurate map of the constellation star pattern as seen from the Earth, but no constellation figure. The stars were identified by the same letters given in the catalog, and they were indicated by different symbols according to their magnitude, with larger symbols representing brighter stars. At the bottom was a scale of degrees that allowed the reader to determine the size of the constellation. Although there was no true coordinate system whereby individual stars could be located in the heavens, each plate had written on it an indication of the direction to the celestial pole ("parte verso il polo"). Also, the constellation's direction of rotation as it revolved around the sky was mentioned: leading edge ("verso dove") and trailing edge ("donde"), although Gingerich (1981) has stated that this was only accurate for the plate containing Ursa Minor and the Pole Star.

De le Stelle Fisse was a true star atlas. The catalog, plates, tables, and text were linked; the constellation star patterns were geocentrically presented as they appeared in the sky; and information was given about stellar magnitudes and locations. Although Piccolomini's constellation plates lacked the artistic beauty that later atlases would have in showing mythological figures, they possessed their own austere beauty, and in many ways they resembled the star atlases today where constellation figures are rarely included.

5.6 GIOVANNI PAOLO GALLUCCI AND HIS COORDINATE SYSTEM

5.6.1 Life and times

Giovanni Paolo Gallucci was born in Salo, Italy in 1538. After studying in his home town and in Padua, he moved to Venice. His interests stemmed from astronomy and physics to medicine and literary works. He became well known for his scholarship, and he was one of the founding members of the second Academy, which was instituted in Venice on June 21, 1593. He was a gifted teacher and was sought after as a tutor for the nobility. He was a prolific writer. He published several books on astronomy and was especially interested in astronomical instrumentation. He died in Venice around 1621.

5.6.2 *Theatrum Mundi, et Temporis*

In 1588, Gallucci published a star atlas entitled *Theatrum Mundi, et Temporis . . .* The text of this book discussed a number of topics of astronomical and astrological

Figure 5.7. The constellation of Andromeda, reproduced from Gallucci's 1588 *Theatrum Mundi*. Approximately 15.2 × 12.7 cm. Note the trapezoidal projection and the coordinate system keyed to the margins. Courtesy of the Collection of Robert Gordon.

significance. In the 1603 edition of this book in the British Library, Books 1–4 have a number of illustrations, including an armillary sphere, an Earth globe, the cosmological system of Aristotle, and diagrams of the Moon, Sun, planetary orbits, and eclipses. Book 6 consists of a number of tables, and there is a supplement of volvelle pointers that could be cut out and attached to the respective volvelles earlier in the text (see Section 8.2, especially Figure 8.5).

The constellations were found in Book 5. Gallucci included a total of 48 woodcut maps of the classical Greek constellations, each of which was preceded by a catalog of the principal stars, together with their locations and magnitudes. The constellation figures were lively but primitive in design, and the orientation was external. The plate showing Ursa Major measured 15.9 × 12.1 cm, and that depicting Andromeda is shown in Figure 5.7.

Gallucci's book is notable for being the first star atlas to use a coordinate system from which the location of the stars could be read from the margins in both celestial latitude and longitude. The star positions were taken from the catalog that was published by Copernicus only 45 years earlier. Gallucci used a trapezoidal projection system, which was popular among terrestrial cartographers of the time in giving a fairly accurate representation of three-dimensional features on a two-dimensional surface, without major distortion. In his system the parallels of celestial latitude were equidistant straight lines, and the circles of longitude were also straight lines that radiated from the ecliptic pole.

The *Theatrum Mundi* was very popular and went through many editions after the 1588 original, including 1589 and 1603. In addition, it was translated into Spanish as *Theatro del Mundo y del Tiempo* ... in 1606, and this was followed by new editions in 1611, 1614, and 1616.

5.7   THOMAS HOOD: SETTING THE STAGE FOR THE GOLDEN AGE

5.7.1   Life and times

At the close of the 16th Century, two interesting celestial hemispheres appeared that were produced by the Englishman Thomas Hood. He was the son of a merchant tailor who developed an interest in mathematics and astronomy. Hood attended Cambridge University and was a Fellow at Trinity College. He then moved to London, where he actively lectured in the 1580s and 1590s, giving popular talks on mathematics, astronomy, geography, hydrography, and navigation. Some of his lectures were published, including two books on the use of terrestrial and celestial globes in 1590 and 1592. He was a follower of Copernicus, and his lectures helped to popularize Copernican theory in England.

5.7.2   *The Use of the Celestial Globe*

In 1590, Hood wrote a book entitled *The Use of the Celestial Globe in Plano, Set Foorth in Two Hemispheres* ... By taking a look at what is examined in this book, a

Figure 5.8. The northern celestial hemisphere of Hood's 1590 *The Use of the Celestial Globe in Plano* (facsimile). 53.7 × 54.4 cm. Note the accurate latitude and longitude coordinate system centered on the ecliptic pole, the beautiful detail, and the text information written beside each constellation. Photograph taken from a facsimile copy from the British Library and published as *The English Experience* #533 (Theatrum Orbis Terrarum, 1973). Courtesy of the British Library (Maps 184.h.1). *See also* Color Plate Gallery.

picture emerges of some of the issues being discussed in astronomy some 50 years after the death of Copernicus. The text is in the form of a dialogue between a master and student, and it was aimed at helping the reader recognize the stars and their constellations. It included a description of the great circles in the sky, ways to locate a star in celestial latitude and longitude, an alphabetical list of the most notable stars and their positions, and a survey of the constellations and their history and mythology. There was an interesting discussion of the nova of 1572, whose appearance in Cassiopeia provided support for the Copernican system over the Ptolemaic system.

Hood countered some of the explanations proposed by traditionalists of the latter system, such as the nova being a new comet or a star that had been previously hidden from view by an "exhalation" in the sky. The origin of the Milky Way also is discussed, with a refutation of some of the theories of Aristotle (i.e., that it was formed by a meteor that ignited the air or by vapors from the Earth). Hood also described ways of classifying the stars (e.g., according to their magnitudes, color, form, and twinkling).

Bound in many editions of *The Use of the Celestial Globe* were two celestial hemispheres, each centered on an ecliptic pole using polar equidistant projection and external orientation (Figure 5.8). In case they are missing, the reader is told on the title page: "The hemispheres are to be sold in Abchurch-lane at the house of Th. Hood." They were labeled *Augustinus Ryther (Anglus) sculpsit. 1590,* and they were dedicated to Lord Lumley, who had commissioned Hood's lectures. The maps were influenced by the globes of Mercator and were the first hemispheres to be printed in England. Beside each constellation appeared a short commentary in Latin, giving its name in various languages, brief mythological facts, its planetary associations, and information on its stars. According to Hood, 1,025 stars are represented in the two hemispheres. In the northern map, Hood included the constellations of Antinous (associated with Aquila) and Cincinnus (referred to as Coma Berenices in his book), both of which Vopel had introduced in his globe of 1536 (see Section 4.3.3.1). In the southern map near the ecliptic pole, there appeared a dedication with Lord Lumley's coat of arms, next to which was a blank area devoid of constellations (representing the area of the southern sky never seen by the mapmaker).

Hood's hemispheres had a number of useful features that integrated the best of what was being done in celestial mapping, and it is remarkable that they were not better known and more influential. His coordinate system was quite sophisticated for the time, including not only degrees of longitude marked around the periphery but also degrees of latitude marked in his solstice and equinoctial lines. In addition, for reference purposes, he added radiating lines every 30 degrees of longitude and circles representing every 10 degrees of latitude. Thus, it was easier to find the location of the stars in the heavens than in previous maps. Also, he included a star brightness scale with different symbols representing stellar magnitudes from 1 to 6. His constellations were attractive and accompanied by detailed information written in on the map. Granted, he referenced his coordinate system to the ecliptic rather than to the equatorial pole, and he used an external rather than an observer-friendly geocentric orientation. Nevertheless, these were conventions commonly used at the time. In a sense, Hood's maps were the epitome of the star maps in the pre-Golden Age period, and they set the stage for what was to follow.

5.8 BIBLIOGRAPHY

Ashbrook, J. (1984) Chapter 78. Johann Bayer and his star nomenclature. *The Astronomical Scrapbook.* Cambridge, U.K.: Cambridge University Press, pp. 411–418.

Ashworth, W.B. Jr. (1997) *Out of this World—The Golden Age of the Celestial Atlas*. Kansas City, MO: Linda Hall Library.

Brown, B. (1932) *Astronomical Atlases, Maps and Charts*. London: Search Publishing.

Condos, T. (1997) *Star Myths of the Greeks and Romans: A Sourcebook*. Grand Rapids, MI: Phanes Press.

Gingerich, O. (1981) Piccolomini's star atlas. *Sky & Telescope*, December, 532–534.

Hingley, P.D. (1994) *Urania's Mirror*—a 170-year old mystery solved? *Journal of the British Astronomical Association*, **104**, 238–240.

Hood, T. (1973) *The Use of the Celestial Globe. The English Experience #533*. (Facsimile of the original of the 1590 edition in the British Library, London, and the Library of Congress, Washington, D.C.). Amsterdam: Theatrum Orbis Terrarum.

Johnston, P.A. (1985) *Celestial Images—Astronomical Charts from 1500 to 1900*. Boston, MA: Boston University Art Gallery.

Kanas, N. (2002) Mapping the solar system: Depictions from antiquarian star atlases. *Mercator's World*, **7**, 40–46.

Kanas, N. (2003) From Ptolemy to the Renaissance: How classical astronomy survived the Dark Ages. *Sky & Telescope*, January, 50–58.

Kanas, N. (2005) Are celestial maps really maps? *Journal of the International Map Collectors' Society*, **101**, 19–29.

Kanas, N. (2006) Alessandro Piccolomini and the first printed star atlas (1540). *Imago Mundi*, **58**(1), 70–76.

Katzenstein, R. and Savage-Smith, E. (1988) *The Leiden Aratea: Ancient Constellations in a Medieval Manuscript*. Malibu, CA: J. Paul Getty Museum.

Snyder, G.S. (1984) *Maps of the Heavens*. New York: Abbeville Press.

Stott, C. (1995) *Celestial Charts: Antique Maps of the Heavens*. London: Studio Editions.

Warner, D.J. (1979) *The Sky Explored: Celestial Cartography 1500–1800*. Amsterdam: Theatrum Orbis Terrarum.

Whitfield, P. (1995) *The Mapping of the Heavens*. London: The British Library.

6

The "Big Four" of the Golden Age of pictorial star maps

The Golden Age of pictorial celestial cartography took place in Europe, roughly from 1600 to 1800. During this period, grand sky atlases were produced that attempted to accurately place the stars and planets in the heavens in coordinate systems that paralleled those on Earth. These were influenced from increasingly more accurate placement that resulted from new star catalogs that built on that of Ptolemy. Using bigger and better instruments, Tycho Brahe and Johannes Hevelius determined the locations of the stars using naked eye observations. Exemplified by Astronomer Royal John Flamsteed, telescopes and micrometers were added to these instruments, which made their positioning even more accurate. In addition, the rapid advances in printing techniques since the development of movable type procedures in the 1450s led to the ability to depict images with more detail and accuracy. Paralleling developments in terrestrial mapmaking, the use of coarse woodblocks gave way to intaglio processes in celestial maps that allowed fine images on copper and steel plates to be reproduced as fine images on paper. Now, maps of the heavens could be both aesthetically pleasing and technically accurate, and mapmakers increasingly competed with each other to produce bigger and better star atlases. Some of these are truly works of art and will be described in Chapter 7. In this chapter, I will deal with four individuals who are considered to be the most influential during this period for the standards they set and their influence on others. But first, a word about the conventions that will be used.

6.1 STAR MAP CONVENTIONS

In this and subsequent chapters (including the Appendices) where star maps are discussed, basic material will be presented that will help the reader in two ways:

(1) to identify the source of the map, and (2) to understand what it shows. Some of this material will depend on certain conventions that will be used throughout this book, as follows:

- *Source*: This generally will be the creator of the map or the author or publisher of the book from which the map comes.
- *Edition*: The text will mainly refer to the first state of the map or the first edition of the book from which it comes. Many of the illustrations, however, will be from later (and more available) editions. In most cases, the year of the edition used will be clear. The same goes for entire atlases: usually, the first edition will be described. In some cases, subsequent editions will be mentioned, particularly if there are important differences. Readers wanting to research later editions can find these by accessing the on-line search catalog of books: *worldcat.org*
- *Title*: When a map or a book from which it comes has a title, this will be given *in italics*. Usually, the whole title will be given at least once, but in some cases just the key words by which the map or book is commonly known. Many maps do not have a title but have a description of some kind. In these cases, the entire or major part of the descriptive label will be given *in italics*.
- *Year*: The year of publication of the map will be given, or if unknown, its approximate year (e.g., ca. 1750).
- *Area of sky shown*: Many maps are hemispheres, showing half of the sky from an ecliptic (or equatorial) pole in the center to the rim of a circle 90 degrees away that represents the ecliptic (or celestial equator). Other maps are planispheres, showing more than half of the sky extending from a northern pole to a southern region or from a southern pole to a northern region. Areas above or below the ecliptic will be referred to as north or south latitude (e.g., N lat. or S lat.)—see Section 1.5. Areas above or below the celestial equator will be referred to as north or south declination (e.g., N dec. or S dec.) or, as stated in many modern star catalogs, "+" or "−" degrees dec., respectively. Thus, one might see an indication that "the map is a planisphere centered on the north equatorial pole down to 15 degrees S dec.", or "−15 degrees dec." Other maps are segments of the sky whose area is described (e.g., "eight degrees above and below the ecliptic line") or clear from the illustration itself.
- *Orientation*: As discussed in Section 1.5, on some maps the pattern of stars in the constellations is the same as it appears while looking up at the sky from the Earth: this is called a "geocentric" orientation. In other maps, the pattern of stars is left-to-right reversed, as if we are gods looking back at the sphere of stars from the heavens beyond. This orientation developed from conventions used in celestial globes and is called an "external" orientation. Note that although the orientation describes the star pattern of a constellation it does not say anything about the orientation of the figure itself, which may be facing us front-on or from behind, depending on the desire of the map's author and engraver. Many authors followed the conventions used by Ptolemy for his 48 classical constellations, but not all, and in some cases this became problematic in defining a particular star's location, as we will see below.

- *Projection*: Just like their terrestrial counterparts, celestial maps used a variety of projection systems aimed at accurately transferring information from the three-dimensional celestial sphere onto a two-dimensional piece of paper. Terms such as "polar", "trapezoidal", "sinusoidal", and "conic" will be used to describe the particular system used in the map (see Chapter 1 and Glossary for a description of these projection systems).
- *Size*: Oddly enough, one of the most inconsistent ways of categorizing a celestial (or terrestrial) map is by size. Some people measure the page size, others the size of the total plate area, still others the size of the "true map" part of the print (i.e., excluding the scale markings along the border). Some people use centimeters, others inches. In addition, images may vary in size from plate to plate in a given book due to idiosyncrasies in the construction of the plates (e.g., northern and southern celestial hemispheres in the same atlas might differ in diameter by several millimeters, and even one hemisphere might have a slightly different diameter if measured horizontally rather than vertically). For maps tightly bound in books, the area in the binding might take up some of the distance along the horizontal dimension. Also, later editions of a book might use slightly larger or smaller plates than the original edition. Finally, humidity and other ravages of time might cause some crinkling of the pages, thereby shortening the length of a map in one book but not in another published at the same time. In the main text of this book and in Appendix B, I have measured the dimensions obtained from an identified source using the following conventions: (1) measurements are in centimeters; (2) for hemispheres and planispheres, the least distorted horizontal or vertical diameter is given as measured from the northern plate (e.g., "the Northern Hemisphere map was 15 cm in diameter"); (3) for rectangular maps, unless otherwise indicated, the distances between the innermost borders are given, first for the vertical then for the horizontal dimension (i.e., the dimensions of the map that exclude any grid marks that might be going around its rim). Where possible for comparison purposes, the map that is measured is the one that contains the constellation of Ursa Major (e.g., "the Ursa Major map was 15 [vertical] × 20 [horizontal] cm). In a few places in the main text of the book, and in most of the cases in the catalog of Appendix B, I have been unable to measure a map size directly and have used the same measurements cited from another source (although it has not always been clear to me if the source used the image or plate size).
- *Other information*: For most of the maps described in the text, additional information is given to help identify the map or to put a map's importance in context with the other maps.

6.2 JOHANN BAYER

6.2.1 Life and times

Johann Bayer was born in 1572 in Germany in the village of Rain, located about 80 miles northeast of Augsburg. Little is known about his life, in part because several distinguished individuals in southern Germany had the same family name. He attended the local Latin school and possibly continued his education in Augsburg. His family was probably Catholic. In 1592 he matriculated at the University of Ingolstadt, where he studied philosophy and law. After leaving school, he practiced law in Augsburg. Bayer had other interests as well, including mathematics and archeology. However, his passion and later claim to fame was astronomy, and he was especially interested in the locations of the heavenly bodies in the sky. This passion resulted in his publishing a star atlas, which he dedicated to two prominent citizens of Augsburg, who subsequently rewarded him with an honorarium and promoted his civic career. Basil Brown (1932) has written that on December 18, 1612, Bayer was appointed municipal counsel in Augsburg (in whose records his name was spelled "Bayr"). His epitaph in the local Dominican cemetery states that he was never married and that he died in Augsburg on March 7, 1625.

Figure 6.1. A later-colorized image of the constellation Bootes, from a mid-1600 edition of Bayer's *Uranometria.* 27.5 × 37.8 cm. Note the geocentric orientation, the use of Greek letters to indicate stellar magnitude, the use of an accurate grid system in the margins, a bundle of wheat instead of Coma Berenices to the lower right of the image, and the five bright stars of the handle of the Big Dipper to the upper right. *See also* Color Plate Gallery.

6.2.2   *Uranometria*

Bayer's atlas was entitled *Uranometria Omnium Asterismorum* ... and was a major
step forward in celestial cartography. First published in 1603 in Augsburg, it consisted of 51 star maps engraved on copper by Alexander Mair. There were 48 plates
for each of Ptolemy's classical Greek constellations. The constellation name did not
appear on the map itself but was included in a separate constellation catalog that
accompanied each map. The 1603 edition had catalog information written on the
verso of the plates that sometimes showed through; this was not the case in subsequent editions, where the verso was blank. Each constellation map had a carefully
engraved grid using a trapezoidal projection and geocentric orientation, and the
margins were calibrated for each degree (see Figure 6.1). This permitted star positions
to be read off to a fraction of a degree from the margins using a simple straightedge.
The Ursa Major map was 27.3 × 37.3 cm in size. Map number 49 (see Appendix C)
was less precise but depicted the Magellanic Clouds and 12 new post-Ptolemy southern constellations charted by various sources (e.g., Vespucci, Corsali, Keyser, De
Houtman). The final two maps were celestial planispheres that were labeled on the
page preceding the plate as *Synopsis Coeli Superioris Borea* (or *Inferioris Austrina*).
The maps were centered on an ecliptic pole down to 15 degrees S lat. using a polar
equidistant projection and an external orientation. Neither had constellation figures.

On his maps, Bayer generally categorized each star according to its magnitude by
using Greek letters, with "alpha" being the brightest star in the constellation, "beta"
the second, etc. Astronomer and science writer Joseph Ashbrook (1984b) has pointed
out that the ordering of stars from "a" to "g" in the constellation Ursa Minor in
Piccolomini's atlas is the same as the ordering from "alpha" to "eta" in Bayer's atlas,
suggesting that Bayer borrowed from Piccolomini while developing his star atlas.
However, it should be noted that when Bayer ran out of Greek letters, he used
Roman letters for the less bright stars that he labeled.

Unlike Piccolomini, Bayer included mythological figures in his plates, which were
seen as being an aid to locating the stars in the constellation. Curiously, many of the
constellation figures were shown from the back, even though the star patterns
themselves were depicted geocentrically as seen from the Earth. This rendered the
classical descriptions of the stars in such constellations inaccurate. For example,
the star Rigel, described by Ptolemy as being in the left foot of Orion, appeared
in the right foot of Bayer's Orion figure.

The basis of this atlas was the catalog of 1,005 stars observed by Tycho Brahe,
although Bayer revised some of the magnitudes and added an additional 1,000 stars
of his own. In the separate catalogs that accompanied each of Bayer's constellations,
the stars were listed according to their Ptolemaic numbers in the first column. This
was followed in subsequent columns by their corresponding Greek (or Roman) letter;
a description of where the star was in the constellation figure, and finally by further
magnitude information.

Bayer's atlas was very popular and went through several editions without the
accompanying text. In her classic catalog of astronomical maps, astronomy historian
Deborah Warner (1979) lists these as occurring in 1624, 1639, 1641, 1648, 1655, 1661,

1666, and 1689. In addition, the text without the maps was published as *Explicatio characterum aeneis Uranometria imaginum* ... in 1624, 1640, 1654, 1697, and 1723.

A list of the plates from my examination of the 1603 edition of the *Uranometria* in the British Library is given in Appendix C. Brown (1932) also gives a list of the constellations as well as the numbers of stars in each magnitude range (for a total of 1,706) that are found in this atlas.

6.2.3   Derivative atlases

6.2.3.1   *Julius Schiller*

The *Uranometria* set the standard for future celestial atlases due to its beauty and accuracy. One such atlas that was influenced by Bayer was Julius Schiller's *Coelum Stellatum Christianum* ..., published in 1627 in the same year of his death. Although there are some unknowns about the man, he most likely was a prominent Catholic figure who like Bayer practiced law in Augsburg and was an amateur astronomer. He was interested in renaming all of the constellations according to Biblical references. For example, Pegasus became the Archangel Gabriel, Taurus became St. Andrew, and Argo Navis became Noah's Ark. Schiller persuaded Bayer to help him with this project, and he also obtained advice on observational data from Tycho Brahe and Johannes Kepler.

The resulting atlas (*Coelum Stellatum Christianum*) contained 49 constellation maps and two hemispheres. The constellation maps used a trapezoidal projection with an external orientation, and the Ursa Major map (labeled "Naviculae S. Petri", or the boat of St. Peter) measured 23.8 × 30.8 cm. All of the figures were shown from the front, since Schiller thought it would be disrespectful to show them from the back. Schiller's atlas indicated for the first time the great nebula in Andromeda (now known as M31), which was not reported by Ptolemy or Tycho Brahe and did not appear in Bayer's atlas. The northern constellations depicted figures from the New Testament, the southern constellations depicted figures from the Old Testament, and the zodiac constellations depicted the 12 apostles. The ordering of constellations in Schiller's atlas was the same as that in Bayer's, although the plates were slightly smaller. In addition, Schiller's plates showed faint images for constellations around the featured constellation, whereas Bayer only depicted the image for a main constellation. Compare Schiller's "St. Sylvester" (a.k.a. Bootes) in Figure 6.2 with Bayer's Bootes in Figure 6.1. The two hemisphere maps were not centered on an ecliptic or equatorial pole but on the Sun's location in the zodiac during the vernal or autumnal equinox. They used a stereographic projection with external orientation, and each measured 24.2 cm in dia.

On the left-hand page before each map on the right was a printed table giving the Christianized and Ptolemaic names for the constellation. Included in the table were the Roman number and position of the main stars in the constellation with two descriptions: one with reference to the new Christian figure, and one for the old pagan constellation (see Figure 6.3). Supplementary information on the constellation was also included.

Figure 6.2. The Christianized constellation St. Sylvester (a.k.a. Bootes), from the 1627 edition of Schiller's *Coelum Stellatum Christianum*. 23.8 × 30.6. Note the external orientation, the five bright stars of the Big Dipper to the upper left of the image of the saint, and the wand with ribbons instead of Coma Berenices to the lower left. *See also* Color Plate Gallery.

Although Schiller's atlas was of high quality, his Christianized constellations never achieved wide acceptance. He also published in 1627 a counterproof edition of his atlas (*Coelum Stellatum Christianum Concavum . . .*), which contained constellations with geocentric star patterns but without accompanying figures.

6.2.3.2   Aegidius Strauch

According to Deborah Warner (1979), Aegidius Strauch was born in 1632 and died in 1682. He was a Professor of Mathematics at the University of Wittenberg. Later, he became a Professor of Theology at the Gymnasium in Danzig. In 1659 he published a popular astronomy book entitled *Astrognosia Synoptice et Methodice in Usum Gymnasiorum Academicum Adornata*. Written for students, this popular book went through four editions by 1684. Warner states that it contained 32 celestial maps in geocentric orientation, each of which showed one or two constellations that were modeled after Bayer.

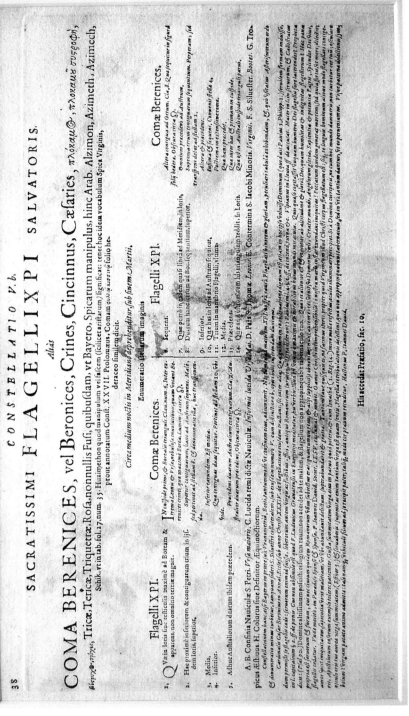

Figure 6.3. The verso of the print shown in Figure 6.2, giving the title and star table for the next constellation in the atlas, *Sacratissimi Flagelli XPI Salvatoris* (a.k.a. Coma Berenices). Note the star table giving information about the location of the star in both the pagan (Ptolemaic) and Christianized versions.

6.2.3.3 *Ignace-Gaston Pardies*

Ignace-Gaston Pardies was born at Pau, France, on September 5, 1636. He entered the Society of Jesus on November 17, 1652. For a time he taught classical literature and wrote a number of short works, both prose and verse. After he was ordained as a Jesuit priest, he taught philosophy and mathematics at the College of Louis-le-Grand in Paris. He published a number of works on the physical sciences, mathematics, and optics in several sources, including the *Philosophical Transactions of the Royal Society*. He also corresponded actively with a number of prominent scientists of his age, such as Newton, Leibniz, and Huygens. He died on April 22, 1673, of a fever he contracted while ministering to the prisoners of Bicetre, near Paris.

His famous atlas of six celestial maps was printed posthumously in 1674 and was entitled *Globi Coelestis in Tabulis Planas Redacti Descriptio, Auctore R. P. Ignatio Gastone Pardies* ... The first and sixth maps were centered on the north or south equatorial poles down to 45 degrees N or S dec., respectively. The middle four maps were centered on the vernal equinox, the summer solstice, the autumnal equinox, and the winter solstice, respectively. All of the maps used a gnomonic projection with a geocentric orientation, and each had explanatory texts in Latin down the left side and French down the right side (except for the first map, which was entirely in Latin). The plate centered on the vernal equinox contained an asterism of the fleur-de-lis just above the constellation Aries; in later French maps this was labeled Lilium. The plate centered on the summer solstice contained Ursa Major and was 47.8 × 48.6 cm.

The constellation images were influenced by those depicted in the atlas of Bayer and the globes of Blaeu. An important feature of Pardies' atlas was the inclusion of the paths through the constellations of a number of historically important comets, which the Linda Hall Library catalog (see Ashworth, 1997) attributes to the influence of an important cometary book by Lubieniecki, published in 1666–1668 (see Section 8.5.4.1). A second edition of the *Globi Coelestis* was published in the early 1690s and included additional comets along with the positions of southern stars as determined by Halley. Brown (1932) mentions a third edition ca. 1700 that was nearly identical to the second. Pardies' maps were popular among Jesuit astronomers and influenced the works of later cartographers.

6.2.3.4 *Augustine Royer*

The French architect and cartographer, Augustin Royer (active mid-1600s), published a set of four celestial maps in 1679 under the title of *Cartes du Ciel Réduites en Quatre Tables, Contenant Toutes les Constellatios* ... Also included were tables of stellar longitudes and latitudes. The first and last maps were centered on the north or south ecliptic poles down to about 23.5 degrees N or S lat., respectively. Each was about 34 cm in diameter and used a polar equidistant projection with a geocentric orientation. The north map contained the new constellation of Sceptrum (see Section 4.4). The other two maps were centered on the summer or winter solstice and covered half of the zodiac from equinox to equinox along the ecliptic in an area ranging from 35 degrees N lat. to 35 degrees S lat. These rectangular maps each measured 17.8 × 45 cm and also had a geocentric orientation.

A French translation of Halley's catalog was also published in 1679 that was labeled *Catalogue des étoiles australes ...* It included a copy of Halley's southern hemisphere map that was centered on the south ecliptic pole, was about 37 cm in diameter, and used a polar stereographic projection with a geocentric orientation. This catalog and map were sometimes bound with Royer's other four celestial maps.

Artistically, Royer's constellation images and geocentric orientation were similar to Bayer, and he was one of the first people to adopt Bayer's Greek-lettering system to identify the brighter stars. But unlike Bayer, all of the constellation figures were shown from the front. Royer also added some new stars and constellations, of which only two survive: Columba and Crux Australis (the Southern Cross). According to Brown (1932), Royer was the first person to clearly show the Southern Cross as a distinct constellation in a flat star map (Bayer's atlas showed the asterism, but it was a part of Centaurus). Copies of Royer's maps are very rare, and they influenced the maps and globes of Coronelli.

6.2.3.5   John Bevis

Another person influenced by Bayer was John Bevis, and his story is one of the most interesting in celestial cartography. He was born to a prosperous family in Old Sarum, Wiltshire, England, on October 31, 1695. He matriculated at Christ Church, Oxford, on April 4, 1712, where he studied medicine. After spending some time in France and Italy, he returned to England by 1728 and by 1730 had a thriving medical practice in London. He was also interested in optics and learned to make lenses.

He gradually found himself pursuing his interest in astronomy until it became a full-time occupation. In 1731 he became the first European to record the Crab Nebula, 27 years before Charles Messier listed it as M1 in his famous catalog. On May 28, 1737, he made the only recorded visual telescopic observation of the occultation of one planet by another when he noted Venus eclipsing Mercury. He engaged in an active program of telescopic observations of the sky, and in 1738 he built an observatory just north of London at Stoke Newington. He communicated with a number of other astronomers, including James Bradley, the Astronomer Royal. He wrote a number of papers on eclipses, comets, and occultations for the *Philosophical Transactions*, confirmed the effects of aberration in right ascension, and was one of the first people to see the Great Comet of 1744. In 1749 he published an edited version of Edmond Halley's posthumous astronomical tables. He also observed Halley's Comet in 1759 and the transits of Venus in 1761 and 1769. He was elected to the Berlin Academy of Sciences in 1750, and he became a Fellow of the Royal Society in 1765, serving as its Foreign Secretary for four years. He died on November 6, 1771, after falling from his telescope while measuring the meridian altitude of the Sun. He never married and was described by his estate's executor as a cheerful and kind-hearted individual.

During the late 1730s, Bevis began to think about compiling a star atlas that would surpass those of Bayer and Flamsteed (see Section 6.4). With financial backing from the instrument maker John Neale, much of Bevis' time from 1746 to 1750 was spent on this project. The plan was to raise funds to support the costs of engraving and to acknowledge the subscribers in the atlas itself. The atlas, which was initially to be called *Uranographia Britannica*, was patterned after Bayer's atlas (see Figure 6.4): it used an ecliptical coordinate system; it employed Bayer's Greek letters to label the brightest stars; it had identical constellation images; and it included the same number of plates that covered the same area of the sky but were slightly smaller (the northern celestial hemisphere measured 30 cm in diameter, and the map of Ursa Major measured 26.8 × 36.2 cm). Plates 1–49 were unlabeled and followed the order and style of Bayer's atlas. Plates 50 and 51 showed hemispheres of the constellations that were labeled as being—according to Ptolemy—north and south of the ecliptic, respectively, and each measured about 29.6 cm in dia. The hemisphere maps used a polar stereographic projection with an external orientation, and the constellation maps used a geocentric orientation.

But there were some differences between the two atlases. Bevis' atlas was updated: stars were added from the catalogs of Hevelius, Flamsteed, Halley, and

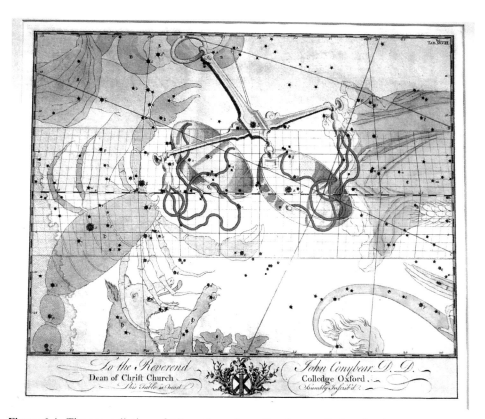

Figure 6.4. The constellation of Libra, the only zodiacal constellation that is not a person or animal, from Bevis' *Atlas Celeste*, ca. 1786. 26.8 × 36.3 cm. Note the central grid area that represents the zodiac and the dedication at the bottom to the Dean of Christ Church, Oxford. *See also* Color Plate Gallery.

Bevis himself (for a total of over 3,500); stellar positions were corrected for the epoch of 1746; there were more constellations (including 10 introduced by Hevelius and patterned after his atlas—see Section 6.3), for a total of 79; a few novae and supernovae were shown; and for the first time in any star atlas a number of "nebulae" were depicted (including the Crab Nebula and what we now know to be the Great Galaxy in Andromeda). In addition, the star images were smaller than those of Bayer, making the plates appear more realistic and less cluttered, and neighboring constellations were shown in each plate, although they were not highlighted. The hemispheres also had constellation figures in Bevis' atlas, and Bootes held the reins of two dogs, Asterion and Chara, in the hand that held his scythe, which as seen in Figure 6.1 do not appear in Bayer's image (see also Figure 6.6). Finally, at the bottom of each constellation map was a 3.4 cm tall dedication strip addressed to a person or organization that gave some financial support for the atlas. For example, the first plate featuring Ursa Minor was dedicated to Oxford, and the second plate featuring Ursa Major was dedicated to Cambridge. The British Library has a proof set of Bevis' atlas that includes blank dedication spaces on the otherwise completed maps.

Some of the star charts began to be printed in 1749, and no expense was spared in making them accurate and artistically beautiful. However, this led to the atlas' demise, in that the next year John Neale became bankrupt and the copper plates were sequestered and later destroyed. Although no formal editions of this atlas exist, a few near-complete sets were compiled, which included star tables for the plates and a star catalog of 3,551 entries that he made. In 1786, bound sets of Bevis' plates (complete with an explanatory advertisement) appeared on the market under the title *Atlas Celeste*, which is the name most associated with the maps. These were assembled anonymously, probably from individual charts resulting from an estate auction in 1785 that included Bevis' work.

6.2.3.6   Philippe de la Hire

In 1705, this famous French astronomer and polymath produced two influential celestial hemispheres whose constellations were based on Bayer. These will be described more fully in Section 7.12.

6.3   JOHANNES HEVELIUS

6.3.1   Life and times

Johannes Hevelius was born in the Hanseatic town of Danzig (later to become Gdansk, Poland) on January 28, 1611, the son of a wealthy brewer and property owner. Wanting his son to follow in his footsteps as a businessman, his father sent him to nearby Poland to study the language when he was 9 to 16 years of age. After he returned, he continued his studies in Danzig, where he became interested in mathematics and astronomy. Under the influence of his teacher Peter Kruger,

Hevelius began observing the heavens. He also learned the scientific language of Latin, became experienced in drawing and copperplate engraving, and became proficient in making astronomical instruments out of wood and metal. In 1630, at the age of 19, Hevelius' parents sent him to Holland to study law, and over the next four years he traveled throughout Europe to expand his education. During this time, he met a number of prominent scientists in London and Paris. He returned to Danzig in 1634, where he became involved in his father's business and city government, later becoming a council member of the old town ("Altstadt"). At the age of 24, he married his wife Katharina, the daughter of a distinguished Danzig businessman, and they settled down to the life of a successful merchant family.

Upon the death of his old mentor, Peter Kruger, Hevelius' interest in astronomy was rekindled, and he observed the June 1, 1639 annular eclipse of the Sun with great diligence. He soon became interested in the Moon and its eclipses, but he realized that an accurate lunar map did not exist. Hevelius decided to remedy this situation by producing an accurate and detailed map of the Moon. This led to his publication in 1647 of *Selenographia*, the first lunar atlas. This monumental work, the first of Hevelius' three great tomes, made Hevelius famous throughout Europe, and it will be described in Section 8.5.2.1.

In 1649, Hevelius' father died, and he assumed sole ownership of the family brewery. Despite being occupied in his business pursuits, he continued to make astronomical instruments and observe the heavens. In March, 1662, Hevelius' wife died, the couple never having had any children. The next year, at the age of 52, Hevelius married Elisabeth, the 16-year-old daughter of a Danzig merchant. Despite the age disparity, Elisabeth, who was keenly interested in astronomy, became a valuable assistant to Hevelius in his astronomical observations. In addition, she maintained the household, and the couple subsequently had four children.

Hevelius' observatory in Danzig's Altstadt, called "Stellaburgum", was the best in Europe until the national observatories were established in Paris and Greenwich in the 1670s. It was visited by King John II Casimir of Poland and other dignitaries, its work was supported by a yearly grant from Louis XIV of France, and in 1664 Hevelius was elected as a Fellow of the Royal Society of London. The observatory instruments were mounted on a huge wooden platform that was built across the roofs of three adjoining houses. Also included were a museum and library; a printing press, which Hevelius used to produce his books; and a shop for making and repairing his astronomical instruments. On September 26, 1679, when Hevelius and his family were out of town, a fire erupted from a burning candle in the adjoining stable, and the observatory was destroyed. At the age of 68, Hevelius had to start over, which he did with the support of many people, including Louis XIV and the new king of Poland, John III Sobieski. Within two years Hevelius was again observing the heavens, and he continued his astronomical activities until his death on January 28, 1687, his 76th birthday. Before he died, Hevelius published a second great book in two parts: *Machinae Coelestis* ... This book described and illustrated the observing instruments and methods he used at his observatory, as well as giving a plethora of his data in terms of tables and figures. It will be described further in Section 8.3.5.2.

6.3.2   The *Prodromus Astronomiae* and the *Catalogus Stellarum Fixarum*

Hevelius' third great book was actually completed and published posthumously in 1690 by his wife Elisabeth, with support from King Sobieski. It was entitled *Prodromus Astronomiae . . .*, and it consisted of three parts that had separate cover sheets: a preface (simply labeled *Prodromus*), a catalog of stars (named *Catalogus Stellarum Fixarum*), and an atlas of the constellations (named *Firmamentum Sobiescianum, sive Uranographia*). The preface contained comments on the technology and methodology that were related to the production of the star catalog, including examples of how Hevelius calculated a star's longitude and latitude from observations he made with his sextant and quadrant, as well as tables of the position of the Sun. The page numbers of the preface and the catalog were continuous.

The catalog was printed from Hevelius' hand-written manuscript that had miraculously escaped the 1679 fire that destroyed his observatory, and it is currently at the Clark Library at Brigham Young University in Provo, Utah (where it was acquired as the one-millionth volume!). Measuring 19.2 × 15.5 cm, the manuscript was composed of 183 leaves, 145 of which contained star positions. Also, in the hand-written version, the constellations follow one another alphabetically, with individual stars in each constellation listed according to decreasing magnitude. For each star, there was a series of columns containing such information as the location of the star in the constellation; its reference number and magnitude according to Tycho Brahe's catalog; Hevelius' own magnitude estimation; the star's ecliptic coordinates in longitude and latitude as measured by its angular distances (given in degrees/minutes/seconds) from reference stars using Hevelius' sextant; the star's longitude and latitude as calculated from meridian altitudes that were measured using Hevelius' azimuthal quadrant; corresponding measurements from other observers, such as Tycho Brahe, Ulugh Beg, and Ptolemy; and the star's equatorial coordinates in right ascension and declination as calculated from its ecliptic coordinates using spherical trigonometry.

The printed *Catalogus* found in the *Prodromus Astronomiae* had 107 pages, each of which measured 35.5 × 22.0 cm in total page size. It was similar to the manuscript, except that the two columns describing a star's ecliptic coordinates were combined into one column, where the source of the measurements (sextant or quadrant) was unspecified. In addition, only the best single value for the star's longitude and its latitude was given, rather than several similar values as in the manuscript. The *Catalogus* contained more than 600 new stars, for a total of 1,564, and 12 new constellations first recorded by Hevelius. Despite being constructed from naked eye observations, it was an important advance over previous star catalogs, and its measurements were used in the making of many celestial globes in the early 18th Century.

6.3.3   *Firmamentum Sobiescianum*

Hevelius' star atlas, *Firmamentum Sobiescianum, sive Uranographia*, was also part of the *Prodromus Astronomiae*, although it was likely published separately in limited

Figure 6.5. The northern celestial hemisphere from Hevelius' *Firmamentum Sobiescianum*, first published in 1687. Approximately 50.8 × 49.5 cm, 46.5-cm dia. hemisphere. Note the external orientation and the nicely drawn constellation images. Courtesy of the Collection of Robert Gordon. *See also* Color Plate Gallery.

circulation in 1687, and it had its own cover page and pagination. The atlas consisted of two hemispheres and 54 double-page plates of 73 constellations, including the new ones that he invented. The northern and southern celestial hemispheres were centered on an ecliptic pole and used a polar stereographic projection with external orientation, and the northern map was 46.5 cm in diameter. Most of the star locations were based on Hevelius' own measurements. However, the positions of the southern polar stars were based on the catalog and map published in 1679 by Edmond Halley, which resulted from the observations he made of 341 southern stars at St. Helena in 1676 (see Section 4.3.2.4). In the corners of the hemispheric maps were angelic putti, those at the top holding the labels and those at the bottom working with astronomical instruments like those in Hevelius' observatory (see Figure 6.5).

Figure 6.6. "Fig. F." constellations from Hevelius' *Firmamentum Sobiescianum, sive Urano-graphia,* first published in 1687. 29.2 × 36.3 cm. Note the external orientation, Bootes (shown from the back) standing on Mons Maenalus and holding his two hunting dogs (now called Canes Venatici), Corona to his right, Coma Berenices to the lower left, and the handle of the Big Dipper in Ursa Major to the upper left.

Each of the remaining 54 maps was labeled and featured one or a few constellations. As in Bayer's atlas, the stars were plotted using a trapezoidal projection and ecliptic-centered coordinates. However, unlike Bayer, the orientation was external, and the stars were not labeled with Greek or Roman letters. The constellation images were very pleasing, and being an engraver Hevelius managed to produce clear, beautiful images. Figure 6.6 shows Bootes and the constellations Hevelius introduced: the hunting dogs (Canes Venatici) and Mons Maenalus (now obsolete). See list in Appendix C and another example in Figure 4.8.

Despite the artistry and accuracy of Hevelius' catalog and atlas, and the fine instruments he used in his work, he was still a throwback to earlier times. Harkening back to Dürer, his star patterns and constellation figures were mirror-reversed, as on a celestial globe. His stars were unlabeled, reminiscent of the Ptolemaic system of identifying stellar location primarily by position in a constellation (which is how they were identified in his star catalog). And he refused to mount telescopic sights on his quadrants and sextants, unlike many of his contemporaries, claiming that the naked eye was more accurate for positional work. These bows to tradition were typical of Hevelius. Nevertheless, his work influenced a number of other people, as will be shown below. Curiously, many of them were German, who might have been expected to support the work of their fellow countryman, Johannes Bayer, than that of the Polish astronomer Hevelius.

6.3.4   Derivative atlases

6.3.4.1   *Johann Zahn*

Johann Zahn was a German mathematician and mystic who was born in 1641 and died in 1707. In 1696 he published a natural history book entitled: *Specula Physico-Mathematico-Historica Notabilium ac Mirabilium Sciendorum* ... It was an

Figure 6.7. The northern celestial hemisphere from Zahn's *Specula Physico-Mathematico-Historica* ..., published in 1696. 35.2 × 40.7 cm, 31.9 cm dia. hemisphere. Note that this is a near-perfect copy of Hevelius' northern hemisphere map (compare with Figure 6.5). *See also* Color Plate Gallery.

Figure 6.8. The southern celestial hemisphere from Zahn's *Specula Physico-Mathematico-Historica* ..., published in 1696. 35.2 × 40.9 cm, 31.9 cm dia. hemisphere. Note that this is a near-perfect copy of Hevelius' southern hemisphere map and likewise incorporates the work of Halley. *See also* Color Plate Gallery.

encyclopedic work dealing with a number of topics in astronomy, cosmography, astrology, botany, and zoology (involving both real and fantasied animals), along with an early history of the world. It was profusely illustrated and included maps of the Moon and its phases, the Sun, planets, comets, meteors, and a variety of terrestrial maps (including one showing California as an island).

Also included were two celestial maps that were derived from the hemispheres found in Hevelius' *Firmamentum Sobiescianum*. They were labeled *Facies Hemisphaerii Coelestis Superior Borealis* (or *Inferior Australis*) and *Secundum restitutionem noviss: Ioan. Heveli Gedan*, with the southern map adding the statement *cum Supplemento Egmundi Halleji Angli* (reflecting that—as in Hevelius' maps—Zahn's hemispheres incorporated the work of Halley's catalog of the southern stars). Each hemisphere measured 31.9 cm in diameter, was centered on an ecliptic pole, and used a polar stereographic projection with external orientation. These maps are almost perfect copies of the hemispheres of Hevelius (see Figures 6.7 and 6.8).

Figure 6.9. The double-hemisphere map from Schenck's *Atlas Contractus*, published around 1705. 48.3 × 56 cm, each hemisphere 26.3 cm dia. Note that the constellations are taken from Hevelius. The format of this map is nearly identical to maps by Eimmart and Lotter and quite similar to the double-hemisphere map by Seutter. *See also* Color Plate Gallery.

6.3.4.2 *Petrus Schenck*

Petrus Schenck was a map engraver and publisher, with shops in Amsterdam and Leipzig. He was born in 1660 and died in 1718 or 1719. In 1705 he published a book entitled *Atlas Contractus* that contained a 48.3 × 56 cm plate that featured two side-by-side celestial hemispheres, each 26.3 cm in diameter and centered on the ecliptic poles (see Figure 6.9). The projection was polar stereographic, and the orientation was external. The maps were close copies of the hemispheres of Hevelius. A large downward-curving ribbon at the top gave the name of the map: *Planisphaerium Coeleste*, and a smaller upward-curving ribbon at the bottom added: *Secundum Restitutionem Hevelianam et Hallejanam*. Across the top of the map were three circular diagrams illustrating the cosmologies of Tycho Brahe and Ptolemy and the effect of the Moon on the Earth's tides. Across the bottom were three additional circular diagrams illustrating the phases of the Moon, the cosmology of Copernicus, and the Earth's orientation as it circled the Sun. Outside the bottom border was an attribution to "Petri Schenck". According to Warner (1979), this map was also bound in R. and J. Ottens' *Atlas Sive Geographia Compendia*, published around 1756.

Together with his partner Gerard Valk, Schenck republished in 1708 Cellarius' *Harmonia Macrocosmica*, which will be referred to in Section 7.1.2.

6.3.4.3   Georg Christoph Eimmart

Georg Eimmart was a German astronomer living in Nuremberg who was born in 1638 and died in 1705. Around the time of his death, he produced a 48.4 × 56.3 cm plate that featured two side-by-side celestial hemispheres, each 26.4 cm in diameter and centered on the ecliptic poles. The constellation images followed those of Hevelius, and the format of the map (including the three upper and three lower circular diagrams) was nearly identical to that of Schenck (see Section 6.3.4.2). The only major differences were that the second word in the title was spelled with an "a" (*Caeleste*), and the attribution at the bottom margin was to "G. C. Eimmarti". This map was quite popular, and according to Warner it was copied by both Schenck and Lotter (see Section 6.3.4.7).

6.3.4.4   Johann Leonhard Rost

Johann Rost was an astronomer living in Nuremberg who was born in 1688 and died in 1727. In 1723 he published a general astronomy text entitled: *Atlas Portatilis Coelestis*. This book contained 14 plates that were divided into panels showing one or more constellations, for a total of 41 celestial images. According to Warner, the plates measured 16 × 8.4 cm, and the orientation was external. However, the star positions were relatively inconsequential, and the images generally had a non-existent coordinate system. The figures were strongly influenced by Hevelius.

6.3.4.5   Mattheus Seutter

Mattheus Seutter was born in 1678 in Augsburg. He studied engraving with the great Johann Baptist Homann in Nuremburg (see Section 7.8.2). He settled in his native Augsburg, where he established himself as a successful cartographer and instrument maker. By 1727, he was appointed as the Imperial Geographer. His daughter married the cartographer Tobias Lotter (see Section 6.3.4.7). He died in 1756.

Around 1730, Seutter published a book in two volumes entitled *Atlas Novus Sive Tabulae Geographicae*. In the first volume, there was a 48.5 × 56 cm print with two side-by-side celestial hemispheres that was entitled *Planisphaerium Coeleste. Secundum Restitutionem Hevelianam et Hallejanam*. Each hemisphere was 26.5 cm in diameter and was centered on an ecliptic pole using a polar stereographic projection with external orientation. The constellations followed Hevelius, and the print was formatted similar to that of Eimmart/Schenck/Lotter. However, these names did not appear as an attribution in the bottom margin, and the ribbon containing the title of the map at the top curved upward and was topped by heavenly figures. In addition, instead of the three top and three bottom circular diagrams that were found in the other maps, Seutter had two diagrams at the top (of day and night on the Earth and the phases of the Moon) and five diagrams at the bottom (of the phases of the Moon, the cosmologies of Brahe, Copernicus, and Ptolemy, and the Earth's orientation as it circled the Sun). According to Warner (1979), this print was copied in a later ca. 1745 edition of the *Atlas Novus* ..., along with another print showing an armillary sphere and the path of a comet entitled *Cometa qui Anno Christi 1742 Apparuit* ..., which was 34.5 × 24 cm with geocentric constellations and a north circumpolar view.

6.3.4.6 *Christoph Semler*

Christoph Semler was a Protestant clergyman living in Halle who was born in 1669 and died in 1740. He had a great interest in astronomy, mathematics, and mechanics. In 1731 he published at atlas entitled: *Coelum Stellatum*, which consisted of 35 maps, each with one or more constellation figures. The maps varied in size up to 10 × 13 cm, and the images were geocentric. The stars and constellations were printed in white against a black background to give a sense of seeing the stellar patterns in the night sky. According to the Linda Hall Library celestial catalog (Ashworth, 1997) each of Semler's plates was printed from a woodblock and was meant only to outline the constellations and pinpoint the stars; there essentially was no coordinate system. But, reminiscent of Piccolomini (see Section 5.5), the direction to celestial north was indicated by an arrow appearing on each plate. Semler derived all of his constellations and star positions from Hevelius' *Firmamentum Sobiescianum*, although his images were not externally oriented.

6.3.4.7 *Tobias Conrad Lotter*

Tobias Lotter was a cartographer from Augsburg who was born in 1717 and died in 1777. He was the son-in-law of the famous cartographer and instrument maker Mattheus Seutter, and he succeeded him in his business. Around 1772, Lotter published a book entitled *Atlas Novus*, in which was bound a double hemisphere entitled *Planisphaerium Coeleste. Secundum Restitutionem Hevelianam et Hallejanam.* The print was 47.5 × 56.6 cm, and each hemisphere was 25.7 cm in diameter. The constellations were derived from Hevelius, and the print pretty much followed the format of the maps of Eimmart (and Schenck). However, the upper ribbon with the title was more waving than arcing downward, and the attibution in the bottom margin was to Lotter. In addition, between the three upper circular diagrams were pictured two small armillary spheres, and between the three lower circular diagrams were pictured two small globes.

6.3.4.8 *Johann Doppelmayr and Antonio Zatta*

The star charts produced by the famous German astronomer, geographer, and mathematician, Johann Doppelmayr, and the prolific Venetian printer and map publisher, Antonio Zatta, were influenced by Hevelius' *Firmamentum Sobiescianum.* See Sections 7.8 and 7.9 for more details.

6.4   JOHN FLAMSTEED

6.4.1   Life and times

John Flamsteed was born in Denby, England, on August 19, 1646. He was the son of a well-to-do maltster. His mother died when he was three, and his stepmother died when he was eight. At the age of 14, Flamsteed contracted a chronic rheumatic condition, and he was sickly throughout the rest of his life. Although he attended a university preparatory school in nearby Derby until age 16, his father decided not to send him to university owing to his medical problems. However, he was self-taught, showing an affinity for mathematics and astronomy. He obtained and read a copy of Sacrobosco's *De Sphaera* (see Section 3.5.2), as well as other astronomical texts. Flamsteed made himself a telescope and a quadrant, and he began systematic observations of the sky in his early teens. His interests included plotting variations in the Sun's altitude, predicting solar eclipses, and calculating the positions of celestial bodies. He published his findings, corresponded with scientists, and soon came to the attention of the Royal Society. In 1670, he was introduced to Sir Jonas Moore (1627–1679), the Surveyor-General of the Royal Ordnance, who became his patron. Moore persuaded King Charles II to issue a warrant allowing Jesus College, Cambridge, to grant Flamsteed an M.A. in 1674.

At that time, there was great interest in using astronomical events to calculate the location of a ship at sea. The method for calculating one's location in latitude had been known since the time of the ancient Greeks. This was based on measuring the elevation above the horizon of the Sun at noon or of the Pole Star at night. However, no reliable method existed for determining longitude. One solution was to use the changing position of the Moon with reference to the background stars as a sort of clock hand, but this required having an accurate catalog of stellar positions. The best catalog available at the time, that of Tycho Brahe, was not accurate enough. With Moore's help, Flamsteed convinced Charles II that data from existing star catalogs had insufficient accuracy to calculate longitude, and the King decided to do something about this. In 1675 he appointed Flamsteed as his Astronomer Royal and tasked him with supervising the building of a Royal Observatory at Greenwich. This was completed in 1676, and Flamsteed took up residence and began his observing program, which included accurately determining stellar positions and making observations of the Sun, Moon, planets, and comets. He was elected a Fellow of the Royal Society in 1677. Like many scientists of his day, he subsequently took his holy orders to become a reverend in the English church. He married his wife Margaret on October 23, 1692. He died on December 31, 1719.

6.4.2   *Historiae Coelestis Britannicae*

One of his major duties was to complete a telescopic catalog of the magnitudes and positions of the stars, a project that involved him for the next 43 years until his death. The making of an accurate star catalog would not only assist in the determination of longitude, but it was also desired by such scientists as Isaac Newton to test some of their theoretical ideas. Flamsteed made measurements of stellar positions from Greenwich using a variety of instruments, most of which were specially constructed and purchased with his own money. But unlike Hevelius, Flamsteed used these instruments in conjunction with a telescope, which increased the accuracy of his measurements.

He carefully made and checked his measurements over the years, refusing to publish any of his findings until he was completely satisfied with them. This perfectionist quality, together with Flamsteed's tendency to be somewhat irritable, contributed to an unfortunate occurrence involving the catalog and some of his equally strong-minded colleagues. Responding to pressure from some of the members of the Royal Society, particularly Edmond Halley and Isaac Newton (who needed accurate stellar positions to test some of his ideas regarding the Moon's motion and gravity), Flamsteed finally sent them a preliminary manuscript copy of his catalog and gave them permission to publish some of his accompanying observations (but not the catalog). Instead, in 1712 they published this material as a completed catalog, with Halley acting as editor. Halley and Newton believed they had a right to do this because the English government was paying Flamsteed as a civil servant, and English ships were being lost at sea due to inaccurate stellar positions that were being used for navigation. Even though Prince George, Queen Anne's consort, supported this publication, Flamsteed was furious, and he subsequently obtained and burned about 300 remaining copies of the 400 published. He continued to work on his catalog and had completed two of the three volumes at the time of his death.

The final three-volume folio, called the *Historiae Coelestis Britannicae*, was completed and published in 1725 by Flamsteed's wife and two former observatory assistants, Joseph Crosthwait and Abraham Sharp. Volumes I and II included tables of his observations of the stars and Moon. Volume III included the observations of others (e.g., Ptolemy, Ulugh Beg, Tycho Brahe, Hevelius), as well as Flamsteed's own catalog of nearly 3,000 stars. Unlike previous catalogs, the stars in each constellation were arranged according to increasing right ascension and included information on their location in the constellation, their longitudes and latitudes, and their magnitudes. The brighter stars used Bayer's Greek letters and not the so-called "Flamsteed numbers", which were actually introduced in 1783 by Lalande in his catalog and do not appear in any of Flamsteed's work. Flamsteed's catalog was a landmark in accuracy and was used for many decades. Interestingly, the original reason for its development disappeared within 70 years, when the invention of an accurate maritime clock by John Harrison and others made the lunar method of determining longitude at sea obsolete.

6.4.3   *Atlas Coelestis*

Flamsteed's wife and former assistants were also responsible for the posthumous publication in 1729 of his great star atlas, the *Atlas Coelestis*, which had been in development during his lifetime. When it appeared, it was the largest star atlas ever published, it contained more stars than the atlases of Bayer and Hevelius, it used

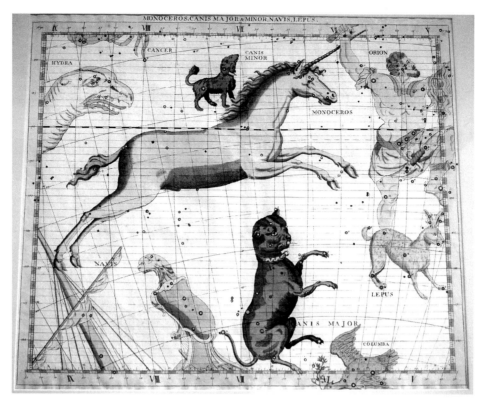

Figure 6.10. The map labeled "Monoceros, Canis Major & Minor, Navis, Lepus", from Flamsteed's *Atlas Coelestis*, published in 1729. 47.3 × 58.2 cm. Note the absence of constellations (e.g., Lacaille's "Pyxis") between Monoceros and Navis and the double-grid system, the major one centered on the celestial equator and the other centered on the ecliptic. *See also* Color Plate Gallery.

a very detailed grid system, and its stellar positions were accurately based on Flamsteed's exceptional star catalog (see Figure 6.10). The individual constellations were depicted in 25 maps using a geocentric orientation. The maps themselves were quite large: Ursa Major was 44.9 × 58.9 cm, and the double-page Hydra was 44.5 × 97.1 cm. All of the constellation figures were drawn as seen from the front according to the classical Ptolemaic descriptions (e.g., Rigel appeared in the left foot of Flamsteed's Orion), a point made explicit in the introductory comments to the atlas as an important advantage over Bayer's work. All of these maps were labeled according to their predominant constellations. In addition to these maps, there were two unlabeled celestial hemispheres added by Abraham Sharp that were centered on the south or north equatorial poles, were 50.6 cm in diameter, used a polar stereographic projection, and differed somewhat from the constellation maps in style (e.g., they used an external orientation). A list of the maps in Flamsteed's atlas is given in Appendix C.

Flamsteed's was the first major celestial atlas to use maps that emphasized the newer equatorial grid system that was centered on the projection of the Earth's equator into the sky. This more practical system corresponded to the apparent rotation of the heavens as seen from the Earth, and it could be used with greater ease with meridian circle telescopes (see Section 8.4). But as a nod to tradition, Flamsteed's plates also included a secondary grid system based on the older ecliptic system, a trait to be found in subsequent atlases for many years. As indicated in Section 1.5, the equatorial grid systems found in today's modern celestial atlases speak in terms of declination (measured in degrees above or below the celestial equator and analogous to the celestial latitude of the ecliptic system) and right ascension (measured in hours of time along the celestial equator and analogous to the celestial longitude of the ecliptic system).

Another feature of Flamsteed's atlas was the use of the sinusoidal projection (also known as the Sanson–Flamsteed projection—see Section 1.6). This system attempted to improve the distortions in constellation star patterns found in earlier trapezoidal projections used by Bayer, Hevelius, and others by more accurately reproducing the situation that exists on a sphere. But despite this more sphere-like feature, the sinusoidal projection only marginally improved star pattern distortions, which are inherent when one tries to represent three-dimensional figures on a two-dimensional piece of paper.

A major drawback of Flamsteed's atlas was the large size of the maps, which made them unwieldy to use during telescopic observations. What was needed was a smaller version of the atlas. In addition, the constellation figures were not as graceful or aesthetically pleasing as those found in other atlases. Nevertheless, in terms of its accuracy of stellar positions and the many advances listed above, it was widely used for decades and influenced a number of subsequent atlases, which will be mentioned below. Warner (1979) states that Flamsteed's atlas went through two subsequent printings, in 1753 and 1781.

6.4.4   Derivative atlases

6.4.4.1   *John Hill*

John Hill was a physician, naturalist, and apothecary who worked in London. In addition, he was an active writer, both literary and scientific. He was born around 1707 and died in 1775.

In 1754, he published a book entitled *Urania, Or a Compleat View of the Heavens* ..., which was in fact a dictionary of astronomy from A to Z (see also Section 4.4). It also contained 13 celestial plates. One of these was an 18.9 cm diameter hemisphere centered on the S equatorial pole using a geocentric orientation. It was entitled *The Southern Hemisphere*. The other maps varied in size from 11.8–19.6 cm high and 13.8–32 cm wide; most (but not all) were wider than higher. The plate with Ursa Major was 16.6 × 19.9 cm. They contained one or a few constellations, and the main constellation was titled in the lower right corner. Most of these maps showed lines of latitude that were parallel to the upper border and were

5 degrees apart; there were no lines of longitude. The orientation of the constellations was geocentric.

The traditional constellations were derived from Flamsteed. In addition, Hill added a number of constellations of his own invention that were taken from the animal world, such as a scaly lizard, a black snail, an earthworm, a leach, and different kinds of shellfish. Although these were not taken up by any other astronomers, the book was popular enough for a second edition to be published in 1768.

6.4.4.2   Jean Fortin

Jean Fortin was an artisan and globe maker for the French king and royal family. He was born in 1750 and died in 1831. It is unclear why he was commissioned to produce a revised edition of Flamsteed's atlas in France, but the enterprise was very successful. The resulting *Atlas Céleste de Flamsteed ...* was first published in 1776 (and labeled the second edition of Flamsteed's original) and again in 1795 (labeled the third edition). Both used smaller re-engravings of Flamsteed's plates, which were about $\frac{1}{3}$ the original size. As in the original, the perspective was geocentric, the constellation figures were seen from the front, and a sinusoidal or Sanson–Flamsteed projection was used (see Figure 1.5).

There were a total of 30 maps, 26 centered on the major constellations as seen from Greenwich, one new map showing Corvus and the rear part of Hydra (omitted from Flamsteed's original), a northern and southern hemisphere (*Hemisphere Boreal* or *Austral*—see Figures 4.5 and 4.6) made by Le Monnier (see Section 7.13), a map for the alignments of the principal northern stars, and a map of the southern stars devised by the French astronomer Nicolas Lacaille (see Section 4.3.2.5). The map featuring Cygnus (see Figure 1.1) measured 15.7 × 20.6 cm, and the hemisphere maps centering on the equatorial poles each measured 16.9 cm in diameter. The maps were labeled, and a list of them from the index of the 1776 edition is given in Appendix C.

The 1795 edition included some additional stars and eight new constellations (Le Mural, Le Solitaire, Le Messier, Le Taureau Royal de Poniatowski, Trophée, Le Harpe de George, and Le Petit and Le Grand Telescopes de Herschel). A Portuguese translation was published in 1804.

All of the above features made this atlas very popular, and even today one can purchase the entire atlas at a reasonable price. Notably, the great comet hunter Charles Messier used the plates from the *Atlas Céleste de Flamsteed* to illustrate the paths of comets and the location of nebulae he discovered in the late 18th Century.

6.4.4.3   Johann Bode

In 1782, Johann Bode published a German version of Fortin's atlas entitled *Vorstellung der Gestirne*. He made a number of improvements over the French work, however, and these will be described in Section 6.5 on the great German astronomer. The *Vorstellung* in turn inspired an atlas by Jamieson and a set of star cards called

Urania's Mirror, both of which were quite popular in England and will be described in Section 6.5.5. Although Bode gets the credit for directly influencing these two English works, the lineage also goes back to Flamsteed and Fortin, and their influence can also be seen in the star maps.

6.4.4.4 Maximilian Hell

Maximilian Hell was a Jesuit astronomer and first Director of the Universitäts-Sternwarte in Vienna who was born in 1720 and died in 1792. In 1789 he introduced some new constellations in a series of four maps whose title began *Monumenta aere perenniora, inter Astra ponenda ...* These appeared in the 1790 Vienna issue of *Ephemerides Astronomicae ad Meridianum Vindobonensem ...* According to Warner (1979), the maps were derived from Fortin's 1776 atlas and measured 16 × 21 cm, except for the fourth, which was 28.5 cm square. They used a Sanson–Flamsteed projection with a geocentric orientation.

The new constellations honored William Herschel and his patron, King George III. In honor of the latter was the constellation Psalterium Georgianum, and in honor of the former's telescopes were the constellations Tubus Herschelii Minor and Tubus Herschelii Major. These instruments were placed on either side of the location of Uranus in 1781 when Herschel discovered it. These constellations appeared in Fortin's 1795 atlas, Bode's monumental 1801 atlas (see below), and several 19th-Century atlases, although they no longer exist today (see Section 4.4).

6.4.4.5 Kornelius Reissig

Little is known about Reissig, except that he was an honorary member and correspondent of the Russian Imperial Science Academy. However, in 1829 he published in St. Petersburg a star atlas, the name of which is translated as *Constellations Represented on XXX Tables*. Although the constellations had Cyrillic names, the basic format and figure design of these geocentric maps owed much to the Flamsteed tradition, as can be seen by comparing Figure 6.11 with Figure 6.10. The presence of "Typographia" below the unicorn's rear legs suggests Bode's influence as well (see Sections 4.4 and 6.5).

6.4.4.6 Society for the Diffusion of Useful Knowledge

The Society for the Diffusion of Useful Knowledge (S.D.U.K.) was founded in 1826 in England. Its purpose was to provide information for the common man at affordable prices. Since many of its writers and scholars worked for free or for little pay, the Society was able to publish a great deal during its existence, and for a while its works became quite popular. Eventually, however, it began to experience financial problems, and in 1846 it suspended its operations.

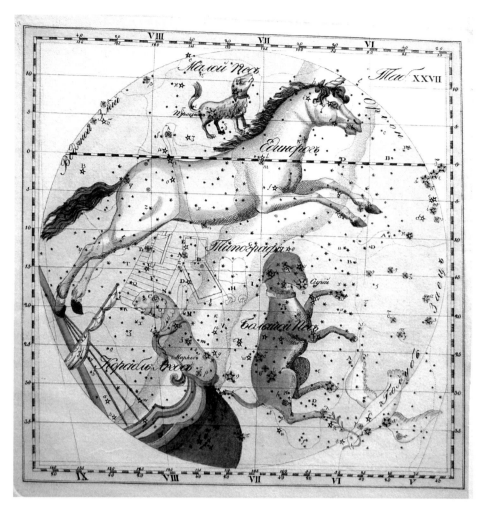

Figure 6.11. The map showing the region around Monoceros (the unicorn), from Reissig's *Constellations Represented on XXX Tables*, published in 1829. 19.1 × 19.2 cm. Note the similarity to the map shown in Figure 6.10 from Flamsteed's atlas, although the constellations are labeled in Cyrillic. *See also* Color Plate Gallery.

From 1829 to 1843, the Society published a series of cheap and accurate maps. These included a celestial section labeled *The Constellations* that consisted of a title page, an explanation, and six star charts: two circumpolar and four centered on the solstices and equinoxes. The constellations were geocentric, the projection was gnomonic, and there was a double-grid coordinate system that was centered on the celestial equator. The chart showing the constellations around the autumnal equinox (in Virgo) was 25.2 × 24.9 cm (see Figure 6.12). Larger maps also were produced. The maps showed the influence of both Pardies and Flamsteed.

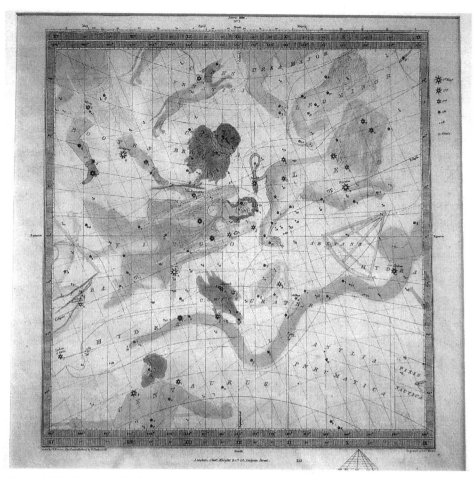

Figure 6.12. The constellations around the location of the Sun in the zodiac at the time of the autumnal equinox, from the 1833 edition of *The Constellations*, published by the Society for the Diffusion of Useful Knowledge. 25.2 × 24.9 cm. Note the influence of Flamsteed and the original pastel colors on this inexpensive, practical English celestial map. *See also* Color Plate Gallery.

6.5 JOHANN BODE

6.5.1 Life and times

Johann Elert Bode was born in Hamburg, Germany on January 19, 1747. He was the son of a Hamburg merchant. As a child, he was gifted in mathematics and largely taught himself astronomy, and he published his first treatise while still a teenager. He continued to write and soon became a prolific popularizer of astronomy and observer of the skies. For a time Bode was employed as a calculator at the Berlin Academy of Sciences.

In 1774 he co-founded their *Astronomisches Jahrbuch (Astronomical Yearbook)*, serving as its editor for over 50 years. The *Jahrbuch* improved and then flourished under his tutelage, ultimately being circulated worldwide. Also in 1774 and for the next few years, Bode observed a number of nebulae and star clusters, some of which he discovered (using Messier's nomenclature: M81, M82, M53, and M92). Many of these, along with sky objects noted by other observers, were reported in the pages of the *Jahrbuch*. In 1777 he published a deep-sky catalog of 75 objects, which appeared in the 1779 *Jahrbuch*. Most of the objects were copied from other sources and were not directly observed by Bode, and as a result his catalog contained a number of errors and non-existent entries. On January 6, 1779, he discovered a comet that bears his name, and in subsequent years he observed a number of other comets and calculated their orbits.

In 1786, Bode was elected as a member of the Berlin Academy of Sciences and also became the director of the Academy's Berlin Observatory, where he served for nearly 40 years. Over the years, he improved the observing equipment there, and he continued to observe the sky and record his findings. Bode retired from the Observatory in 1825. He continued to work on the *Jahrbuch* until his death in Berlin on November 23, 1826. But before his death, Bode published three important books that contained a series of star maps.

6.5.2 *Anleitung zur Kenntniss des Gestirnten Himmels* and Bode's Law

The first of Bode's books appeared in 1768 and was entitled *Anleitung zur Kenntniss des Gestirnten Himmels (Instruction for the Knowledge of the Starry Heavens)*. Warner (1979) states that this was the most popular introductory book on astronomy of its time, and that by 1867 it had gone through 15 German editions. It also went through two Austrian editions and was translated into Dutch and Danish. It contained a number of celestial figures. In the 1777 edition I saw in the British Library, these included a clear polar stereographic grid (labeled "Tab. I" in the upper right corner); a page with two circular charts of the solar system with insets of constellations, lunar phases, and the rings of Saturn ("Tab. IV"); and a picture of the full Moon with illustrations of the new and half Moon ("Tab. V"). There was also a 23.5 cm diameter planisphere centered on the north equatorial pole down to 38 degrees S dec. in polar stereographic projection with geocentric orientation ("Tab. II"). According to Warner, this planisphere gave rise to a similarly oriented but larger (60 cm dia.) planisphere that was published in 1787 and labelled *Stereographischer Entwurf des Gestirnten Himmels* ... Subsequent printings of this added additional constellations.

There was no celestial map labeled "Tab. III", but in its place was a series of 12 monthly star charts labeled at the top as *Vorstellung einer Gegend des gestirnten Himmels* ..., followed by the compass direction of the area of the sky represented. The name of the month for best viewing was given in the upper right margin (Figure 6.13). The charts were referenced to the celestial equator and used a polar projection with geocentric orientation. The map with Ursa Major measured 13.5 × 19 cm. The constellation images were fairly primitive, and many were drawn as seen from behind, even though the star patterns were geocentric.

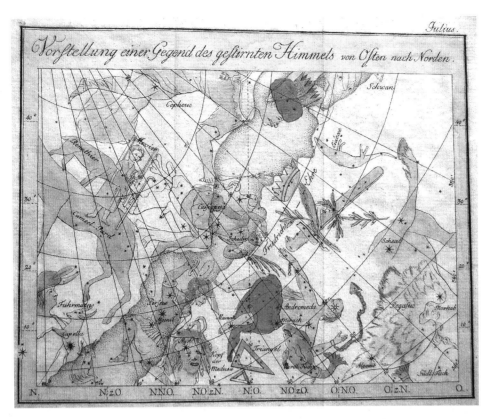

Figure 6.13. The constellations for "Julius" (July), from a ca. 1787 edition of Bode's *Anleitung zur Kenntniss des Gestirnten Himmels*. 15.5 × 19.3 cm. Note the crudely drawn constellation figures; the now obsolete constellations of the Reindeer and Messier; and the new constellation of "Friedrichs Ehre" (e.g., Frederick's glories), introduced by Bode in 1787 to honor the death of his patron, Frederick the Great. *See also* Color Plate Gallery.

In this book, Bode put forth an idea on planetary distances that was originally proposed in 1766 by Johann Titius (1729–1796) of Wittenberg. Sometimes referred to as the "Titius–Bode Law", it is usually shortened to simply "Bode's Law". This law consisted of a formula that described the relative distances of the then-known planets from the Sun. When Uranus was discovered in 1781 by William Herschel (1738–1822), its distance from the Sun seemed to fit the formula of Bode's Law. The law also predicted that a planet should exist in the space between Mars and Jupiter. A search of the area in the early 1800s did not find a planet but instead discovered a group of asteroids (see Section 8.5.1) that were thought to represent a planet that had broken up (today, we think that the objects in the asteroid belt are instead material left over from the early solar system that never coalesced). However, Bode's Law failed to predict the correct location of Neptune when it was discovered in 1846 and similarly did not account for the distance of Pluto when it was discovered in 1930, so most modern astronomers see it as a fluke that coincidentally explained several of the planetary distances of the inner planets.

6.5.3   *Vorstellung der Gestirne*

As mentioned in Section 6.4.4.3, in 1782 Bode published a smaller version of Flamsteed's atlas in German entitled *Vorstellung der Gestirne auf XXXIV Kupfertafeln nach der Pariser Ausgabe Des Flamsteadschen Himmels Atlas*. As indicated in the title, the format and constellation figures and order were similar to those found in Fortin's 1776 French Flamsteed atlas. However, the plates were newly engraved, and according to Warner 1,520 additional stars from a variety of sources were added to the 2,919 charted by Flamsteed and Fortin. There were a total of 34 charts that were numbered consecutively using Roman numerals. Two were contemporary 17 cm diameter celestial hemispheres centered on the northern and southern equatorial poles using a polar stereographic projection with geocentric orientation; two were 17.2 cm northern and southern hemispheres drawn "according to the ancients" (i.e., showing Greek and

Figure 6.14. Plate III showing Perseus and Andromeda, from the 1782 edition of Bode's *Vorstellung der Gestirne* ... 15.8 × 20.6 cm. Note the similarity of this map to the maps in the atlases of Flamsteed and Fortin. *See also* Color Plate Gallery.

Roman constellations); one was an 18.5 cm diameter hemisphere of the southern stars according to Lacaille; two were maps of nebulae and star clusters; and one was a map of the alignment of the principal stars.

The remaining 26 plates were star maps that were numbered consecutively (II–XXVII) in the upper right corner but were unlabeled. They were centered on the principal constellations that were visible from Berlin (see Figure 6.14). Each used a Sanson–Flamsteed projection with geocentric orientation and was aligned to the celestial equator. The map containing Ursa Major measured 15.8 × 21 cm. One new feature over the Flamsteed and Fortin atlases was the addition of constellation boundaries that were drawn according to the parameters of Flamsteed's star catalog. Since some of these stars carried erroneous constellation identifications, Bode's literal interpretation sometimes led him to overlap his boundary lines into another constellation or to "gerrymander" the boundaries in eccentric ways.

A second edition of this atlas was published in 1805. This contained a number of new stars and constellations, and the text was written in both German and French. A star catalog accompanied it.

6.5.4 *Uranographia*

In 1801, Bode published his great star atlas, *Uranographia sive Astrorum Descriptio* ..., by any standard a work of superlatives. In size, it was the largest celestial atlas ever produced: the celestial planisphere that was centered on the vernal equinox measured 56.7 cm in diameter, and the constellation chart showing Ursa Major was 52.5 × 73 cm. The accompanying star catalog, labeled *Allgemeine Beschreibung und Nachweisung der Gestirne ... von 17240 Sternen ...*, listed 17,240 stars down to magnitude 8, thousands more than in any other atlas. The star maps depicted more than 100 constellations and asterisms, many of which were added by Bode himself, which again set a record for total number. Finally, some 2,500 nebulae that had been catalogued by William Herschel in the late 1700s were shown in this truly monumental work.

In all, there were 20 engraved plates: two geocentric hemispheres centered on the vernal and autumnal equinoxes, and 18 star maps centered on the major constellations. A list of these maps is given in Appendix C. The star patterns were geocentric, and most of the constellation figures were seen from the front. Bode used a conic projection aligned to the celestial equator, with the hour circles being straight lines and the circles of declination being curves (see Figures 4.9 and 6.15). According to Warner (1979), Bode was encouraged to use this projection by Lalande, who thought that it showed less constellation distortion than the trapezoidal or sinusoidal projections. The brighter stars were identified with Bayer letters. Like his earlier "Flamsteed" atlas, the constellations had boundaries drawn on the plates. There were a number of original constellation designs, some diverging from conventional representations from Bayer, Hevelius, and Flamsteed, and others being original to this atlas, as we have discussed in Section 4.4. But the *Uranographia* represented the end of the era of the great pictorial atlases, since after Bode celestial atlases became

Figure 6.15. The hemisphere "Coelum Stellatum Hemisphaerium Arietis", centered around the location of the Sun in the zodiac at the time of the vernal equinox, from Bode's 1801 *Uranographia.* 56.7 cm dia. hemisphere. Note the large number of stars and constellations, more than in any previous atlas. Note also the added colorization (see Color Plate Gallery), whose poor quality suggests that it was done by non-professionals, perhaps children.

less artistic and more utilitarian, gradually dispensing with constellation figures (as will be discussed in Chapter 10).

6.5.5   Derivative atlases

6.5.5.1   *Christian Friedrich Goldbach*

Christian Goldbach was a German astronomer who was born in 1763 and died in 1811. He taught astronomy in Moscow. In 1799 he published a celestial atlas derived from Bode's *Vorstellung der Gestirne ...* which was entitled *Neuster Himmels-Atlas zum Gebrauche für Schul- und Akademischen Unterricht.* It consisted of 56 prints, 26 of which were white-on-black constellation maps centered on the major constellations north of the Tropic of Capricorn. Each map used a Sanson–Flamsteed projection with a geocentric orientation, and the Ursa Major map was 15.8 × 20.6 cm. Facing each of these maps was a print of the same area showing the white stars on a black background but without coordinates, constellations, or other markings. These were intended to simulate the view of the night sky. This pairing was well received by the astronomical community. There were also two 16.8 cm diameter black-on-white hemispheres centered on the equatorial poles using a polar projection and geocentric orientation. The final two maps were also black-on-white hemispheres: one an 18.3 cm diameter Lacaille map of the southern constellations, and the other a 16.6 cm diameter map centered on Polaris which showed no constellations but labeled the major stars.

6.5.5.2   Alexander Jamieson

Bode inspired two popular English works and an American work in the early 1800s. One of the English books was the atlas published by Alexander Jamieson, who according to Brown (1932) resided at Kensington and was interested in astronomy but did little observing himself. Jamieson also wrote a book with the interesting title: *A Grammar of Logic and Intellectual Philosophy on Didactic Principles*, as well as other books on language and science. Brown states that Jamieson prepared a series of large star maps in 1820 but never published them due to the expense.

In 1822, Jamieson published a book entitled *A Celestial Atlas*. This book contained a catalog of stars and constellations, astronomical exercises, and 30 celestial maps, some of which were hand-colored. Two of these were celestial hemispheres that were 16.6 cm in diameter and were centered on the equatorial poles using a polar projection and geocentric orientation. One additional plate showed the principal stars seen from Great Britain, and another showed the Moon and planets. The remaining 26 plates were constellation maps that used a sinusoidal projection and geocentric orientation centered on the equatorial poles (see Figure 6.16). The overall design was similar to that of Flamsteed/Fortin and Bode's *Vorstellung der Gestirne*. But in addition, Jamieson included a number of constellations developed by Bode and appearing in his *Uranographia* (e.g., Felis, Honores Friederici, Globus Aerostaticus, Machina Electrica, and Officina Typographica), as well as the new constellation Noctua the owl (Figure 9.5), which he placed on the tip of Hydra's tail. However, Jamieson's atlas included far fewer stars, since it only had those that were visible to the naked eye. This made it appear less cluttered. The book was popular and had the honor of being allowed to be dedicated to King George IV.

6.5.5.3   Urania's Mirror

A second English work inspired by Bode via Jamieson's atlas was an interesting set of star charts entitled *Urania's Mirror*. This consisted of a boxed set of 32 hand-colored constellation cards that was first published by Samuel Leigh & Co. in London in 1825

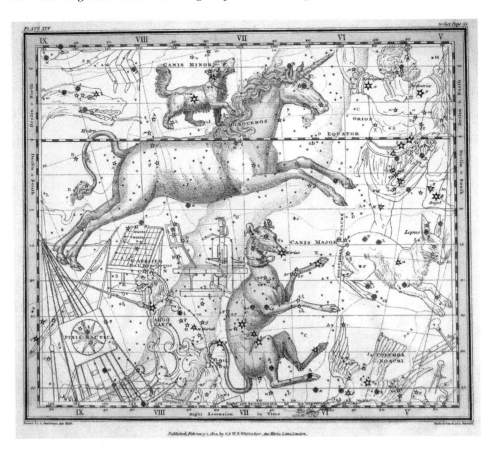

Figure 6.16. Plate XXV featuring the region of Monoceros, Canis Major and Minor, from Jamieson's 1822 *A Celestial Atlas.* 17 × 20.1 cm. Note the stylistic similarity to Bode's *Vorstellung*. Note also the constellation below Monoceros' rear legs labeled here (in French) as *l'Atelier de l'Imprimeur*, which Bode introduced in his 1801 *Uranographia* as "Officina Typographia", the printing press, to honor Gutenberg. *See also* Color Plate Gallery.

(both a first and second edition). Also included was a cover illustration of Urania, the Muse of Astronomy. The authorship of *Urania's Mirror* was mysteriously attributed to "a lady" and has remained a mystery, although the librarian of the Royal Astronomical Society, Peter Hingley (1994), has provided evidence that it was produced by Dr. Richard Rouse Bloxam (ca. 1765–1849), who was the Rector of Brinklow in England and later moved to Rugby, where he served as Assistant Master of Rugby School for some 38 years.

The cards were on stiff paper, and they measured 20 × 14 cm in total card dimension (see Figure 6.17). The constellations were clearly inspired by Jamieson's atlas, even including Noctua the owl, which was first introduced by Jamieson (see Section 6.5.5.2). However, the cards had no coordinate system, and the number of

Figure 6.17. An original hand-painted card of the constellation "Orion", from a later edition (ca. 1840) of Leigh's *Urania's Mirror*. 18.3 × 12.4 cm (image size). Note the holes punched in the brightest stars that give a representation of the constellation pattern when the card is held up to the light. *See also* Color Plate Gallery.

stars was limited. In the first edition, only the stars of the featured constellation or constellations were present; in the second and subsequent editions, a few stars outside of these constellation boundaries (which were drawn in) were added.

There were small holes punched in for each star in the main constellations, allowing the viewer to hold the card up in front of a light source and directly view the star pattern. Thus, it is apparent that one of the uses of the cards was as a self-instruction aide in learning the constellation patterns of the night sky. A total of 80 constellations were featured on the cards, not all of which are still recognized today. A facsimile of the cards was produced by Catherine Tennant in 1993 and published in London by Chatto & Windus under the title *The Box of Stars*.

6.5.5.4   Elijah H. Burritt

Bode's influence also crossed the Atlantic. In 1833 the American Elijah H. Burritt wrote a popular celestial book entitled *The Geography of the Heavens*, whose accompanying *Atlas* borrowed heavily from English sources, especially the atlas of Alexander Jamieson. Not only is the style of the constellations similar, but Burritt even included Jamieson's Noctua the owl in his atlas. More will be said about Burritt in Section 9.4.

6.5.5.5   M.C.G. Riedig

Another imitator of Bode's work was M.C.G. Riedeg, an instrument maker and cartographer who lived in Germany. In 1832, in Leipzig, he published an atlas entitled *Himmels Atlas*, and in 1841, also in Leipzig, he published a set of 20 star maps entitled *Stern-Karten in 20 Blattern*. Both of these were patterned after Bode's *Uranographia*, although they were more modest in scope. For example, the plates that featured Orion in the 1841 work measured 14.4×20.5 cm in size and had far fewer stars. But the constellation images were quite crisp, and their style was similar to that of Bode's great work (see Figure 6.18).

6.5.5.6   Joseph J. von Littrow

Joseph von Littrow was a German astronomer who in 1839 produced his *Atlas of the Starry Heavens*, which emphasized the stars (complete with Bayer letters) and reduced the constellations to faint outlines. However, he generally followed the tradition of Bode and included all of his constellations. In fact, according to Ashworth (1997), this was the last time that all of Bode's constellations would appear in a professional celestial atlas. See Section 10.1.2.3 for more on von Littrow.

6.5.5.7   G. Rubie

Rubie was a British educator who in 1830 published *The British Celestial Atlas* in London. Notably, this atlas was organized using a series of monthly star charts. According to Ashworth, Bode was the first person to show how the sky changes month by month in an astronomy textbook he wrote in 1768. Rubie was one of the first persons to pick up on this idea in his own atlas.

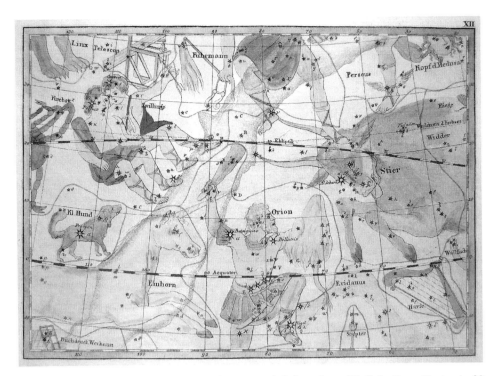

Figure 6.18. Plate XII featuring the region around Orion, from Riedig's *Stern-Karten in 20 Blattern.* 14.4 × 20.5 cm. Note the attractive constellation images, the ecliptic and celestial equator lines, and the constellation of the harp, depicted in Bode's *Uranographia* as Harpa Georgii (honoring King George III of England). *See also* Color Plate Gallery.

6.5.5.8   *Kornelius Reissig* (see Section 6.4.4.5)

6.5.5.9   *Karl Friedrich Vollrath Hoffmann*

Hoffmann (1796–1842) was a German geographer and cartographer who managed a geographical institute in Stuttgart and lectured at the University of Munich. He wrote several books, including a celestial atlas in 1835 entitled *Vollständiger Himmels-Atlas . . . (Complete Atlas of the Heavens . . .).* It contained 32 lithographed star charts with charming images and constellations (e.g., Felis, Officina Typographica, Friedrichs Ehre) introduced by Bode.

6.6   BIBLIOGRAPHY

Ashbrook, J. (1984a) Chapter 41. The first observed occultation of one planet by another. *The Astronomical Scrapbook.* Cambridge, U.K.: Cambridge University Press, pp. 209–212.

Ashbrook, J. (1984b) Chapter 78. Johann Bayer and his star nomenclature. *The Astronomical Scrapbook*. Cambridge, U.K.: Cambridge University Press, pp. 411–418.

Ashbrook, J. (1984c) Chapter 79. Hevelius: his star catalogue and atlas. *The Astronomical Scrapbook*. Cambridge, U.K.: Cambridge University Press, pp. 419–427.

Ashworth, W.B. Jr. (1997) *Out of this World—The Golden Age of the Celestial Atlas*. Kansas City, MO: Linda Hall Library.

Brown, B. (1932) *Astronomical Atlases, Maps and Charts*. London: Search Publishing.

Hingley, P.D. (1994) Urania's Mirror—a 170-year old mystery solved? *Journal of the British Astronomical Association*, **104**, 238–240.

Johnston, P.A. (1985) *Celestial Images—Astronomical Charts from 1500 to 1900*. Boston, MA: Boston University Art Gallery.

Kanas, N. (2002) Mapping the solar system: Depictions from antiquarian star atlases. *Mercator's World*, **7**, 40–46.

Kanas, N. (2005) Are celestial maps really maps? *Journal of the International Map Collectors' Society*, **101**, 19–29.

Kanas, N. (2006) The cartographic legacy of John Flamsteed. *Sky & Telescope*, July, 30–36.

McCarroll, S. (2005) *Celestial Images: Antiquarian Astronomical Charts and Maps from the Mendillo Collection*. Boston, MA: Boston University Art Gallery.

Ritter, M. (2001) Seutter, Probst and Lotter: An eighteenth-century map publishing house in Germany. *Imago Mundi*, **53**, 130–135.

Volkoff, I.; Franzgrote, E.; and Larsen, A.D. (1971) *Johannes Hevelius and His Catalog of Stars*. Provo, UT: Brigham Young University Press.

Warner, D.J. (1979) *The Sky Explored: Celestial Cartography 1500–1800*. Amsterdam: Theatrum Orbis Terrarum.

7

Other important star maps of the Golden Age

The four people highlighted in Chapter 6 were very important for their innovations and influences on others. Each made contributions that added to the accuracy of stellar location in the sky and included more stars in the process. Although attractive visually, their atlases were not necessarily the most beautiful of the Golden Age. In fact, in terms of sheer beauty, some of the works from people discussed in this chapter were superior. Other atlases had wide appeal for different reasons and were quite popular with the public.

In the maps described below, the same guidelines and conventions are followed as were described in Section 6.1. We will begin with one of the most beautiful celestial atlases ever produced, that by Andreas Cellarius.

7.1 ANDREAS CELLARIUS

7.1.1 Life and times

Andreas Cellarius was born around 1596 in the town of Neuhausen, near Worms, Germany. His father was a pastor. Sometime after 1599 the family moved to Heidelberg, where Andreas subsequently attended the Sapierzkolleg and the University of Heidelberg. Since he later wrote publications on the art of fortification and on the country of Poland, it is thought that he may have spent time in the military and traveled through Poland. He subsequently settled in Holland, and in 1625 he was identified as a 30-year-old schoolmaster (rector) in the marriage records of Amsterdam. His wife was named Catharina, and the couple had a son the next year (followed by two more in 1631 and 1635). He was employed as the rector of the Latin School in The Hague in 1630 and at Hoorn in 1637. During his rectorship at Hoorn, he published works on designing impregnable city fortifications (1645) and on a description of Poland (1652), as well as some poems praising his assistant rector. He resigned his position from the Latin School in early 1665, and he died in Hoorn in November of the same year.

7.1.2   *Harmonia Macrocosmica*

His best known publication was his celestial atlas *Harmonia Macrocosmica seu Atlas Universalis et Novus ...*, also sometimes known from its frontispiece as *Atlas Coelestis, seu Harmonia Macrocosmica*. This was intended to be part of a two-volume work on cosmography, but the second part never appeared. The *Harmonia Macrocosmica* was initially published by the Amsterdam publisher Johannes Janssonius in

Figure 7.1. A diagram showing the orbit of the Moon around the Earth according to Ptolemy, from Cellarius' *Harmonia Macrocosmica*, ca. 1661. 42 × 50.4 cm, 36.8-cm dia. hemisphere. Note the locations of the epicycle of the Moon as it revolves along its deferent orbit. Note also that the Moon's epicycle is pulled in toward the Earth by a "crank" mechanism hinged on a central epicycle (dotted line), which would correctly account for its location in the sky ("save the phenomena") but would result in an erroneous increase in its apparent diameter, especially at the quarter phase. *See also* Color Plate Gallery.

Figure 7.2. A diagram showing a planisphere centered on the north ecliptic pole down to about 20 degrees in the southern hemisphere, from Cellarius' 1660 edition of *Harmonia Macrocosmica*. Approximately 43.2 × 50.8 cm, 40.7 cm dia. planisphere. Note the beautiful constellation images and the Baroque elements (clouds, putti) in the periphery. Courtesy of the Collection of Robert Gordon. *See also* Color Plate Gallery.

1660, with Warner (1979) citing subsequent reprints by Janssonius in 1661 and 1666. The atlas was reissued without the Latin text in 1708 by the Amsterdam publishers Gerard Valk and Petrus Schenk, who engraved their names on the plates before printing the volume.

The atlas began with a prologue that discussed the ideas of various astronomers throughout history and the physical nature of the universe. This was followed by chapters dealing with the specific theories of Ptolemy, Copernicus, and Tycho Brahe. Then appeared discussions of general issues, such as the climatic zones of the Earth; stellar magnitudes; the nature of the Sun, Moon, and planets; and the zodiac.

There were a total of 29 double-page plates, which are listed in Appendix C. Cellarius' plates were artistically superb and done in colorful French Baroque style. The Latin title of the plate was given in two cartouches at the top left and right corners. Around the periphery, there frequently were acanthus leaves, shells, or putti floating in clouds among astronomers and gods. Rarely have the art and science of celestial images been more beautifully depicted, which has accounted for the great popularity of these plates today in interior decorator offices and corporate boardrooms.

The first 21 plates consisted of elaborate illustrations of the heavens and diagrams of the orbits of the Sun, Moon, and planets according to different cosmological theories, primarily from Ptolemy, Copernicus, and Tycho Brahe (Figures 3.2 and 7.1). These often were engraved using a central archetypical circular image, which contained scientific diagrams of planetary motions and projections of the Earth's great circles in the sky. Smaller diagrams were often added in the corners below the main diagram to illustrate specific points. Many were quite detailed, especially those relating to theoretical issues, and they served to enhance and clarify the text.

The last eight plates were constellation maps. Because these expensive plates often appear on the open market, I will describe them here in some detail. Two were labeled *Coeli Stellati Christiani Haemisphaerium Prius* or *Posterius* and were celestial hemispheres centered on the vernal or autumnal equinox, respectively. These were copies of the Judeo-Christian hemispheres from Schiller's atlas (see Section 6.2.3.1). The autumnal map measured 39 cm in diameter. Two additional maps were labeled *Haemisphaerium Stellatum Boreale* (or *Australe*) *Antiquum* and were planispheres centered on an ecliptic pole down to about 20 degrees in the opposite hemisphere using a polar stereographic projection and external orientation. The northern planisphere was 40.7 cm in diameter (see Figure 7.2). The third pair were labeled *Haemisphaerium Stellatum (Boreale cum Subiecto Haemisphaerio Terrestri* or *Australe Aequali Sphaerarum Proportione)* and were 41.6 cm diameter planispheres giving an oblique view of the constellations, which were externally oriented. The final pair were labeled *Haemisphaerii Borealis Coeli et Terrae Sphaerica Scenographia* and *Haemisphaerium Scenographicum Australe Coeli Stellati et Terrae* and showed the constellations in an external orientation between the north or south celestial poles, respectively, and the vernal equinox; they measured 41.5 cm in diameter. Both Ashworth (1997) and Warner (1979) trace the style of the constellations in these star maps to the work of Petrus Plancius.

In the 1600s and 1700s, Dutch cartography reigned supreme, and Cellarius' atlas was one of the highlights of this tradition. Admittedly, it did not contain anything new in terms of celestial mapping, nor did it influence the subsequent cartographic development of the field in any significant manner. However, the prints remain popular today with decorators and collectors due to their great beauty and Baroque drama, and this is reflected in the high price, particularly for the constellation plates.

7.2   ATHANASIUS KIRCHER

7.2.1   Life and times

Perhaps the most interesting and colorful polymath of all time, Athanasius Kircher was born in the town of Geisa in what is now Germany on May 2, 1602, the last of nine children. His father had a doctoral degree and was a lecturer in theology, and Athanasius was exposed early to intellectual pursuits. He studied humanities at the Jesuit College in Fulda, and on October 2, 1618 he entered the Society of Jesus. After his novitiate, he left for Cologne in 1622 to escape advancing Protestant soldiers as a result of the Thirty Years' War. There he studied philosophy and soon became interested in the natural sciences and classical languages, which he taught at Jesuit colleges in Koblenz and Heiligenstadt. In 1625 he began theological studies in Mainz and was ordained a priest in 1628. Although he subsequently was given a position teaching mathematics, philosophy, and Oriental languages at the University of Wurzburg, he was driven out of the area by the troubles of the Thirty Years' War and went to Avignon, France in 1631. While in France, he pursued an interest in deciphering Egyptian hieroglyphic writing, and this activity brought him to the attention of a prominent French senator who recommended to the General of the Jesuits that Kircher be allowed to pursue his interests at the Collegio Romano. When this was finally approved, he was already in transit to Vienna to replace astronomer Johannes Kepler as the Imperial Mathematician to the Hapsburg court. Fortuitously, a shipwreck landed him near Rome in 1635, where he found out about the change in his assignment.

At the College, Kircher lectured on a variety of subjects, including physics, mathematics, and Oriental languages. However, due to his great intellectual promise, he was released from his teaching duties after several years and was allowed to pursue areas of interest to him. Travels to southern Italy peaked his interest in earthquakes and volcanoes, and in 1638 he climbed the rumbling Mt. Vesuvius to observe the crater first-hand, and this began a life-long interest in subterranean phenomena. Back in Rome, he began collecting a variety of antiquities and ethnologically important items, which soon formed the core of his famous museum at the College which attracted scholars worldwide (and still exists today). He continued with his interest in deciphering hieroglyphics. Although his writings in this area proved to be largely incorrect (especially after Champollion's work on the Rosetta Stone in the 1820s), he was one of the first scholars to make the attempt, and he stimulated great interest in this area. He also stimulated work in the relationships between different languages, including a number of ancient languages such as Persian and Coptic. He made forays into such diverse areas as medicine, microbiology, harmonics, archaeology, fossils, geology, astronomy, physics, magnetism, and Chinese and Egyptian culture. He also did pioneering work on counting machines and speaking tubes, and he was the inventor of the magic lantern. He published over 30 books and papers, and he corresponded with a number of scholars.

Despite his great creativity and productivity, many of his ideas proved to be wrong. Partly, this was because he took speculative risks in his work that went beyond what the data suggested. Also, he looked for grains of truth in questionable (but at the time legitimate) subjects such as alchemy and astrology. In addition, he was bound by his faith to reject the Copernican cosmology in favor of the model put forth by Tycho Brahe. In fact, he was a pious man and hoped to do missionary work in China, but this never transpired, so to compensate he erected a sanctuary near

Rome for pilgrims to come to renew their faith. He died in this city on November 28, 1680.

7.2.2   Books of astronomical interest

Kircher's writings transcended traditional boundaries between disciplines, and his books reflected this syncretic approach. Along with ancient Christian and Hebrew concepts, they included the Eastern mysticism that had been popularized by neo-platonic philosophers. For example, Kircher was a believer in the hermetic tradition, which stated that a mythical syncretic figure and possible contemporary of Moses, Hermes Trismegistus, had discovered the true wisdom and religion of the patriarchs going back to Adam, and that this ancient wisdom had been recorded in Egyptian hieroglyphics. Some of the illustrations for Kircher's books carried symbolic representations of this and other ancient traditions (e.g., Zoroaster, Orpheus, the Kabbalah), as well as the unique associations that were a product of his creative mind. Due to his fame and reputation, many of the images from his books were used as illustrations of other books and prints, and astronomical sources were no exception.

Two of his more famous books are especially noteworthy in this regard. The first was *Oedipus Aegyptiacus* (Egyptian Oedipus), which was published in Latin from 1652 to 1654. This encyclopedic work consisted of three parts in four volumes totaling more than 2,000 pages, and it included dozens of exotic fonts and hundreds of illustrations. It summarized many of his notions concerning hieroglyphics and Egyptology. He published other books on Egypt as well, and his ideas were later used to produce syncretic images of Egyptian constellations that reflected some of his mystical views of this society (Figure 2.10).

Another book that influenced astronomical thinking was *Mundus Subterraneus (Underground World)*, which was published in 1665. It was another magnificent work, consisting of two immense folios written in Latin, with fascinating and sometimes fanciful illustrations. It dealt with a variety of topics and hypotheses, combining ancient and medieval ideas of natural philosophy with observations of his own. He subscribed to the idea that there was a central fire in the middle of the Earth and that chimneys and fissures from this source ran to other pockets of fire and to the surface, exiting from volcanoes. Paralleling this system was another one filled with water and consisting of subterranean pools and canals. Giant whirlpools in the oceans forced water underground, where it circulated and emerged as streams and rivers. Where a water channel happened to run close to a pocket of fire, it was warmed and emerged as a hot spring. He imagined the Earth to be a living organism, and these two systems were like the human circulatory system (Figure 7.3).

The *Mundus* also dealt with related topics, such as geology, paleontology, mining, metallurgy, and alchemy (the legitimacy of which he described ambivalently). He put forth the notion that some fossils were inorganic products of natural stony processes that made them look like plants and animals. His ideas also included the solar system, where he discussed issues such as the placement of craters on the Moon and the eruption and movement of sunspots. Regarding the Sun, Kircher took the non-Aristotelian view that its surface was rough and uneven and contained

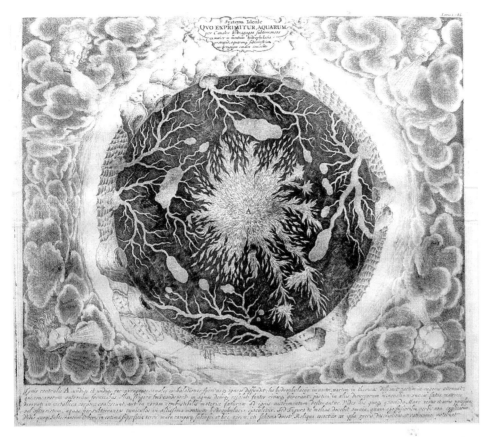

Figure 7.3. An engraved plate from the first Dutch edition (1682) of *Mundus Subterraneus*, first published in 1665 by Athanasius Kircher. 32.7 × 40.3 cm. Note his eccentric ideas concerning the center of the Earth, including the central fire, underground lakes, and passages and whirlpools through which water enters from the surface.

mountains and volcanoes. One of his illustrations that resulted from this notion became the basis for illustrations of the solar surface that were used by later writers (see Figure 8.17).

7.3 ALAIN MANNESON MALLET

7.3.1 Life and times

Alain Manneson Mallet was born in 1630 in France. He was originally a foot soldier under Louis XIV, rising to the rank of sergeant-major of artillery and then inspector of fortifications. For a time he served under the King of Portugal. When he returned to France, his military engineering and mathematical background led to

Figure 7.4. A print from Mallet's 1683 *Description de l'Univers*. 14.5 × 10.1 cm. Note the five constellation images, in which are shown the locations of supposed novae (see text). Also, note the accompanying rural scene at the bottom, meant to aesthetically enhance the astronomical part of the print. *See also* Color Plate Gallery.

his appointment to the court of Louis XIV, and he wrote books on both of these areas of expertise. He died in 1706.

7.3.2 *Description de l'Univers*

Mallet was widely traveled and was interested in surveying and geography. In 1683 he published in Paris his encyclopedic description of the world (*Description de l'Univers*) in five volumes, dealing, respectively, with world maps and celestial charts, Asia, Africa, Europe, and America. Although the maps were based on earlier prototypes, Mallet drew many of the plates himself. The text of this massive undertaking dealt with astronomical, geographical, and socio-cultural knowledge. The *Description* was quite popular and went through many editions, including one in German.

Most of the astronomical plates were found in Volume I of Book I: *De La Sphère en Général*. They were meant to be illustrations for the text and included depictions of cosmological systems, constellations, planets, and other heavenly bodies. In Chapter 3 there were two hemispheres labeled *Planisphère des Constellations Septentrionalles* (or *Meridionalles*) that were centered on the north or south ecliptic poles using a polar projection and external orientation. There were two additional hemispheres similarly labeled and presented but with a geocentric orientation. Another two hemispheres labeled *La Voye Laictée* showed crudely drawn constellations in an external orientation, but the purpose of these plates was to illustrate the path of the Milky Way. These hemispheres measured 9.4 cm in diameter.

Warner (1979) cites a particularly beautiful star map labeled *Etoilles Nouvelles* that shows five constellations in which supposed novae appeared: Cassiopée (with the nova of 1572); Cygne (with the novae of 1600 and 1670/71); Andromède (with a "new star" at her belt, which is actually the Andromeda galaxy); Baleine (with a variable star in its neck); and Fleuve Eridan et le Lièvre (with several novae that were seen by Cassini). This plate measures 14.5×10.1 cm (see Figure 7.4).

Mallet often showed a scientific diagram at the top of his plates and accompanied this at the bottom with a beautiful rural scene to provide aesthetic value and stimulate the interest of the reader. As a result, the plates from *Description de l'Univers* are commonly found today and make for beautiful displays (Figure 8.24).

7.4 VINCENZO MARIA CORONELLI

7.4.1 Life and times

Vincenzo Maria Coronelli was born in Venice on August 16, 1650. He took Holy Orders as a Minorite friar in Venice, living first at San Nicolo della Lattuga (from 1665 to 1671), then at Santa Maria Gloriosa dei Frari (from 1671 to 1674). He also received a Doctor of Theology degree at the Collegium Santo Bonaventura in Rome in 1673. Despite his religious activities, he became famous for his developing skills as a globe maker, cosmographer, cartographer, and encyclopedist.

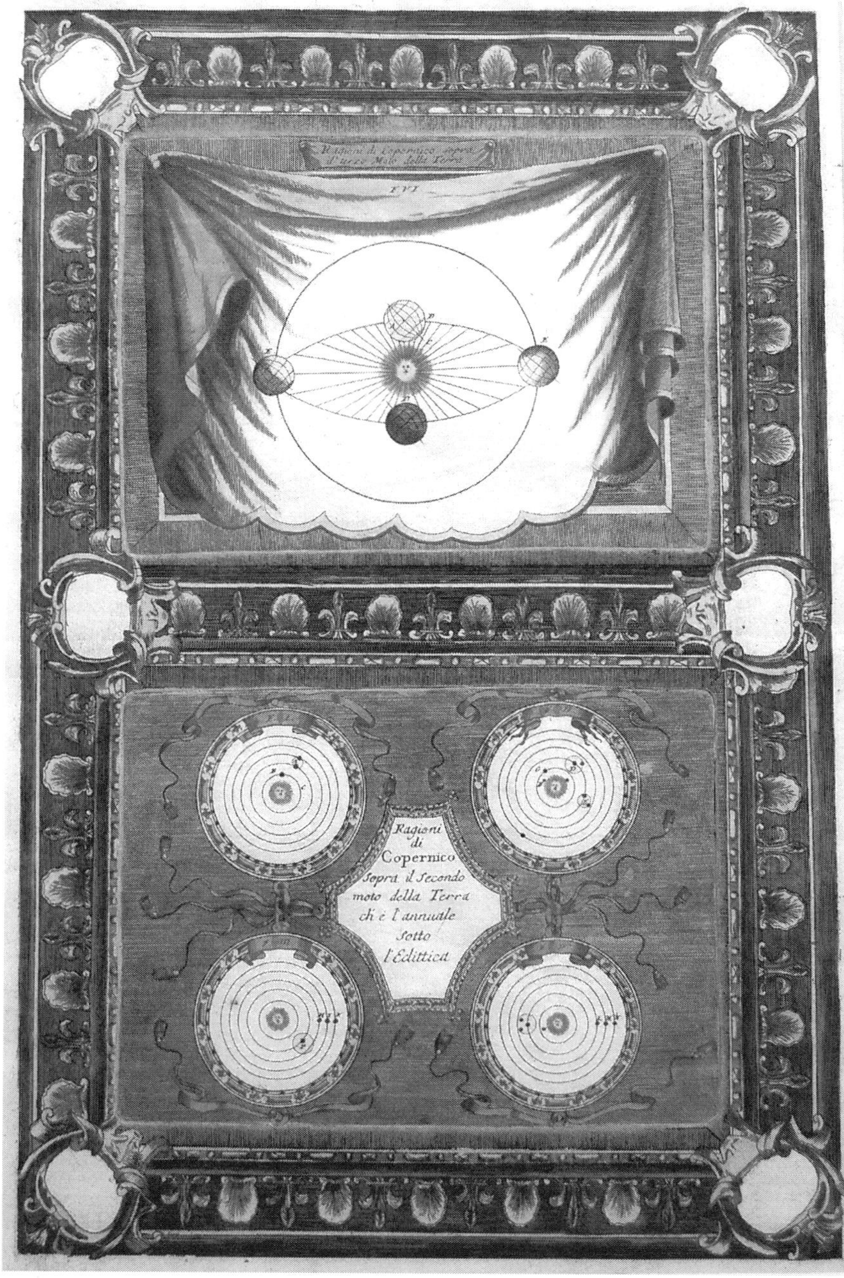

Figure 7.5. A double print from Coronelli's *Corso Geografico Universale*, published in 1692. 38.4 × 25.5 cm. Note the diagram of the third motion of the Earth at the top, the second motion of the Earth at the bottom, and the beautiful Baroque margin uniting the diagrams into one print. *See also* Color Plate Gallery.

In 1678 he created two finely crafted globes of diameter 175 cm for the Duke of Parma, and this brought him to the attention of the French ambassador. In 1681 he moved to Paris, where he lived for two years and became Geographer Royal to King Louis XIV. His fame spread, and he worked in several European countries during the following years, continuing to make globes and writing books related to geography and cartography. He returned to Venice for good in 1705, where he founded the first geographical society (the *Accademia Cosmografica degli Argonauti*) and held the position of Cosmographer of the Republic of Venice. He also published the first six volumes of the first encyclopedia to be ordered alphabetically, although this was never completed. He died in Vienna on December 9, 1718.

7.4.2 Books with celestial plates

During his lifetime, Coronelli is said to have engraved over 400 maps in his own hand. His celestial maps were influenced by the star catalogs of others, such as Bayer, Hevelius, and Halley. Many of these first appeared in his *Atlante Veneto* in 1690 and were subsequently reproduced in follow-up books (Mosley, 2011).

For example, in his 1692 *Corso Geografico Universale*, there was a similar set of world-system diagrams showing the cosmologies of Ptolemy, Copernicus, and Tycho Brahe; the different movements of the Earth around the Sun; etc. They were arranged two to a page, one above the other. However, the beautiful Baroque marginal details provided a border that integrated them into one print. The top of one such print showed the third motion of the Earth, which supposedly accounted for the stability of its axial direction as it moved around its orbit, and the bottom showed the second motion of the Earth (and other planets) in revolving around the Sun. Both illustrations were in reference to the heliocentric system of Copernicus (Figure 7.5). The print measured 38.4 × 25.5 cm.

These same diagrams appeared in the 1693 *Epitome Cosmografica*, except that they were one to a page rather than two, and they were smaller. For example, the third motion illustration was 13 × 18.6 cm. Also in this book was a star catalog in which the stars were organized by constellation.

Another book, the 1701 *Globi Differenti del P. Coronelli*, contained a number of planispheres, which Warner (1979) states were also in the *Corso* and *Epitome*, but which I could not find in the British Library editions. One of these prints had side-by-side hemispheres centered on the ecliptic poles using a polar equidistant projection with geocentric orientation. The title was *Planisferii Celesti, Calcolati per l'Anno MDCC* ... Around the periphery of the plate was a prominent dedication to "Marco Bembo" and tables showing stellar magnitudes and positions (Figure 7.6). The northern hemisphere measured 28.3 cm in diameter, and the total print size was 44.1 × 59.4 cm. There were no constellation figures. Each hemisphere had an equatorial grid system superimposed on the ecliptic one, showing that Flamsteed was not the first person to employ a double-grid system in a single chart. However, Flamsteed emphasized the equatorial system, and he was the first to use this kind of double grid in a major atlas (see Section 6.4.3).

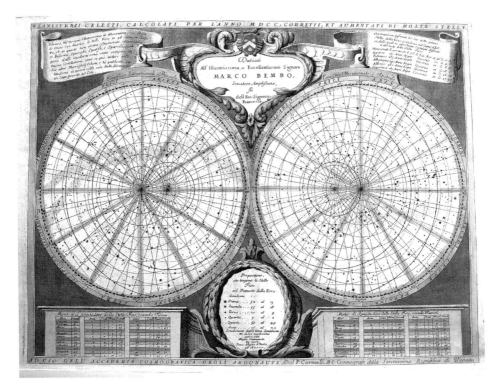

Figure 7.6. A double hemisphere from Coronelli's *Globi Differenti del P. Coronelli*, published in 1701. 44.1 × 59.4 cm, each hemisphere 28.3 cm dia. Note the lack of constellation figures and the double-grid system, showing that Flamsteed was not the first person to use this in a single chart. *See also* Color Plate Gallery.

This book also contained two separate hemisphere prints labeled *Planisfero Settentrionale* (or *Meridionale*) ... centered on the ecliptic poles using a polar equidistant projection with geocentric orientation. There were constellation figures and a superimposed equatorial grid system, and each hemisphere was 40.5 cm in diameter. Surrounding the hemispheres were tables of stellar magnitudes and a dedication to the *Republica di Venetia*.

7.5   JOHN SELLER

7.5.1   Life and times

John Seller was a pioneering English cartographer and instrument maker who was born around 1630. He established a famous shop in Wapping, near the Tower of London, that was frequented by sailors and sold many items related to navigation, including nautical charts, terrestrial and celestial maps, globes, and instruments. He was a Baptist and a bit of a gadfly, and at one point he was found guilty of conspiring to kill King Charles II; however, he was let off. In 1671 he was appointed Hydrographer to the King (under Charles II, James II, and William III), which gave him a monopoly on publishing nautical atlases in England. Also in 1671, he published the first volume of a maritime atlas entitled *The English Pilot*, to which he added several additional volumes before financial problems caused him to sell off the rights to this popular publication in 1680. He continued to publish atlases and maps, including some of the earliest British maps of the American colonies. In the 1680s he participated in a project to survey England and Wales and publish an atlas called *Atlas Anglicanus*, but this came to a premature end after only six maps were completed. In 1695 he published a small and somewhat plain atlas of the counties of England and Wales called *Anglia Contracta*, which was based on the work of John Speed and was accompanied by text. This popular book went through several editions, and its maps were used to illustrate other atlases of the time. Seller died in 1697.

7.5.2 *Atlas Maritimus*

In 1675, Seller published his famous nautical atlas, the *Atlas Maritimus*, or the *Sea-Atlas* ..., which went through many editions into the 1700s. Although there was a standard core to this folio atlas, individual copies were assembled according to the wishes of the purchasers, which meant that their content varied. Like other such composite atlases, maps were sometimes added from other sources as well (given the relative lack of copyright protection at the time). In the late 1670s, Seller produced some interesting celestial charts, which were then bound in various editions of this atlas (even including copies with the original 1675 title page).

Warner (1979) describes several of these. One was labeled *A Mapp of the two Hemispheres of the Heavens* ... and consisted of two celestial hemispheres, each of which was 21.1 cm in diameter and was centered on an equatorial pole using a polar stereographic projection with geocentric orientation. The example in the British Library measured 21.6×44.4 cm in plate size and included six zodiac pictures above and six below the hemispheres. There was also a planisphere labeled *A Coelestial Planisphere* ... that was 35.2 cm in diameter and was centered on the north equatorial pole down to 38 degrees S dec. using a polar equidistant projection with external orientation. Instructions for its use were given in the left and right margins. A third map was published in 1679 and depicted the zodiac constellations in four long strips arranged vertically on a map that measured 44.5×57.2 cm in total size. It was labeled *Zodiacus Stellatus Cujus Limitibus Planetarum Omnium* ... Each strip was 10.8×57.2 cm in size and was centered 8 degrees above and below the ecliptic using a cylindrical projection with geocentric orientation. According to Warner, Halley laid down the star positions, and this was the first published map solely of the zodiac. Halley also helped produce two hemispheres published by Seller in 1679 which were labeled *Planisphaerium Coeleste, contained in two Hemispheres of the Heaven* ... Each was centered on an ecliptic pole using a polar stereographic projection with geocentric orientation. The maps were 46.6 cm in diameter, although a similar southern map exists labeled *S: Coelestiall Hemisphere* that was 33 cm in diameter. Finally, there was a double hemisphere of the world labeled *Novissima Totius Terrarum Orbis Tabula* that included three celestial images at the top (seasons, Hevelius lunar map, Tycho Brahe cosmology) and three at the bottom (Ptolemy cosmology, Hevelius lunar map, Copernicus cosmology).

7.5.3 *Atlas Coelestis*

In 1680, Seller produced a small atlas entitled *Atlas Coelestis Containing the Systems and Theoryes of the Planets* . . . that was well received and revised around 1700. It was one of the first "pocket" astronomical atlases. It contained a text, tables, and a number of astronomical diagrams, including the cosmological systems of Ptolemy, Copernicus, and Tycho Brahe; Kircher's map of the Sun; planetary images; and Hevelius' lunar maps. There was also a series of 31 double-page maps of the

Figure 7.7. A double hemisphere from Seller's *Atlas Coelestis*, published ca. 1680. Each hemisphere is approximately 7 cm in diameter. Note the traditional constellation images, the signs of the zodiac above and below, and outlines of the planets with their symbols at the bottom. Courtesy of Jonathan Potter Ltd. *See also* Color Plate Gallery.

constellations that were oriented to the north ecliptic pole using a polar stereographic projection and geocentric orientation. The map with Ursa Major was 11 × 13.8 cm.

The *Atlas Coelestis* also had a plate entitled *A Mapp of the Constellations about ye North Pole* that measured 9.7 cm in diameter, was oriented to the northern equatorial pole, and used a polar stereographic projection with geocentric orientation. Another plate was labeled *A Mapp of ye two Hemispheres of ye Heavens* (Figure 7.7). Each celestial hemisphere measured 7 cm in diameter, was centered on an equatorial pole, and used a polar stereographic projection with geocentric orientation. These two maps also appeared in Volume 2 of an atlas published by Seller in 1700 that was entitled *Atlas Terrestris*.

7.6   JOHN SENEX

7.6.1   Life and times

John Senex was born at Ludlow, England, in November 1678, the son of a Shropshire gentleman. From 1695 to 1702, he was apprenticed to the London bookseller Robert Clavell of the Stationer's Company. He then began what was to become a very successful career as an engraver, cartographer, and map and book publisher, working at a number of locations in London. Early in his career, he produced a series of plates for London almanacs and for a book on microscopes and telescopes, but from 1706 he became involved with map production. From 1710 to 1740, he was located on Fleet Street, first at the Globe in Salisbury Court (until 1721), then opposite St. Dunstan's Church. He published a number of fine terrestrial maps of countries and continents, some loose and some in atlases, and celestial and terrestrial globes. Some of his maps and globes were produced in partnership with Charles Price and James Maxwell. He had an interest in road maps and in 1719 issued a revised edition of Ogilby's *Britannia* in miniature form, which was quite popular and went through many editions. He was also interested in astronomy, collaborated with leading scientists (including Halley), and was elected as a Fellow of the Royal Society in 1728. Although he died in London in 1740, his business was continued by his widow, Mary.

7.6.2   Celestial maps

According to Warner (1979), there was no reliable and extensive star atlas in England during the first quarter of the 18th Century. The charts of Seller were not satisfactory for high-quality astronomical work, those of Hevelius were hard to find, and Flamsteed had not yet published his great atlas. In part through his collaboration with Halley, Senex produced a number of celestial maps that helped fill this void which were quite popular with astronomers and navigators. All of the maps mentioned below are collected together in the British Library, along with a 1763 advertisement that states: *Atlas Coelestis: Containing the following Hemispheres. Wherein are*

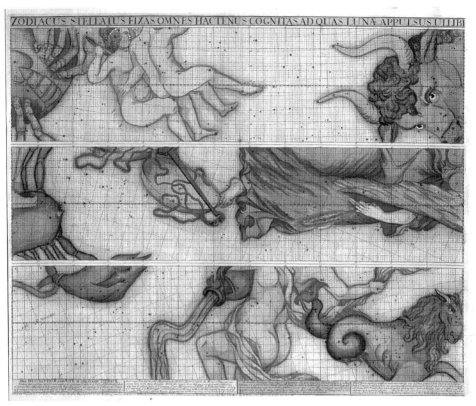

Figure 7.8. One of two pages depicting the zodiac in three long horizontal strips, produced by Senex in 1718. Approximately 53 × 64.5 cm (each page). Note the prominent grid system. Courtesy of Jonathan Potter Ltd. *See also* Color Plate Gallery.

carefully laid down all the Stars in Mr. Flamsteed's Catalogue, as published by Dr. Halley ... It is not known if an actual *Atlas Coelestis* was finally published.

One of these maps was originally published in 1718 and showed the zodiac constellations in three long strips arranged vertically over two pages on a map labeled *Zodiacus Stellatus Fixas Omnes Hactenus Cognitas* ... (Figure 7.8). Each strip was centered 8 degrees above and below the ecliptic using a cylindrical projection with geocentric orientation. The copy in the British Library measures 53 × 129 cm (both pages side by side). Since the stars were derived from Halley's unauthorized edition of Flamsteed's catalog (see Section 6.4.2), the appearance of this and other maps by Senex created great consternation at Greenwich.

Around 1721, Senex produced two celestial hemispheres that later were bound in Knapton's 1728 *Atlas Maritimus Commercialis*. The maps were labeled *Stellarum Fixarum Hemisphaerium Boreale* (or *Australe*). Each was centered on an equatorial pole using a polar stereographic projection with geocentric orientation, and the northern map was 59.5 cm in diameter. There was an overlying ecliptic grid system as well. Again, Senex used Flamsteed's pirated catalog for stellar positions and prior work by Halley for the locations of novae and nebulae, which included the "nubecula maj" and nubecula min" (i.e., the Magellanic clouds) and the "nebula" in Andromeda (i.e., the Andromeda Galaxy, M31). At about the same time, similar-sized hemispheres (the northern was 60.5 cm in diameter) with the same main title were published, but these were centered on the ecliptic poles. Their constellation images were cruder and less well defined than in the equatorial versions.

In 1746, Senex produced a planisphere centered on the north equatorial pole down to 38 degrees S dec. that used a polar equidistant projection with geocentric orientation. This was 60 cm in diameter and was labeled *The Planisphere on which the stars visible in the Latitude* ... Warner states that Charles Messier used Senex's planisphere image as background for the route of Halley's comet during its 1758/59 return, in anticipation of its 1682 return.

There is also an illustration in the British Library collection entitled *A Scheme of the Solar System with the Orbits of the Planets and Comets belonging thereto*. This is 57.5 cm in diameter and shows the relative sizes of all the known planets.

7.7 CORBINIANUS THOMAS

7.7.1 Life and times

Corbinianus Thomas was a Benedictine monk who was born in 1694 and lived in Salzburg. Little is known about him, except that he was a Professor of Mathematics and Theology at the University of Salzburg. He died in 1767.

7.7.2 *Mercurii Philosophici Firmamentum*

In 1730, Thomas published a star atlas entitled *Mercurii Philosophici Firmamentum Firmianum Descriptionem* ... The text included information on astronomy and the use of globes and other astronomical instruments. It also had a number of beautiful illustrations, including an armillary sphere; views of the planets; two side-by-side lunar maps whose features were named by Hevelius and Riccioli (see Section 8.5.2.1 for more on this famous "double Moon" image); the cosmological systems of Ptolemy, Tycho Brahe, Riccioli, and Copernicus; and a diagram of two celestial hemispheres containing Egyptian images of the constellations entitled *Coelum Aegyptiacum* ... In addition, there were two 12 × 13 cm hemispheres of the north and south constellations that were entitled *Haemisphaerium Coeli Boreale* or *Australe*.

There were also 54 small constellation maps measuring 11 to 13.3 cm high and 12.5 to 23.2 cm wide. The map containing Ursa Major was 13.2 × 19.4 cm. Most contained an individual major constellation or two smaller neighboring constellations (see Figure 7.9). The orientation was geocentric, and the attractive constellation images had contemporary hand coloring. Unlike Bayer, who used a uniformly darkened zodiac, Thomas showed a more pleasing band. He had an unusual nomenclature system: Bayer Greek letter for the star, Roman numeral for the magnitude, and

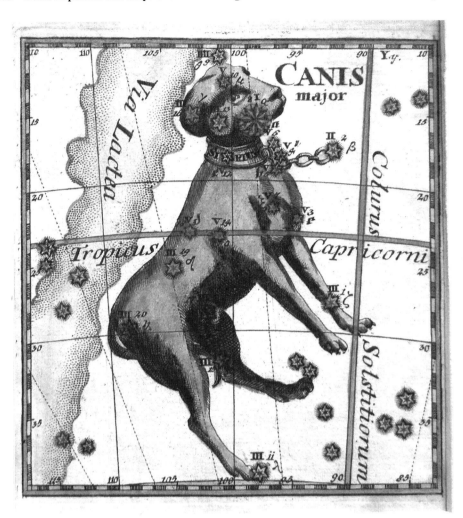

Figure 7.9. The constellation of Canis Major, from Thomas' 1730 *Mercurii Philosophici Firmamentum*. 11.9 × 11.2 cm. Note the use of Bayer Greek letters for stars, Roman numerals for stellar magnitudes, and Arabic numerals for references to the star catalog. Note the Via Lactea (Milky Way) to the left of the image, and the crossing lines representing the Tropic of Capricorn and the Colurus Solstitiorum. *See also* Color Plate Gallery.

Arabic numeral for the reference to a star catalog. The artistic style and choice of constellations were derived from Coronelli, although some of the images were taken from Hevelius (but were not right-to-left reversed). Warner (1979) states that Thomas invented one new constellation, Corona Firmiana, to honor his patron, the Archbishop of Salzburg, but it was never picked up by anyone else. He was also one of the first cartographers to provide individual plates for southern constellations (rather than including several of them on one plate).

7.8   JOHANN DOPPELMAYR

7.8.1   Life and times

Johann Doppelmayr was born on September 27, 1677 in Nuremberg. He was the son of a merchant who also had an interest in science. He studied at the Aegidien Gymnasium in Nuremberg from 1689 to 1696. After graduating, he attended the University of Altdorf, where he studied law, mathematics, and natural philosophy from 1696 to 1699. He continued his studies in mathematics and physics at the University of Halle until 1700, when he began a two-year period of travel in Germany, Holland, and England. During this time, he learned to speak French, Italian, and English. He also studied astronomy and learned to grind and figure his own telescope lenses.

In August 1702 he returned to Nuremberg, and in 1704 he was appointed Professor of Mathematics at his alma mater, the Aegidien Gymnasium, where he remained the rest of his life. He married in February 1716 and had four children. He wrote on a number of topics, including astronomy, geography, cartography, spherical trigonometry, and scientific instruments. He also collaborated in the production of terrestrial and celestial globes. He was a member of several scientific societies, including the Royal Society of London and the Berlin and St. Petersburg Academy of Sciences. He died on December 1, 1750 in Nuremberg.

7.8.2   Johann Baptist Homann

Doppelmayr often collaborated with the cartographer Johann Baptist Homann (1664–1724), who was a former Dominican monk who came to Nuremberg and learned to be a map engraver. In 1702 he founded a famous cartographic publishing firm that continued through his heirs until 1848. According to Warner (1979), there were a few celestial charts published by Homann prior to his association with Doppelmayr. One of these was a terrestrial map in two hemispheres entitled *Planiglobii Terrestris cum Utroq Hemisphaerio Caelesti Generalis Exhibitio*, which was bound in Homann's 1707 *Neuer Atlas* ... The map included two celestial hemisphere inserts: *Hemisphaerium Boreale* or *Australe*. Each was 13 cm in diameter and was centered on an ecliptic pole using a polar stereographic projection with external orientation. They were derived from Hevelius. Slightly smaller versions of this print were also bound in Homann's *Atlas Scholasticus* ... (ca. 1732) and *Atlas Novus* ... (1702–1750).

7.8.3   *Atlas Coelestis*

In the early 1700s, Doppelmayr prepared a number of astronomical and cosmological plates that had appeared in several of Homann's general atlases. In 1742 these were collected and issued as the *Atlas Coelestis in quo Mundus Spectabilis* ... There were a total of 30 plates, which are listed in Appendix C. Twenty dealt with various astronomical themes, such as the cosmological systems of Copernicus (Figure 1.2) and

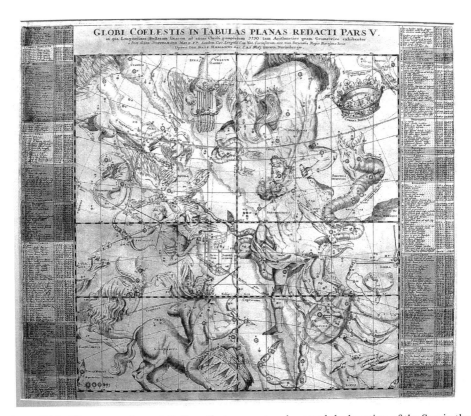

Figure 7.10. The stars and constellations that are centered around the location of the Sun in the zodiac during the winter solstice, according to Doppelmayr (from his *Atlas Coelestis*, 1742). 48.5 × 57.4 cm. Note the influences of Hevelius (including the presence of his constellation Scutum Sobiescianum) and the paths of several comets, including Kepler's 1607 viewing of what was later to be called Halley's comet (to the right of Serpens Ophiuchi). *See also* Color Plate Gallery.

Tycho Brahe (Figure 3.18), planetary motions, the solar system, the Moon's surface (Figure 8.20), and the theory of eclipses. There were also two hemispheres illustrating the passage of comets in the sky labeled *Motus Cometarum in Hemisphaerio Boreale* or *Australe* that were centered on the ecliptic poles using a polar stereographic projection with external orientation, but which did not show stars (just the constellation figures). The plates measured 47.3 × 56.7 cm, and the hemispheres were 43.7 cm in diameter.

The remaining 10 plates were actual star charts. Two were hemispheres centered on the equatorial poles using a polar equidistant projection and geocentric orientation that were labeled *Hemisphaerium Coeli Boreale* (or *Australe*), *in qua loca stellarum fixarum* ... In the four corners were putti holding various instruments, and on the sides were top-to-bottom tables listing the constellations and the names, magnitudes, and locations of their main stars. The northern plate measured 48.5 × 58 cm, and the hemisphere was 41.5 cm in diameter. Two other plates were also hemispheres, but they were centered on the ecliptic poles, used a polar stereographic projection, and had an external orientation. They were labeled *Hemisphaerium Coeli Boreale* (or *Australe*) *in quo Fixarum* …, had pictures of observatories in the corners, and had relatively small side tables of stars. The northern plate measured 48.7 × 58 cm, and the hemisphere was 45.8 cm in diameter. The remaining six plates were part of a series labeled *Globi Coelestis in Tabulas Planas Redacti Pars I* (or *II* … *VI*) that used a gnomonic projection with geocentric orientation. They were respectively centered on the north equatorial pole to 45 degrees N dec., the vernal equinox, the summer solstice, the autumnal equinox, the winter solstice (see Figure 7.10), and the south equatorial pole to 45 degrees S dec. The map with Ursa Major (*Pars I*) was 48 × 58.8 cm. The maps included the paths of important comets. There were no corner pictures, but there were top-to-bottom side tables of the constellations and their stars. Warner points out that the choice and style of the constellation images suggest that Doppelmayr's maps were strongly influenced by Hevelius.

Although still listing 1742 on the title page, later editions of this atlas have an altered title (*Atlas Novus Coelestis* …) and often contain additional plates.

7.9 ANTONIO ZATTA

7.9.1 Life and times

Antonio Zatta was a prolific Venetian printer and map publisher who was born in 1757. He produced a large number of atlases and maps that included various countries throughout the world. Generally, these were of high quality in terms of their accuracy and artistic style. He died in 1797.

7.9.2 *Atlante Novissimo*

His best known work was *Atlante Novissimo*, which was published in Venice in 1779. It included 218 maps in four volumes. In Vol. I were two celestial hemispheres centered on a north or south equatorial pole in folio size which were labeled *Planisferio Celeste Settentrionale* (or *Meridionale*) … (Figure 7.11). Each used a polar stereographic projection with geocentric orientation. The plates measured 31 × 40.3 cm, and the hemispheres were 24 cm in diameter. In the corners were beautiful drawings of famous observatories: Pisa, Bologna, Padua, and Milan in the northern plate; and Paris, Greenwich, Cassel, and Copenhagen in the southern plate. Warner (1979) comments that the style of these maps followed Hevelius and ignored recent observations by Lacaille. Also included in this volume was a print labeled *Mappa Dell'Universo* that depicted the cosmologies of Ptolemy and Copernicus in two large 18.8 cm diameter diagrams.

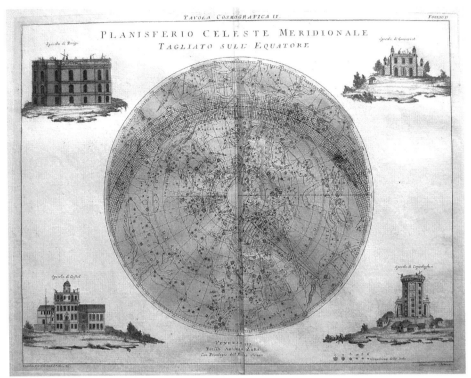

Figure 7.11. The southern hemisphere constellations, from Zatta's 1779 *Atlante Novissimo.* 31 × 40.3 cm, 24-cm dia. hemisphere. Note the Magellanic Clouds and the beautiful drawings of famous European observatories in the corners (at Paris, Greenwich, Cassel, and Copenhagen). *See also* Color Plate Gallery.

7.10　SAMUEL DUNN

7.10.1　Life and times

Samuel Dunn was a cartographer who lived in London, although according to Basil Brown (1932) he was born and worked as a school principal for many years in Crediton, Devon. From the 1750s to the early 1790s, he was quite prolific, authoring a number of books, maps, and charts. Some of these found their way into other publications, including encyclopedias and atlases. On many of his works, he identified himself as a teacher of mathematical sciences. He also was a member of the American Philosophical Society. He died in 1794.

7.10.2　*A New Atlas of the Mundane System*

One of Dunn's most popular contributions to celestial cartography was in *A New Atlas of the Mundane System; Or, Of Geography and Cosmography . . .* , printed for the

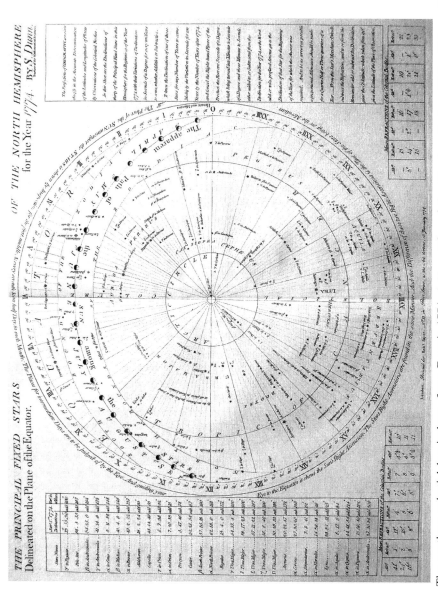

Figure 7.12. The northern celestial hemisphere, from Dunn's 1774 *A New Atlas of the Mundane System*. 34.2 × 45.1 cm, 29.9-cm dia. hemisphere. Note the path of the Sun through the zodiacal constellations and the tables allowing for the calculation of stellar declinations due to precession.

London map and print seller Robert Sayer in 1774. This book contained six celestial charts. Four of these were cosmographical diagrams dealing with such issues as the relative sizes and distances of the Sun and planets, time as it relates to latitude, and the great circles. The other two charts were northern and southern equatorial celestial hemispheres that used a polar projection with external orientation entitled *The Principal Fixed Stars of the North* (or *South*) *Hemisphere* (Figure 7.12). Each image was 29.9 cm in diameter, and there were no constellation figures. Subsequent editions were published in 1788, 1789, 1796, and 1800. According to Warner (1979), the maps were used to illustrate the article on Astronomy in the *Encylopedia Britannica*, both the 3rd British edition (1797) and the 1st American edition (1798).

Warner describes several of Dunn's other celestial productions, including three gore charts labeled *A Chart of the Zodiacal Stars Used in finding the Longitude at Sea* ... These extended from the north to the south equatorial poles and were without constellation figures: two had geocentric star patterns and were published in 1772, while a third had an external orientation and was published in 1782. Dunn also produced a terrestrial map in two hemispheres labeled *Scientia Terrarum et Coelorum* ... that was surrounded by a number of astronomical diagrams and two 17.8 cm diameter celestial hemispheres labeled *The Northern* (or *Southern*) *Hemisphere*. Each of these was centered on an equatorial pole using a polar stereographic projection with external orientation. Warner states that this map was bound in various editions of Thomas Kitchin's *A General Atlas* ... from around 1772 to the mid-1780s, and under the title *A General Map of the World* ... from 1787 into the 19th Century.

7.11 ANTOINE AND NICOLAS DE FER

7.11.1 Life and times

Antoine (also Anthoine) de Fer was a map publisher in Paris who was active in the mid-17th Century. He married in 1633, and he and his wife Geneviève had three sons, the youngest of whom was named Nicolas. Antoine's business focused on re-publishing prints from plates that he purchased from others. According to Warner (1979), in 1650 he produced two celestial hemispheres indicated to be from *A Paris, Chez Anthoine de Fer* ... Each was 26.5 cm in diameter and was centered on an ecliptic pole using a polar stereographic projection with geocentric orientation. They were similar to maps issued by his neighbor Melchior Tavernier. Antoine died in June 1673, and after some legal proceedings the business was retained by Geneviève and the family.

Nicolas de Fer was born in 1646. At the age of 12, he was apprenticed to Parisian engraver Louis Spirinx, and at age 23 he produced his first map. He took over the family business in 1687, which by that time had begun to decline. However, he built it up to the point where in 1690 he became the official geographer to the French Dauphin. Over the next three decades, he acquired additional support from both the French and Spanish royal families, becoming the official geographer to both

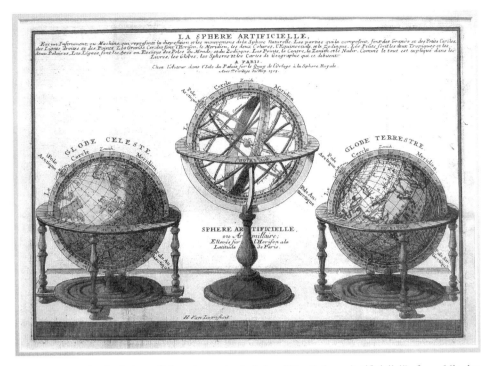

Figure 7.13. The depiction of three important globes ("La Sphere Artificielle"), from Nicolas de Fer's 1703 *L'Atlas Curieux*. 22.5 × 32.1 cm. Note the central armillary sphere and the celestial and terrestrial globes to either side. *See also* Color Plate Gallery.

kings. The company prospered, publishing over 600 sheet maps, some 26 large wall maps, town plans, and a number of atlases. Nicolas died on October 25, 1720. His estate was passed on to his three daughters, whose husbands were also involved in engraving and publishing and continued the business until the mid-1700s.

7.11.2 *L'Atlas Curieux*

In 1700, Nicholas de Fer published an atlas entitled *L'Atlas Curieux ou le Monde . . .* that became very popular, going through several editions in subsequent years. By 1705, this atlas contained three celestial prints that preceded a series of terrestrial maps. One of these prints was entitled *La Sphere Artificielle* (Figure 7.13). It showed three globes, labeled from left to right: *Globe Celeste*, *Sphere Artificielle ou Armillaire*, and *Globe Terrestre*, with descriptive text at the top. This image measured 22.5 × 32.1 cm. The next print showed two celestial hemispheres produced by Philippe de la Hire; this will be described in Section 7.12. The final print was unlabeled and showed the cosmological systems of Ptolemy and Copernicus (top), and Descartes and Brahe (bottom). It measured 23.5 × 33 cm.

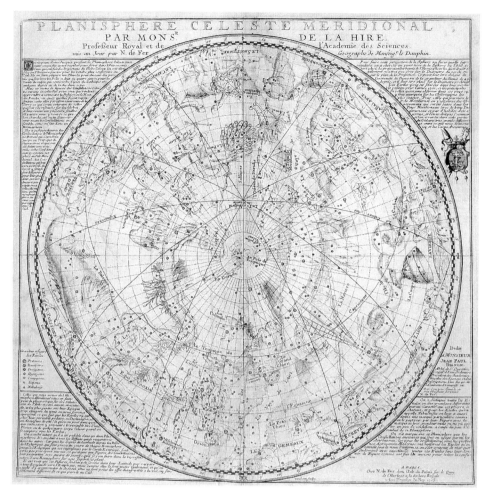

Figure 7.14. The southern celestial skies, from a famous print by Philippe de la Hire and published by Nicolas de Fer in 1705. 46.3 × 46.3 cm, 42.6-cm dia. hemisphere. Note the attention to detail and the lines for the solstices and equinoxes.

7.12 PHILIPPE DE LA HIRE

7.12.1 Life and times

Philippe de la Hire was born in Paris on March 18, 1640. His father was a distinguished artist who was a professor at the Académie Royale de Peinture and Sculpture. Philippe initially studied drawing, painting, and perspective in Italy, but he returned to Paris to take up mathematics and science. He studied under the mathematician Desargues and quickly distinguished himself in this field, especially in the area of geometry. He published important treatises on conics, where he improved several well-known theorems and established a number of new ones. He also was interested in geography, making observations that contributed to a new governmental map of France, working with Cassini to measure the north–south meridian going through Paris, and directing surveying operations that provided water to Versailles. He won professorships in mathematics at the Collège Royale and in architecture at the Académie Royale. He died in Paris on April 21, 1718.

La Hire also became interested in astronomy, and in 1678 he was made a member of the astronomy section of the Académie Royale des Sciences. He installed the first transit instrument at the Paris Observatory, and he produced tables of the movements of the Sun, Moon, and planets. He published in the *Mémoires, Académie Royale des Sciences*, where in 1692 he contributed a 21.7 × 23.3 cm map of the path of Mars through the Praesepe star cluster, and in 1693 he produced a 19.2 × 22.5 cm map of the Pleiades. Both used a geocentric orientation, although reversed as seen through an inverting telescope lens. In association with the famous de Fer publishing house, he produced several influential celestial hemispheres.

7.12.2 Celestial maps

In 1702, La Hire produced a print with two small celestial hemispheres labeled *Planisphères Céleste. Par Mr. De la Hire* ... that were published in Nicolas de Fer's *L'Atlas Curieux*. Each was 16 cm in diameter and was centered on an ecliptic pole using a polar equidistant projection with geocentric orientation. The plate measured 22.8 × 33 cm and contained images of the Sun and Moon at the top and the planets at the bottom. This print was reissued in 1705.

Also in 1705, La Hire produced two larger and more elaborate celestial hemispheres, which were issued by the publishing house of Nicolas de Fer. Entitled *Planisphere Celeste Septentrional* or *Meridional*, each was 42.6 cm in diameter and was centered on an ecliptic pole using a polar equidistant projection with geocentric orientation (Figure 7.14). According to Warner (1979), the descriptive text surrounding the hemispheres also promoted them over smaller celestial maps that focused on individual constellations, and the southern map suggested influences by Halley. The constellations themselves were derived from Bayer, although they did not use Bayer numbers. These two hemispheres were reissued later by de Fer's son-in-law Danet, and they appeared again in Desnos' *Atlas Général* in 1760 and beyond. La Hire's hemispheres were imitated throughout the 1700s (especially in France) by a variety of people, including Le Rouge, Crepy, La Feuille, Chiquet, and De Guignes.

7.13 PIERRE-CHARLES LE MONNIER

7.13.1 Life and times

Pierre-Charles Le Monnier was a famous and influential astronomer who was born in Paris on November 23, 1715. His father was an astronomer and professor of philosophy. The young Pierre-Charles began making and recording serious astronomical observations as a teenager. He also constructed an elaborate lunar map, and this resulted in his admission to the Académie Royale des Sciences on April 21, 1736. He was chosen to participate in the 1736–37 academy-sponsored expedition to Lapland to measure the length of a degree of terrestrial latitude near the pole. To commemorate this work, he created the circumpolar constellation Le Renne (the reindeer), which appeared in his later celestial maps (see Section 4.4 and Figure 6.13). He corresponded with British astronomers and was admitted to the Royal Society in 1739. He was a favorite of King Louis XV, and this helped him procure funds to upgrade the instruments at the Paris Observatory (some of which were based on British models). He made important observations of the Moon and planets and conducted research into terrestrial magnetism and electricity. He was also a professor at the Collège de France, where he publicly lectured on gravitation and other astronomical topics. He died on May 31, 1799.

7.13.2 Celestial maps

Le Monnier produced several influential celestial maps that were copied by others. Two of his hemispheres entitled *Hemisphère Boréal* or *Austral* appeared in 1743 in the book *La Théorie des Comètes* ... , published by Chez G. Martin. Each measured 17.3 cm in diameter, was centered on the equatorial poles, used a polar stereographic projection, and had a geocentric orientation. The plates measured 18×17.9 cm. The constellations were influenced by Senex (except for a reversal of some of the figures),

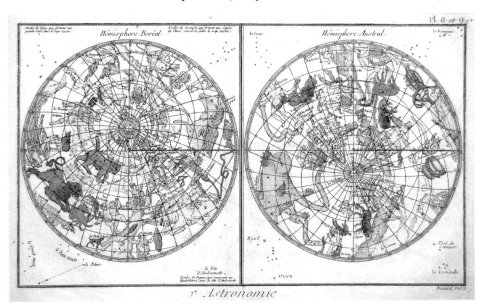

Figure 7.15. Copies of Le Monnier's famous celestial hemispheres, from side-by-side pages of Diderot and d'Alembert's *Encyclopédie*, published in Lausanne and Berne ca. 1780. 19.9×35.1 cm, each hemisphere 16.7 cm dia. Note the beautiful detail in the hemispheres and the constellation star patterns in the corners. *See also* Color Plate Gallery.

and they included Le Monnier's constellation Le Renne. In the corners of the northern map were labeled star patterns of the constellations Cygnus and Cassiopeia (top), and Ursa Major/Ursa Minor and Andromeda (bottom). Similarly placed in the southern map were the star patterns of Grus and Scorpius (top), and Orion and Centaurus/Crux (bottom).

These hemispheres were widely copied, such as in editions of the *Encylopédie, ou Dictionnaire Raisonné des Sciences des Arts et des Métiers*, by Diderot and d'Alembert. In the example shown (Figure 7.15), the two plates are on adjacent pages bound next to each other from a small-sized edition published ca. 1780 in Lausanne and Berne. Each hemisphere measures 16.7 cm in diameter, and the entire double-hemisphere print size measures 19.9 × 35.1 cm. Warner (1979) also cites two sources in which these hemispheres appear: Keill's 1746 French edition of *Institutions Astronomiques*, and Fortin's 1776 and 1795 editions of *Atlas Céleste de Flamsteed* (see Section 6.4.4.2).

In 1755, Le Monnier wrote a book entitled *Nouveau Zodiaque ...*, which contained three influential star maps. According to Warner, the first had the same beginning words in its label as the book, and it showed geocentric views of the zodiacal constellations in three rows of four constellations, each of which measured 14 × 21 cm. Sponsored by the Académie Royale des Sciences, this map was accompanied by a star catalog that updated the positions of Flamsteed's zodiacal stars. The revised catalog and map were designed to help navigators determine terrestrial longitude by means of lunar location in the sky (see also Section 6.4.1). In the British Library copy of this book, this map does not appear. However, there are small 2.5 × 8 cm images of each constellation above its corresponding description in the star catalog.

The other maps mentioned by Warner (1979) in Le Monnier's book were geocentric prints of two star clusters labeled *Carte des Hyades*, measuring 14.7 × 17.2 cm, and *Carte des Pleiades*, measuring 14.7 × 9.5 cm. The latter essentially replicated an earlier map of this cluster that appeared in the 1748 issue of the *Mémoires, Académie Royale des Sciences*, but which was not published until 1752.

Le Monnier had other celestial maps that appeared in the *Mémoires*. Warner describes two of these. One was a 1789 map of 31 members of the Praesepe star cluster, published in 1789. The other was a 10.4 × 18.5 cm geocentric map of the stars of the constellation Solitaire, published in 1776. Le Monnier introduced this tropical bird constellation in this map, which was copied by several subsequent French cartographers. British cartographers, however, redesigned it as the Mockingbird (Thomas Young—see Section 4.4 and Appendix B), and Noctua the owl (Alexander Jamieson—see Section 6.5.5.2 and Figure 9.5).

7.14 THE FRENCH CASSINI FAMILY

7.14.1 Life and times

There was a remarkable family in France that spawned four consecutive generations of astronomers who all were the directors of the Paris Observatory. The grand patriarch was Giovanni Domenico (or Jean Dominique) Cassini (a.k.a. Cassini I), who was born in the Italian town of Perinaldo (near what is now Nice, France) on June 8, 1625. As a youth, he was attracted to the heavens and to astrology, which he later rejected in favor of astronomy. After studying astronomy, mathematics, and engineering at Jesuit College in Genoa, and being tutored by Riccioli and Grimaldi, he became Professor of Astronomy at Bologna in 1650. He became quite active and famous for his work. For example, he determined the obliquity of the ecliptic, studied refraction and solar parallax, observed comets, monitored the phases of Venus, characterized the oblateness of Jupiter, and refined the surface markings of Jupiter and Mars (using these to determine the rotation periods of these two planets). His mathematical and engineering skills also contributed to his work on fortifying the citadel at Urbino.

In 1669, Cassini moved to Paris at the invitation of King Louis XIV to be court astronomer. He became a member of the new Académie Royale des Sciences, and in 1671 he was appointed the first Director of the Paris Observatory. In 1673 he became a French citizen and married a French woman. While at the Observatory, he made a number of important contributions to astronomy. These included observing a number of comets, measuring the distance of Mars and refining the value of the astronomical unit (i.e., the Earth–Sun distance), discovering four moons of Saturn and the large gap in the planet's rings that was named in his honor, characterizing the zodiacal light as a cloud of particles around the Sun, and discovering the "nebula" that later became the star cluster M50. He made ephemerides for the Galilean moons of Jupiter, which helped with the measurement of the speed of light by his co-worker Ole Roemer. He also participated in geographical and surveying projects, including one to determine the dimension of France, and he developed cartographic skills that included celestial and lunar mapping. Later in life, he became blind, and he died in Paris on September 14, 1712.

Jacques Cassini (a.k.a. Cassini II), the son of Giovanni Domenico, was born in Paris on February 18, 1677. He studied optics at the Collège Mazarin and worked with his father and also independently on various astronomical, geographical, and surveying projects. On the death of his father in 1712, he succeeded him as the Director of the Paris Observatory. His astronomical activities included observations of comets, the planets and their satellites, and the proper motion of stars. He achieved memberships in the Académie Royale des Sciences, the Royal Society, and the Berlin Academy. In 1740 he wrote a book entitled *Elements of Astronomy*. He died in Thury, France on April 16, 1756 in a carriage accident.

César François Cassini (a.k.a. Cassini de Thury or Cassini III), the son of Jacques, was born in Thury, France on June 17, 1714. He learned astronomy, geography, and surveying while growing up, and he assisted his father with various geodesic surveys. He became the Director of the Paris Observatory in 1771, but he was less distinguished as an astronomer than as a geodesist and cartographer. For example, he determined the longitude difference between Paris and Vienna, and he was the first of his family to give up the Cartesian view of a north–south elongated Earth for the Newtonian view that the Earth is flattened at the poles. He also made a survey of France using triangulation, and he directed work on a new topographic map of France that was published posthumously by his son in 1789 that is referred to as *Carte de Cassini*. He died in Paris on September 4, 1784.

Jean Dominique Cassini (a.k.a. Cassini IV), the son of César who had the French name of his famous great-grandfather, was born in Paris on June 30, 1748. He continued the family tradition of studying astronomy, geography, and geodesic surveying. He also studied the behavior of chronometers at sea and published his results in 1770. Together with his father, he published an account of the first 100 years of the Paris Observatory in 1771, and he succeeded his father as Director in 1784. He also helped to complete and publish his father's great map of France. He was imprisoned for his royalist sympathies during the French Revolution, but upon his release he retired to the family estate in Thury in 1794. Later, Napoleon made him a senator and count. He died on October 18, 1845.

7.14.2   Celestial maps of Giovanni Domenico (or Jean Dominique) Cassini

Warner (1979) describes a number of celestial maps produced by Giovanni Domenico Cassini (Cassini I). One was a 9.6×18.8 cm illustration for an article entitled *Observations d'une nouvelle comète* ... in the *Journal des Scavans* (1672–1674) that showed the path of the comet of 1672 in a geocentric orientation. The text was republished the same year with a similar 15.5 cm square illustration in the 1666–1699 *Mémoires, Académie Royale des Sciences*. In 1681, Cassini published a booklet on the comet of 1680–1681 entitled *Abrège Des Observations & Des Réflexions Sur La Comète* ... that showed its path on a 51 cm diameter planisphere that was centered on the north equatorial pole down to 41 degrees S dec. using a polar stereographic projection with geocentric orientation. This map also appeared in his *Description et Usage du Planisphère dont la Figure est Inserée* ... He also designed a beautiful silver planisphere with an overlying rete of celestial circles that is now lost but probably resembled his 1681 paper planisphere. In 1708, he published a map in the *Mémoires, Académie Royale des Sciences* with the title *Figure des Pleyades avec le passage de la Lune* ... This 21×29.5 cm map showed 56 stars of the Pleiades measured by Cassini in a geocentric orientation. Finally, in 1692 he produced a detailed map of the Moon that will be described in Section 8.5.2.2.

7.15   GIOVANNI MARIA CASSINI

7.15.1   Life and times

Giovanni Maria Cassini was a fine Italian engraver, globe maker, and painter who was born in 1745 and died around 1824. He did most of his work in Rome and was not a member of the French Cassini family. Hence, he is included in this chapter to avoid confusion with Giovanni Domenico Cassini (who was born 120 years earlier) and his offspring, who did their work in France.

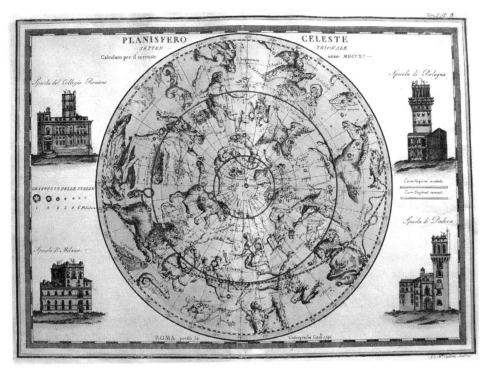

Figure 7.16. The northern celestial hemisphere, from Giovanni Maria Cassini's 1792 atlas *Nuovo Atlante Geografico Universale*, published in Rome. 29.1 × 41.3 cm, 27.3-cm dia. hemisphere. Note the fine images of four Italian observatories in the corners. *See also* Color Plate Gallery.

7.15.2   Celestial maps

Johnston (1985) describes an engraving produced by Giovanni Maria Cassini in 1788 entitled *Tavola Sferica*, which she believes was copied from the work of Antonio Zatta. It measured 17.50 × 23.75 inches (or roughly 44.5 × 60.3 cm) in total page size. It depicted five spheres, a large one in the center, and four smaller ones in the corners. The central and left spheres showed various Earth circles projected into the sky, and the two right spheres depicted the winds of the oceans and the Mediterranean. Cassini also published some fine celestial globe gores, which will be described in Section 8.1.3.

In 1792, Cassini published in Rome Vol. 1 of his atlas *Nuovo Atlante Geografico Universale*. This contained two celestial hemispheres printed in 1790, which were labeled *Planisfero Celeste Settentrionale* (or *Meridionale*) and were centered on the equatorial poles using a polar stereographic projection with geocentric orientation (Figure 7.16). The northern hemisphere measured 27.3 cm in diameter, and the plate size was 29.1 × 41.3 cm. Similar to Zatta's hemispheric prints (see Section 7.9), in the corners were beautiful drawings of famous observatories: Collegio Romano, Bologna, Milan, and Padua in the northern plate; and Paris, Cassel, Greenwich, and Copenhagen in the southern plate. Warner (1979) attributes the style of the constellations to Le Monnier. Vol. 2 of this atlas was published in 1797, and Vol. 3 was published in 1801.

7.16 BIBLIOGRAPHY

Ashworth, W.B. Jr. (1997) *Out of this World—The Golden Age of the Celestial Atlas*. Kansas City, MO: Linda Hall Library.

Brown, B. (1932) *Astronomical Atlases, Maps and Charts*. London: Search Publishing.

Heinz, M. (1997) A programme for map publishing: The Homann firm in the Eighteenth Century. *Imago Mundi*, **49**, 104–115.

Johnston, P.A. (1985) *Celestial Images—Astronomical Charts from 1500 to 1900*. Boston, MA: Boston University Art Gallery.

Kanas, N. (2002) Mapping the solar system: Depictions from antiquarian star atlases. *Mercator's World*, **7**, 40–46.

Kanas, N. (2005) Are celestial maps really maps? *Journal of the International Map Collectors' Society*, **101**, 19–29.

McCarroll, S. (2005) *Celestial Images: Antiquarian Astronomical Charts and Maps from the Mendillo Collection*. Boston, MA: Boston University Art Gallery.

Mosley, A. (2011) Vincenzo Maria Coronelli's *Atlante Veneto* and the diagrammatic tradition of cosmography. *Journal for the History of Astronomy*, **42**, 27–53.

Rowland, I.D. (2000) *The Ecstatic Journey: Athanasius Kircher in Baroque Room*. Chicago: University of Chicago Library.

Stolzenberg, D. (2001) *The Great Art of Knowing: The Baroque Encyclopedia of Athanasius Kircher*. Stanford, CA: Stanford University Libraries.

Warner, D.J. (1979) *The Sky Explored: Celestial Cartography 1500–1800*. Amsterdam: Theatrum Orbis Terrarum.

8

Special topics

In this chapter we will discuss a number of special topics of relevance to celestial maps and prints. Some of these topics relate to objects that have been used in observing and mapping the heavens, such as celestial globes and gores, volvelles, and early astronomical instruments and telescopes. Other topics pertain to non-stellar bodies in the heavens that are mapped in antiquarian prints, such as the components of our solar system and deep-sky objects. Finally, we will discuss frontispieces and title pages.

8.1 CELESTIAL GLOBES AND GORES

As we have seen, the ancients pictured the universe as a series of concentric spheres around a central Earth, with the innermost spheres carrying the Moon, Sun, and planets. However, one of the outer spheres carried the fixed stars, which did not move relative to each other and were the same distance away. Indeed, looking up at the stars at night with our naked eye, it does seem as if they are tacked onto a dark sphere moving around the Earth on a daily basis. This perception led to the idea of representing the heavens as a spherical globe on which the stars and constellations could be imaged. Since such a celestial globe would be viewed from the outside (like God looking down on the stars from the empyreal heavens), the star patterns would be mirror-image reversed from the way we see them looking up from the Earth.

Celestial globes have been produced since ancient times and may be older than their terrestrial counterparts. These globes had several functions. One was educational: to show the constellations in three-dimensional space, to simulate the risings and settings of stars, to demonstrate relative stellar positional distances without resorting to spherical trigonometry, to visualize the important circles such as the ecliptic and celestial equator, and to show the positions and movements of heavenly bodies such as the planets. Another function was to use such information for

Figure 8.1. Nine inch celestial globe on a Bakelite stand, produced by Rand McNally in 1936. Note the stars in yellow (see Color Plate Gallery), the mirror-reversed constellation images in light blue, and the dark blue background, which simulates the night sky. Note also the movable, vertical calibrated meridian ring on which the globe is attached; the hour ring near the north equatorial pole of the globe; and the fixed horizon circle that is connected to the base. Copyright Rand McNally. Reproduced with permission, R.L.07-S-22.

navigational purposes in order to calculate location at sea. Celestial globes also carried decorative and symbolic value, appearing on ancient coins and in mosaic pavements, and later on plates and in paintings. Finally, for some people they were status symbols, whose appearance in a library suggested that the owner was scholarly and worldly. The globes often were accompanied by an instruction manual on how to use them. They also varied greatly in size, from large floor-mounted versions to tabletop pieces to miniature "pocket" globes.

Typically, globes were placed in a frame that was attached to a stand (Figures 8.1 and 8.2). The frame was generally composed of three components: a calibrated meridian ring, in which the globe could rotate on its polar axis; an hour ring with a pointer attached near the north pole; and a fixed horizon circle, on which paper with

Figure 8.2. Close-up view of a 9 inch celestial globe on a Bakelite stand, produced by Rand McNally in 1936. Note the vertical gore lines. Note also the place where the lines representing the ecliptic and the celestial equator cross (i.e., the place in the zodiac where the Sun is located at an equinox, in this case the vernal or spring equinox). The vertical calibrated declination line that passes through this point connects the north and south equatorial poles. Copyright Rand McNally. Reproduced with permission, R.L.07-S-22. *See also* Color Plate Gallery.

printed signs of the zodiac or the months was often glued. The meridian ring could turn through grooves in the horizon circle, allowing the axis of the globe to be adjusted to different angles with respect to the horizon (such as the angle corresponding to the latitude of the observer, where the part of the globe above the horizon circle would correspond to the part of the sky that is visible to the observer).

8.1.1 Early examples

Astrophysicists and scientific writers Marc Lachieze-Rey and Jean-Pierre Luminet (2001) state that celestial globes were probably used to map the heavens by the Chaldeans and Egyptians, who in turn influenced the ancient Greeks. They cite a legend where Atlas was perceived as carrying a globe of the heavens (not the Earth) on his shoulders as punishment by Zeus for his role in the Titan revolt, and Homer credits Atlas for introducing the celestial sphere to man. In another legend the centaur Chiron gave Jason and his Argonauts a celestial globe to help them navigate their way to the Golden Fleece. There is evidence that Anaximander of Miletus created such a globe in the 6th Century BC, followed by other Greek astronomers such as Eudoxus and Hipparchus. London researcher J.V. Field (1996) cites evidence that in the 3rd Century BC, Archimedes may have made a movable device that included a celestial globe that was meant to illustrate the motion of the heavens. Later, in the *Almagest*, Ptolemy described in some detail how to make a celestial globe.

The earliest existing celestial globe is the Farnese Atlas, which once belonged to Cardinal Alessandro Farnese and is now in the National Museum in Naples. It dates back to the 2nd Century AD but was copied from earlier Greek models (probably based on Hipparchus' catalog). It consists of a marble statue of Atlas holding a 65 cm diameter globe on his back, on which are carved a number of heavenly circles (equator, tropics, poles, colures), the ecliptic, and 43 of the 48 classical Greek constellations (Figure 8.3). Since there were no individual stars depicted, its function was probably decorative. The images are oriented left to right from the way we see them when looking up at the sky.

Another early example is the Mainz globe, which was probably made in the 2nd to 3rd Century AD. According to cartographer Edward Dahl and Stewart Museum curator Jean-François Gauvin (2000), it was probably made in Asia Minor, measures 11 cm in diameter, and depicts all 48 of Ptolemy's constellations. It originally served as the crown of a gnomon on a sundial.

Celestial globes were also produced in China. Science writer Colin Ronan (1967) describes a bronze globe cast around 310 AD by the Astronomer Royal Qian Luozhi that had stars colored in red, black, and white to distinguish the separate stellar

Figure 8.3. Paper transcription of two hemispheres drawn from the celestial globe that is part of the 2nd Century AD Farnese Atlas. 25.9 × 52.2 cm. This print appeared in Richard Bentley's 1739 edition of Manilius' *Astronomicon ex Recensione*. Note the classical Greek constellations, the gaps representing areas of damage or places where Atlas holds the globe, and a picture of the statue in the lower center. *See also* Color Plate Gallery.

listings by three earlier astronomers: Shi Shen, Gan De, and Wu Xian. This celestial globe no longer survives.

Islamic astronomers also made celestial globes. According to science historian David A. King (1996), in the 9th Century Qusta ibn Luqa wrote a treatise in Arabic on the celestial globe that remained popular until the 1800s. One of the earliest surviving Islamic globes dates to around 1080 AD and is in the Bibliothèque Nationale de France. It is made of bronze and is mounted on a four-legged bronze stand. Lachieze-Rey and Luminet (2001) state that it is 18.3 cm in diameter. The stars are depicted as small disks and are grouped into the classical 48 constellations. Labels are in old Arabic script. Both the ecliptic and the celestial equator are shown as calibrated bands. King cites another similar example in brass measuring 24 cm in diameter, which was made by Muhammad ibn Hilal in Mosul around 1275 AD. It shows the 48 constellations and some 1,000 stars in inlaid silver dots. It is now in the British Museum.

The earliest celestial globe produced in post-classical Europe was owned by the philosopher and theologian, Nicolaus of Cusa (1401–1464). It was made of wood and glued cloth, and the constellations were drawn by hand in ink (a so-called one-of-a-kind "manuscript" globe). During the Age of Exploration, celestial globes became more widely used in Europe and were often paired with their terrestrial counterparts (Figure 7.13), especially as explorers discovered new lands and circumnavigated the Earth, which required more sophisticated navigational techniques. An early example was the pair of 37 cm diameter engraved globes produced by the mathematician Gemma Frisius in Belgium in 1536–1537. One large interesting and famous celestial globe was housed in the library at Tycho Brahe's Uraniborg. It was constructed out of wood but surfaced with brass sheets.

8.1.2   Celestial gores

As we have seen, celestial globes initially were made of materials such as wood, metal, ivory, marble, or glass, on which the stars and constellation figures were drawn, painted, or engraved. They usually were labor-intensive and one-of-a-kind. But in the 16th Century, the need to make celestial globes cheaper, more standardized, and reproducible led to the use of gores to illustrate the stars and constellations. Like their terrestrial counterparts, celestial gores were consecutive segments whose images were produced on a woodblock or copper plate and then were printed on some material, usually paper. Typically, there were a total of 12 gores (or 24 half gores), each representing 30 degrees of celestial longitude (Figures 8.4 and 8.5). The gores were then cut out, glued onto spheres made out of wood or papier-mâché and plaster, and then covered with varnish for protection. Sometimes, the points of the segments were capped where they met at the poles by circular pieces of paper (called *calottes*) that continued the images to the extreme north and south. Through the use of these printed gores, several identical globes showing the same stellar and constellation material could be produced. Their accuracy was assured, since once the woodblock or engraving was approved, multiple paper copies could be made and placed onto standard spherical cores.

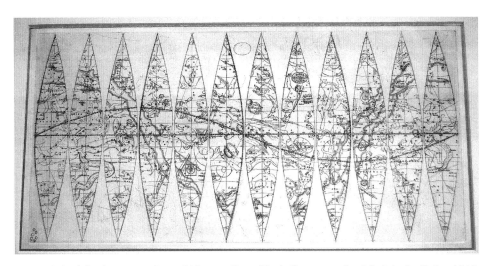

Figure 8.4. A finely engraved set of 12 gores for a 5 inch diameter celestial globe by Bales, 1845. 20.3 × 41.2 cm. Note the ecliptic and equator lines and the typical constellation figures reversed left to right. *See also* Color Plate Gallery.

8.1.3   Important globe makers in Europe

From the 16th to the 18th Centuries, a number of beautiful celestial globes were made, some by people who were also well known for their exceptional maps. Nuremberg became the first major center, especially the workshop of Johann Schöner (1477–1547). By the early 1500s, he was mass-producing both celestial and terrestrial globes. After a period of relative quiescence in the 1600s, Nuremberg again emerged as a center of printed globes in the early 1700s. This stimulated a number of German mapmakers to try their hand at globe-making; examples include Georg Eimmart, Johann Homann, and Johann Doppelmayr.

Another Northern European to make globes was the famous mapmaker Gerard Mercator (1512–1594). Following his production of a terrestrial globe a decade earlier, in 1551 he constructed a celestial globe, which at 41 cm in diameter was the largest globe using printed gores that had been made until that time. It included the latest astronomical information, including the new constellations introduced by Vopel (see Section 4.3.3.1): Antinous and Coma Berenices, which he called Cincinnus. This globe was very popular due to its beauty and precision, and it influenced several subsequent planispheres and globes. For example, Warner (1979) states that the mathematician and instrument maker Emery Molyneux closely copied Mercator in 1592 when he produced the first English globe, which measured 66 cm in diameter. He also copied Plancius' 1589 globe for the new constellations in the southern hemisphere.

In fact, Petrus Plancius (1552–1622), who was mentioned in Section 4.3.2.3, was quite influential as a globe maker. His 1589 celestial globe measured 32.4 cm in diameter and was the first full celestial map to depict a number of new southern stars and constellations by Corsali, Vespucci, and Medina. He also produced a

35.5 cm diameter globe around 1598 that was one of the first to include stars plotted by Tycho Brahe rather than the less accurate and extensive list from Ptolemy, as was typically the case. He also encouraged Keyser to map the southern stars, and he used some of Keyser's observations on this globe. Warner (1979) cites several other globes produced by Plancius: in 1613 (measuring about 9.6 cm in diameter), in 1614 (of which only some gores exist), in 1625 (about 26 cm in diameter), and in 1649 (about 7 cm in diameter). Plancius also included celestial hemispheres as inserts in some of his large terrestrial maps. As we saw in Section 4.3.2.3, he was also responsible for the independent invention of ten constellations (six of which were introduced on his 1613 globe), as well as influencing the selection of several others that were based on the work of Keyser and De Houtman.

Two other famous Dutch mapmakers produced celestial globes: Jodocus Hondius, Sr. (1563–1612) and William Janszoon Blaeu (1571–1638). Before issuing their first celestial globes, both had acquired some practical experiences with the heavens: Hondius had engraved the gores for the globes of Molyneux and Plancius, and Blaeu had spent several months studying with Tycho Brahe and included Brahe's newly positioned stars in his globes. By the early 1600s, both had produced a number of pairs of celestial and terrestrial globes in several sizes, from miniature versions less than 20 cm in diameter to larger versions of over 30 cm. Warner specifically describes Hondius' 34 cm diameter globe, produced in 1600 and patterned after Plancius. Both she and Johnston (1985) describe Blaeu's 34 cm diameter globe and its set of three-gore engravings each measuring 35.6 × 25.4 cm from his *Sphaera Stellifera ...*, first produced in 1598 and again in 1603 in a revised edition that included 14 new southern constellations by Keyser, Plancius, and De Houtman. Blaeu also included celestial hemispheres as inserts in some of his large terrestrial maps, some of which are described in Warner.

Another famous team of globe producers were the Dutch father-and-son Valk family: Gerard (1652–1726) and Leonard (1675–1746). Gerard actually had begun as a map and atlas publisher and had worked for a while with Petrus Schenk (the pair notably producing a version of Cellarius' atlas—see Section 7.1.2) before turning to globes. According to Dahl and Gauvin (2000), the Valk globes were simple in appearance and seem to have been made less for decorative and artistic purposes than for scholarship, containing the latest information on stellar positions and new constellations. During the first 25 years of the 18th Century, they produced seven pairs of terrestrial and celestial globes ranging in diameter from 7.75 to 62 cm.

But perhaps the most famous globe maker was not Dutch but Italian: Vincenzo Maria Coronelli (1650–1718), whose life and cartographic skills we discussed in Section 7.4. Sometime after 1681, during his two years in Paris as Geographer Royal, Coronelli was commissioned by Cardinal d'Estrées to produce a terrestrial and celestial globe for King Louis XIV, and he responded by designing two splendid pieces measuring 384 cm in diameter. They were made in marble and bronze and were very accurate. The constellation figures were lavish and in a Baroque style. The globes were originally displayed in the Château de Marly by order of the King, but now they are at Versailles. Coronelli later made some scaled-down versions of the Marly globes, which can still be found in libraries in Italy and France. Johnston also mentions a celestial globe Coronelli made for the King that measured 15 feet (about 457 cm) in diameter that was the largest in existence until the 1900s.

Coronelli globes are quite scarce today, but printed gores still appear from time to time. Several examples are found in Coronelli's *Libro dei Globi* (Figure 8.5). This book includes gores for celestial and terrestrial globes of different sizes. According to collector Robert Gordon, these sizes are about 5 cm, 10 cm, 15 cm, 46 cm, and 107 cm. Another example that is illustrated by Johnston comes from the 1693 book *Orbis Coelestis Typus*. The gores are described as 20th-Century restrikes made by the Musée du Louvre from the original 1693 plates that are printed on pages measuring 76.2×57.2 cm in dimension. Twelve gores cover the northern celestial hemisphere from equator to 70 degrees N dec. at 30-degree intervals. Twelve other gores cover the south celestial hemisphere from equator to 70 degrees S dec. The last two gores cover the north and south polar areas. The constellation figures are detailed, robust, and theatrical, in keeping with their Baroque style. Printed information can be seen coexisting with the constellation images, but it is discreetly written and placed and does not detract from the beauty of the figures.

Following the lead of Coronelli, other cartographers in France began to produce celestial globes. One such person was Guillaume Delisle (1675–1726), who ran a map-making business. Based on his maps and his fine globes, including a celestial globe produced in 1699 that measured 50 cm in diameter, he was elected to the Académie Royale des Sciences in 1702. In 1718 he was appointed Royal Geographer and tutored the young Louis XV in geography. Unlike Coronelli, Delisle used a less ornamental style in his constellation figures, preferring instead a simpler, less cluttered presentation. Celestial globes continued to be produced in France by people like Didier Robert de Vaugondy (1723–1786), whose father Gilles was a prominent cartographer. In 1750, Didier was commissioned by Louis XV to produce a pair of 45 cm diameter globes for the French Navy.

Another important globe maker was the Italian Giovanni Maria Cassini (1745–1824), who was mentioned in Section 7.15. Johnston discusses a number of gores from Cassini's 1792 atlas *Globo Celeste*, which she says are based on the astronomical observations of Flamsteed and Lacaille. Each engraving of three gores measures approximately 60×43.8 cm in size. Both Ptolemaic and later constellations are depicted, with stars varying in size according to their magnitudes. The figures are shaded to give them a more three-dimensional appearance.

Finally, mention should be made of several English globe producers. The first is cartographer John Senex (1678–1740), who was discussed in Section 7.6. According to Warner, he also produced a number of terrestrial and celestial globes, ranging from 7.6 cm diameter pocket globes (which were popular in England as conversation pieces) to table models that were up to 68 cm in diameter. His larger celestial globes were well known for the skill of their engraving, their durability, and the amount of astronomical material they contained. Many used Latin terms and contained more stars than other globes, suggesting that they were made for scholars in astronomy and geography. His existing globes generally are found in museum collections and rarely come up for sale to collectors. Another important producer of globes was the instrument-making firm of George Adams & Sons. During the last half of the

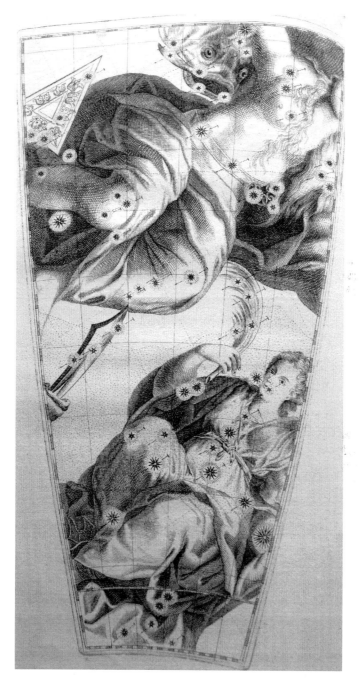

Figure 8.5. A single-gore engraving for a 107 cm celestial globe, from the 1693 edition of Coronelli's *Libro dei Globi*. Approximately 43.2 (vertical) × 27 (top) × 10.2 (bottom) cm. Note the theatrical Baroque style and the rich detail. Courtesy of the Collection of Robert Gordon.

18th Century, they produced a number of globes ranging in diameter from 7 to 71 cm, some of which were based on the star catalog of John Flamsteed. According to Dahl and Gauvin, they also supplied James Cook with instruments that he used to observe the transit of Venus in 1769. Other important globe makers were the two Cary brothers: John (ca. 1754–ca. 1835) and William (ca. 1759–ca. 1825). Little is known of John, but William was an instrument-maker and member of the Astronomical Society. In 1799 the brothers produced *Cary's New and Improved Celestial Globe*, which according to Johnston measured about 53.3 cm in diameter and was made of wood, brass, and paper. It had some 3,500 stars, more than any previous globe. The constellation images were less detailed and dynamic than those found on other globes, however, and Johnston states that the Cary brothers were among the first to completely eliminate constellation designs in their later globes.

During the 1700s, celestial globes increasingly began to be used more as room decorations than for their intrinsic scientific value, giving way to maps and atlases, which were easier to use, could be updated frequently, and were much less expensive. In the 1800s even their aesthetic appeal was lost, and globes began to be mass-produced primarily for educational purposes. Later in the 19th Century, they lost their constellation images completely, reflecting a parallel process in celestial maps, as we shall see in Chapter 10. For a more complete description of celestial and terrestrial globes and their history, the reader is referred to the superb catalog by Dahl and Gauvin (2000).

8.2   VOLVELLES

To determine the position and movement of heavenly bodies in the sky, as well as to tell time and determine one's latitude on the Earth, a number of instruments were developed, such as the armillary sphere (discussed in Section 1.4.2) and the astrolabe (discussed in Section 8.3.2). Such instruments were made of wood or metal, were expensive to produce, and were used mainly by serious astronomers and scholars who had the incentive and funds to purchase them. But with the advent of movable type and the development of printing in Europe in the 1450s, an opportunity arose to incorporate the principles of these instruments into books, which could be produced more cheaply and could reach more people. This led to the popularization of movable devices on the printed page that were called volvelles.

8.2.1   What were volvelles?

According to the 5th edition of *The Shorter Oxford English Dictionary* (OUP, 2002), a volvelle was a "device consisting of one or more moveable circles surrounded by other graduated or figured circles, for calculating the rising and setting of the sun and moon, the state of the tides, etc." (p. 3558). In a sense, they were early analog computers that were used to calculate the time of the day or night, the angular orientation of a heavenly body in the sky, the height of distant objects, or the number of Sundays until Easter. Volvelles were likely used by the ancient Greeks and later by

Arab scholars to illustrate their cosmological theories. Examples from European hand-written manuscripts appear as far back as the 13th Century, but they really became popular with the explosion of printed works during the Renaissance. Printed volvelles were found in a number of prominent 15th and 16th-Century texts. In addition to the sources that will be described later, volvelles appeared in Johannes Regiomontanus' *Calendarium*, Johannes de Sacrobosco's *De Sphaera*, Johann Schöner's *Aequatorium Astronomicum*, Sebastian Münster's *Organum Uranicum*, and Georg Peurbach's *Theoricae Novae Planetarum*. Other examples cited by Warner (1979) were produced by Lucas Cranach the Elder, James Ferguson, Leonhard Thurneysser, and Lucas Janszoon Waghenaer (see Appendix B).

Although a few free-standing volvelles were produced that were made of wood, brass, or other such material, most were incorporated into printed books. The page itself was part of the instrument, and one or more movable paper disks or pointers were attached onto its center by a tab or string. Loose sheets were sometimes inserted that contained the movable parts of the volvelle, which the reader could then cut out and assemble without destroying the book (Figure 8.6). The use of the instrument

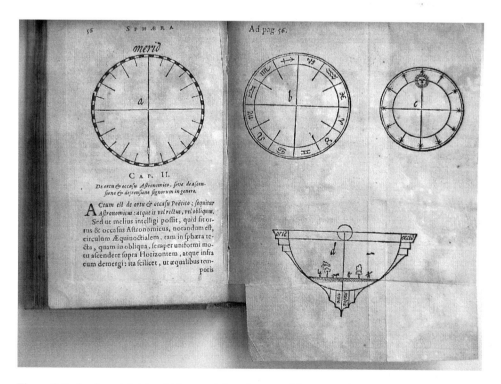

Figure 8.6. An insert (*right side*) containing the movable parts of a volvelle, from the 1647 Leiden edition of Sacrobosco's *De Sphaera*. 17.8 × 13.1 cm insert. Note that these pieces can be cut out and attached to the printed page on the left on to the 5.5 cm dia. printed circular scale. *See also* Color Plate Gallery.

usually was described in the text. Many volvelles incorporated one or more strings that could be used as pointers or could assist with calculations or demonstrations. In some cases, volvelles replaced complicated tables giving the positions of heavenly bodies, since the reader could calculate this information directly by manually operating the instrument. In other cases, simple tables were provided that were designed to be used in conjunction with the volvelle, and what was saved was the need to perform complex mathematical procedures.

Gingerich (1993) has pointed out that some volvelles functioned more as teaching than calculating devices and were used primarily to illustrate the location of a heavenly body or to show its relationship to other heavenly bodies. This became most explicit in a specialized type of volvelle called an "equatorium". According to Evans (1998), this device took into account the non-uniformity of solar, lunar, or planetary movements (due to the eccentricity of their orbits or the particulars of their deferent/epicycle combinations) and allowed one to show the positions of heavenly bodies without the need for complicated mathematical calculations. Thus, it substituted for the "equation" that would otherwise be needed to adjust for the difference between a heavenly body's mean and actual location in the sky. Typically, equatoria used a series of disks to represent the movements of celestial bodies in two dimensions, thus representing the geometry of the heavens according to Ptolemaic concepts. Interestingly, they rarely were used to model other cosmological arrangements, such as those of Copernicus, Tycho Brahe, or Kepler. Their relationship to volvelles can be summed up in the statement: all equatoria are volvelles, but not all volvelles are equatoria. Related paper instruments have recently been described that were used to calculate times when the Moon and Sun would be in syzygy with the Earth during eclipses (Kremer, 2011).

8.2.2 How did volvelles work?

The operations and uses of volvelles were quite varied, and some examples will now be described. The first example is from the 1584 edition of Apian's *Cosmographia* (see Section 3.8.4). This volvelle was used to illustrate the relationship of the Moon and Sun in the sky for different phases of the Moon. As shown in Figure 8.7, the outer circular scale printed on the paper is divided into 12 daylight ("horae diei") and 12 nighttime ("horae noctis") hours of celestial longitude in the sky. The outer movable disk with the lunar symbol on the pointer represents the hour angle of the Moon, whereas the inner movable disk with the solar symbol on the pointer determines the hour angle of the Sun. By positioning the two disks until the current phase of the Moon is shown in the circular window near the center of the volvelle, and by then rotating the outer disk so that the lunar pointer is at the correct longitudinal location of the Moon in the night sky, the inner solar disk is carried around so that its pointer indicates the heavenly position of the Sun.

Another example is from Apian's *Astronomicum Caesareum* (see Section 3.8.4). This impressive tome contained a number of volvelles and equatoria. As befitting its royal patronage, the strings that were used to help position the heavenly bodies with reference to the disks and outer circular scales were made out of silk, and some of

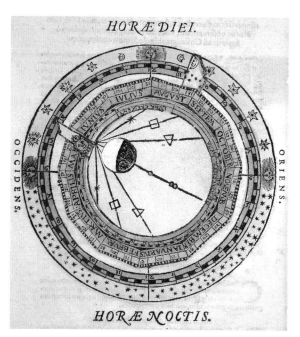

Figure 8.7. A beautifully colored volvelle from the 1584 edition of Peter Apian's *Cosmographia*. 13.4 cm dia. circular scale. Note that by moving the inner and outer disks in the manner described in the text, the relative locations of the Moon and Sun in the sky can be demonstrated for different lunar phases. *See also* Color Plate Gallery.

them supported tiny pearls that could be slid along the string to make it easier to take readings from the various scales. The accuracy was very good, giving planetary longitude positions that Gingerich states were within 15 minutes of arc of the Ptolemaic predictions. In the example shown in Figure 8.8, the position of Mars in the sky can be shown as it revolves in its epicycle along its deferent (labeled "deferens martis"). The various strings align the centers of the movable epicycle and deferent disks with the circular scales. The outermost scale represents the planet's path through the ecliptic in terms of the customary 30 degree increments of each of the 12 zodiacal constellations.

A volvelle that used a pointer instead of an inner disk is shown in Figure 8.9 from the 1614 Spanish edition of Gallucci's *Theatrum Mundi, et Temporis . . .* (see Section 5.6.2). Gallucci's device could be used to calculate the dates related to important calendrical events, such as Easter. On the pointer is a list of the names of the 12 months in order as one approaches the center, and these each line up with a concentric circle printed on the page of the book. As the pointer moves around clockwise, it passes consecutive letters from "A" to "G" that represent the days of the week. For months that do not have 31 days, "ooo" symbols replace the letters and are treated as blank spaces. On the first day of each year (say, it happens to be a Wednesday), that day is given a corresponding letter (e.g., "A"). The letter

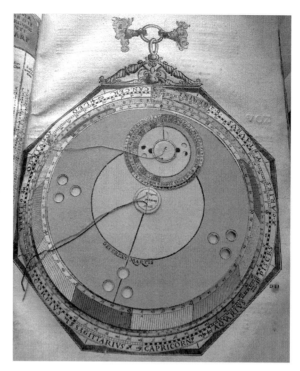

Figure 8.8. An equatorium type of volvelle from Peter Apian's *Astronomicum Caesarium*, published in 1540. Approximately 47×31.8 cm, 29.8 cm dia. hemisphere. Note that by manipulating the various disks and strings, the location and movement of the planet Mars in the sky can be determined as it moves through the ecliptic. Courtesy of the Collection of Robert Gordon. *See also* Color Plate Gallery.

corresponding to Sundays for that year then would be "E", or four letters (days) later. This letter represents the year's so-called "dominical" (for Sunday) letter. By knowing the year's date for Easter Sunday, for example (which in the Catholic Church was the Sunday following the full Moon after the vernal equinox), the pointer could be moved to successive "E's" around each month's circle in order to calculate the number of Sundays until Easter Sunday.

One final example is a volvelle that can be used to calculate the angular position of the Moon's axis. It will be described below (see Section 8.5.2.1 and Figure 8.32).

8.2.3 Planispheres and the demise of volvelles

Paper volvelles offered physical and financial advantages over other wooden or metallic instruments, but they still could be cumbersome to use as part of a book. This especially was true when a sighting was needed that required the whole book to be lifted and positioned with respect to the heavens. In time, new materials were developed, such as heavier paper and cardboard (and later leather, plastic, and light-

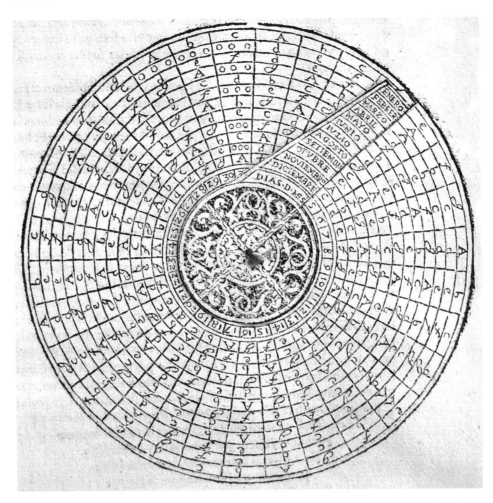

Figure 8.9. A volvelle using a pointer rather than an inner circle, from the 1614 Spanish edition of Giovanni Gallucci's *Theatro del Mundo y del Tiempo*. 13.2 cm dia. hemisphere. Note the outer circles with letters from A to G, which represent the days of the week. As an example of use, for a given year the inner pointer could be set to an important date (e.g., January 1), and from this the number of Sundays (represented by that year's dominical letter) until Easter could be calculated.

weight woods) that were relatively cheap and easier to work with. This allowed for the production of volvelles and other circular devices that were independent of the printed page.

But in the 17th and 18th Centuries, the needs of observational astronomy and the desire for more precision in the location and movement of heavenly bodies led to the publication of increasingly more accurate tables and ephemerides of solar, lunar, and planetary positions. These resulted from finely constructed instruments of measurement, the invention of the telescope, and advances in mathematics, especially spherical trigonometry and calculus. With these changes, the volvelle lost its value as an instrument of calculation.

However, variations continued to exist as teaching and observational tools. An example is the portable celestial planisphere, which is still used today by amateur astronomers to locate the relative position of visible stars, constellations, and (sometimes) selected deep-sky objects. Although described in principle as far back as Roman times, these devices became especially popular after 1700, when heavy paper board versions were made in Europe and the United States.

Portable planispheres typically consist of an outer coated heavy paper jacket that contains a circular or oval window and an inner concentric disk on which is printed the constellations and their stars (Figure 8.10) that are visible in the heavens. By

Figure 8.10. A Victorian planisphere, ca. 1890, constructed under the direction of A. Klippei from Dortmund. Note the 22.2 cm square black heavy paper envelope with four Victorian-decorated pointers. This holds the 25.7 cm dia. revolving wheel, on which are printed the date and the night sky constellations (seen through the window of the envelope). *See also* Color Plate Gallery.

Figure 8.11. A French planisphere, ca. 1900, constructed under the direction of Camille Flammarion. Note the 18 cm square heavy paper base, on which are printed the date and night sky constellations. This is overlayed by a 14.5 cm dia. revolving wheel with clear plastic over the window. *See also* Color Plate Gallery.

aligning the date of observation printed on the disk's edge with the local time of night printed on the jacket, the constellations currently in the sky appear in the window. A volvelle-like variant is shown in Figure 8.11, where the constellations are printed on a cardboard square and the observer rotates an inner movable disk that contains the window. Although still helpful for gross orientation to the night sky, portable planispheres are less detailed than modern-day celestial star charts, and their use for calculation is practically non-existent.

8.3   ASTRONOMICAL INSTRUMENTS BEFORE THE TELESCOPE

We have seen in previous chapters that star maps became progressively more accurate in terms of the positioning of stars in the sky. The invention of the telescope assisted astronomers in this positioning, as well as allowing them to see the surfaces of distant heavenly bodies. But even before the invention of the telescope in the early 1600s, the needs of navigation in the Age of Exploration encouraged the production of ever more accurate instruments that could determine one's location at sea from observations of the sky. In addition, the mapping of new lands required more precise instruments that could be used for surveying purposes. These pre-telescopic instruments typically were made out of wood, brass, or some other metal. In fixed locations such as at observatories, where extreme accuracy was needed, they tended to be quite large, with scales that were subdivided to a fraction of a degree or centimeter. This allowed the positions of heavenly bodies to be determined precisely.

In this section we will focus on some of the more important pre-telescopic instruments that were used by astronomers, navigators, and surveyors to determine accurate stellar and planetary positions in the sky. Two such instruments, the armillary sphere and the celestial globe, had been used since ancient times and have already been discussed in Section 1.4.2 and Section 8.1, respectively.

8.3.1   Antikythera mechanism and the Nebra sky disk

One of the most remarkable of the ancient instruments is the Antikythera mechanism, the use of which has only recently been defined. It was retrieved by sponge divers from the underwater remains of an ancient cargo ship discovered in 1900 off the Greek island of Antikythera (between Crete and the Peloponnese). On May 17, 1902, an archaeologist noted that what was thought to be a rock from the wreck was in fact the encrusted and corroded remains of a gear wheel and other fragments of an ancient bronze mechanism mounted in a wooden frame. Although the cargo ship was Roman and wrecked around 70 BC, the mechanism itself was made in Greece and has been dated back to the end of the 2nd Century BC. It has been studied since its discovery, with findings suggesting that it was a type of analog computer that was used for astronomical and calendrical purposes. It contained over 30 gears and was probably operated by a hand crank. Investigations reported by Freeth and his colleagues (2006) using surface and high-resolution X-ray imaging have suggested that this was a multi-geared device used to calculate and display celestial information, especially cycles such as the phases of the Moon and the luni-solar calendar. Furthermore, it was thought to be useful in predicting lunar and solar eclipses based on Babylonian arithmetic progression cycles. Inscriptions on its surface also have suggested that planetary positions could have been displayed. The mechanism has been associated with Hipparchus, since its gearing was related to his theories involving irregularities of the Moon's motion across the sky.

Although work still is being done on deciphering this amazing instrument (Edmonds, 2011; Evans *et al.*, 2010), its existence suggests that there was a great deal of sophistication in the ancient Mediterranean world in using mechanical devices for astronomical purposes. Cicero wrote about such instruments, and many modern scholars believe that the advanced degree of its construction suggests that similar instruments must have preceded it. Although no such instruments have yet been found, this does not mean there are not other similar devices awaiting us beneath the sea (recalling the old adage: absence of proof is not proof of absence).

Mention should be made of the Nebra sky disk, found in Germany and believed to date back to the 17th century BC. This 30 cm diameter bronze disk is inlaid with gold symbols representing stars (e.g., the Pleiades), the lunar crescent, and the Sun or possibly an actual lunar eclipse. Some believe that golden arcs on the periphery were used as a solar calendar and to mark the solstices. Work continues on interpreting this fascinating instrument or calendar from the Bronze Age.

8.3.2   Astrolabe

Another ancient instrument was the astrolabe (which in Greek meant "star finder")— see Figures 8.12 and 8.13. There were two types. The planispheric astrolabe was a flat, circular instrument consisting of a metal plate called a *tympan*, on whose front was engraved a stereographic projection of the celestial sphere as seen from the south celestial pole. In a sense, this was like crushing a three-dimensional armillary sphere onto a two-dimensional surface. Altitude and azimuth coordinate lines, as well as a line representing the horizon, were inscribed that were relevant for the user's latitude (most astrolabes came with a set of different tympans that could be used interchange-ably for peripatetic users). Overlying this plate was a rotating, cut-away, circular disk called a *rete*. A prominent ring on the rete represented the zodiac, and extending into the cut-away area from this ring were protuberances that indicated the positions and names of important stars. Some astrolabes contained a revolving pointer that extended to the circular, calibrated degree scale at the periphery of the instrument. On the back of the plate was often mounted a rotating sighting device called an *alidade* which, when held up at the eye, could be used to locate the height of a star by reading where it crossed a circular degree scale around the periphery of the plate. By locating the height of the Pole Star, the user could determine his or her latitude as well. In addition, there sometimes was a right-angle coordinate system on the back called a *shadow square* that could be used to calculate the height of a building or some other object on Earth trigonometrically by knowing its distance from the observer and using the astrolabe to sight the angle subtended from its top to the horizon (Figure 3.15).

Planispheric astrolabes had a variety of uses. One could determine the height of a terrestrial object or star, determine his or her latitude, discern the location in the sky of bright stars located above or below the horizon, and tell time during the night according to the positions of the stars. Eastern astrolabes, usually inscribed in Arabic, often had information on the back of the tympan that helped the user find the direction to Mecca. Western astrolabes, usually inscribed in Latin, typically had scales on the back that allowed the user to find the Sun's location in the ecliptic from the current date, and this information could be transferred to the zodiac ring on the rete to find the time during the day.

Figure 8.12. The front of a brass astrolabe made by Badr ibn 'Abdallah ca. 1130 probably in Baghdad. 18.8 × 13.4 cm. Note the rete star map and the tympan engraved with circles of altitude for the latitude of Persia. Courtesy of the Adler Planetarium & Astronomy Museum, Chicago, Illinois, A-84. *See also* Color Plate Gallery.

The other type of astrolabe was the mariner's astrolabe, which was devised late in the 15th Century and was usually made of bronze or brass. It was a heavier and simplified version of the back of a planispheric astrolabe, consisting primarily of an alidade and a degree scale around the rim. Because it was heavy and sometimes weighted at the bottom, it could be used more accurately at sea, since it was less susceptible to movement that would result from the rocking motion of a ship in choppy waters. Portions were cut away in order to reduce resistance from the wind. Its main use was to determine the height of the Sun near the meridian or the height of the Pole Star from the horizon, from which the latitude of the ship at sea could be determined.

Astrolabes were used in classical times, possibly as early as the 2nd Century BC by Hipparchus in compiling his famous star catalog. They became especially popular in the Islamic world, and the oldest surviving example was made in the 9th Century

Figure 8.13. The back of a brass astrolabe made by Badr ibn 'Abdallah ca. 1130 probably in Baghdad. 18.8 × 13.4 cm. Note the Persian writing and the absence of an alidade sighting bar. Courtesy of the Adler Planetarium & Astronomy Museum, Chicago, Illinois, A-84. *See also* Color Plate Gallery.

AD by Ahmad ibn Khalaf. Astrolabes came to Spain in the 10th Century, and in the next century European manuscripts were being written describing how to use this instrument. In the early European universities, astrolabes were used to teach astronomical principles to students, and Geoffrey Chaucer wrote a treatise on their use in 1391. The Portuguese and subsequent explorers used the mariner's version on their travels during the Age of Exploration. However, by the 18th Century their use had been supplanted by newer, more accurate instruments and methods of calculation, as we shall see below.

8.3.3   Nocturnal

Reckoning time has always been important. Early devices have included the water clock or clepsydra, the hourglass, and, during the day, the sundial. At night, a useful instrument was the nocturnal. It consisted of a round metal plate with a calendar scale engraved on it, over which rotated an hour disk marked with two 12-hour periods. On top of this disk was a rotating pointer. It worked on the principle that the stars move around the Pole Star every 24 hours. After setting the hour disk to the correct date on the calendar disk, the observer would then sight the Pole Star through a central hole in the instrument. Next, the pointer would be lined up with prominent nearby stars that point toward the Pole Star, usually *Alpha* and *Beta Ursae Majoris*. The approximate time could then be read by where the pointer cut across the hour disk. This instrument appeared around the beginning of the 14th Century and was in use until the 17th Century, when the invention of the pocket watch made it (and the other time devices mentioned above) obsolete.

8.3.4   Cross-staff and back-staff

The cross-staff was invented around 1300 and consisted of a long wooden staff and one or more perpendicular vanes that moved over it. The end of the staff was held up to the eye, and a vane was moved until its ends corresponded with two points that were being measured, such as the height of a building, the distance between two stars, the diameter of the Moon, or the elevation of the Pole Star from the horizon. In this way, the angular distance between these points could be directly determined by reading the trigonometrically graduated scale on the staff. This simple instrument was quite versatile, being used for surveying as well as for astronomy and navigation. It was also portable and could easily be put together or taken apart.

Latitude at sea could also be determined by measuring the height of the Sun at the meridian, but this would require gazing directly into the solar light. To avoid this, the back-staff was invented around 1595 by an English captain, John Davis. Looking like two large triangles, one on top of the other, the observer would adjust the movable upper vane of the instrument until the Sun's shadow would appear at the far end of the staff. The lower vane was placed at eye level, aligned with the horizon. By reading the number of degrees made by the Sun's rays off the calibrated scales of the instrument, its elevation with respect to the horizon could be determined.

8.3.5   Octants, sextants, and quadrants

Another way of determining the angular distances between two points in the sky was to use an instrument whose main feature was an arc of a circle that was calibrated in units of degrees. Several varieties were made depending on the size of the arc, most typically octants ($\frac{1}{8}$ of a circle), sextants ($\frac{1}{6}$ of a circle), or quadrants ($\frac{1}{4}$ of a circle). These instruments usually were found in observatories and were often of large size (Figures 8.15, 8.16, and 8.17). Some were used to determine the altitude of a heavenly body. By observing such an object in the sky through an eyepiece that moved along the circumference, the height of the object could be read in terms of degrees above the horizon or below the zenith. Other instruments were used to determine the angular distance between two objects. One observer would sight the first object through a movable eyepiece, another observer would sight a second object through another eyepiece, and the circumferential distance in degrees between the two eyepieces would correspond to the angular distance in degrees between the two objects.

Since quadrants were able to measure a separation of up to 90 degrees, they were more versatile and covered a broader expanse of sky, but they were also heavier and more difficult to use than the smaller octants and sextants. In addition, the larger quadrant was more susceptible to metal fatigue and bending, rendering it less accurate; for this reason, quadrants tended to be bolted to a wall in order to make its frame more rigid. Nevertheless, all of these instruments were fairly large, since this allowed for finer graduations of the scale along the arc of circle so that angular distances could be measured to a fraction of a degree.

As an illustration of how these instruments were used, I will now give examples based on descriptions written by two great astronomers who set the tone for observational accuracy, one in the in the 16th Century, Tycho Brahe (see Section 3.9.2), and one in the 17th Century, Johannes Hevelius (see Section 6.3). Both wrote important books describing their observatories and their instruments.

8.3.5.1   Tycho Brahe's *Astronomiae Instauratae Mechanica*

In 1598 (three years before his death), Tycho Brahe published his *Astronomiae Instauratae Mechanica*, which contained a description of the buildings at his observatory on the island of Hven, a description of his instruments, and an account of his scientific contributions to astronomy. It was published after he had left Denmark but before he went to Prague.

Three groups of instrument were described. One group consisted of various armillary instruments and was used to measure right ascensions and declinations with respect to the celestial equator and latitudes and longitudes with respect to the ecliptic (Figure 8.14). A second group was used to determine altitudes and azimuths and included various quadrants and altitude sextants. The third group was used to determine angular distances between heavenly bodies and included various sextants.

One of the more accurate of his instruments was the giant mural quadrant. For stability, it was mounted permanently on a north–south wall facing south, and it covered the sky from the horizon to the zenith. It was used to determine the right ascensions of heavenly bodies based on the times that they transited across the meridian. According to the translated version of Tycho's book edited by Raeder and his colleagues (1946), it was made of brass, and its circumference corresponded to a circle with a radius of nearly 194 cm. The scale on its perimeter was so large as to allow readings to be made to an accuracy of five seconds of arc. In Figure 8.15, the observer who is partially seen at the extreme right is observing a transiting object through the slit on the left wall and calling out its transit appearance and altitude as read from the quadrant's scale. The assistant at the lower right is noting the time of transit from the clocks in front of him, and the assistant at the lower left is recording all of the information. Within the circumference of the quadrant is a picture that was painted on the wall of Tycho Brahe himself pointing at the slit in the left wall, and to his left are images of some of the rooms and instruments that were located at his observatory in Uraniborg.

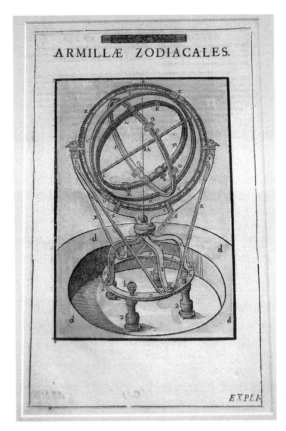

Figure 8.14. Page depicting an armillary sphere from an edition of Tycho Brahe's *Astronomiae Instauratae Mechanica*, first published in 1598. 27.7 × 16.3 cm. The instrument pictured was used by this famous Danish astronomer at his observatory, Uraniborg. Note the letters indicating the parts of the instrument that were keyed to a description in the accompanying text. *See also* Color Plate Gallery.

8.3.5.2   Hevelius' *Machinae Coelestis*

In 1673, Hevelius published a book describing his scientific education and the various astronomical instruments that he used at his observatory. Called *Machinae Coelestis Pars Prior*, it contained 30 plates illustrating these instruments and their use. This was followed in 1679 by *Machinae Coelestis Pars Posterior*, which in two volumes is a synopsis of his scientific contributions and a compilation of some 30,000 pieces of observational data, mostly in tabular form, of distance measurements between stars and planets, lunar and solar eclipses, sunspots, the 1661 transit of Mercury across the Sun's surface, near passes and occultations of planets and stars by the Moon, paths of comets through the heavens, etc. Included are 42 engraved plates illustrating some of these observations. Most copies of this second part were destroyed in the fire,

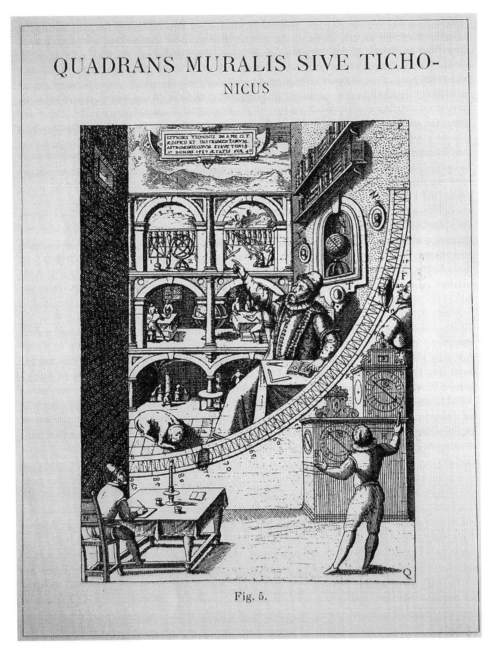

Figure 8.15. Page from a 1946 facsimile of Tycho Brahe's *Astronomiae Instauratae Mechanica*, first published in 1598. 18.8 × 14.2 cm. Note the three observers sighting and recording the meridian transit of a celestial object using the giant mural quadrant at Tycho's observatory, Uraniborg. Note also the painting of Tycho and some of his instruments on the wall to which the quadrant is attached. Courtesy of the Royal Danish Academy of Sciences and Letters.

accounting for its rarity. A perusal of the *Machinae Coelestis* tells us a great deal about the observational techniques of Hevelius.

Many of the observations that Hevelius made were in the service of producing a new star catalog that would improve on Kepler's *Rudolphine Tables*, published in 1627 using the observations of the great 16th-Century Danish astronomer, Tycho Brahe. Hevelius hoped to enhance the accuracy of this earlier work and to record the positions of over 3,000 stars, three times the number found in Kepler's *Tables*. During his schooling, Hevelius had been encouraged by his mentor, Peter Krüger, to be an accurate and persistent observer of the stars and to use the finest instruments in his work, and his role model was Tycho Brahe, who made his observations before the advent of the telescope. Hevelius was not averse to using telescopes in his work (e.g., in observing the Moon and eclipses), but the quality of the lenses of the day resulted in distortions of image and color that necessitated extremely long focal lengths of 150 feet or more. Such telescopes were cumbersome to use, and the quality of their optics was not ideal. Hevelius thought that by using careful observational skills and well-made instruments, he could determine stellar positional accuracy better with his naked eye than with the help of a telescope. Although armillary spheres and astrolabes were still being used to determine the positions of stars and planets in the sky, Hevelius considered these to be inaccurate, and he preferred quadrants and sextants.

In 1644 the city of Danzig presented him with an uncompleted 5-foot radius azimuthal quadrant begun by Krüger but never used. Hevelius completed it and later built 6-foot and 8-foot radius sextants made of wood. These devices were unstable due to flex in the wood over time, and Hevelius later built similar instruments made of brass. His two favorite instruments were the 5-foot quadrant and the 6-foot sextant. The quadrant was used to measure the height in degrees of a star in the sky above a reference point, usually the horizon (called its elevation in altitude) (Figure 8.16). The sextant was used to measure the angular distance in degrees between two stars. It usually required two observers to use, and in later years Hevelius' young wife Elisabeth worked as his assistant (see Figure 8.17).

The publication of *Machinae Coelestis Pars Prior* created some controversy, in that Hevelius' claims that his observations with the naked eye were more accurate than those with the new telescopic sights that were being developed for use with quadrants and sextants raised the ire of many astronomers, chiefly Robert Hooke. To add to the controversy, England Astronomer Royal John Flamsteed reported that some measurements he took with a sextant fitted with telescopic sights agreed fairly well with those of Hevelius. To resolve the dispute, in May 1679 the Royal Society sent Edmond Halley to Danzig to test Hevelius' naked eye observations with those made with instruments taken by Halley that were fitted with telescopic sights. Although Halley believed in the superiority of optically enhanced observations, he was nonetheless impressed with the accuracy of Hevelius' measurements. He wrote a testimonial letter two months later that was subsequently reported in the *Philosophical Transactions*. Although Hevelius was vindicated, in later years improvements in instrumentation showed the superiority of telescopic sights by the time that Flamsteed published his own catalog in 1725.

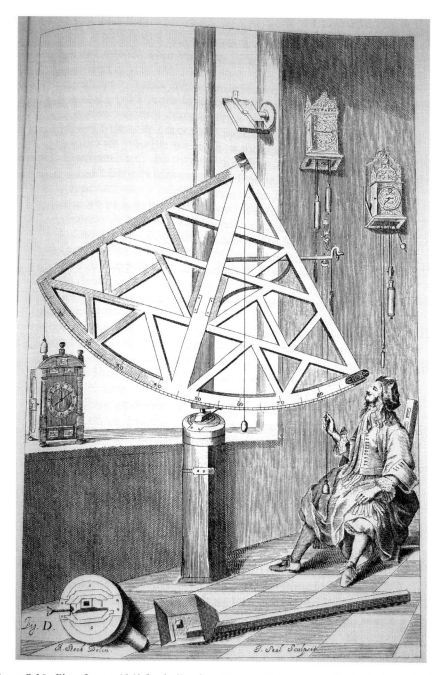

Figure 8.16. Plate from a 1969 facsimile of Part I (*Pars Prior*) of Hevelius' *Machinae Coelestis*, originally published in 1673. 34.6 × 22.5 cm (page size). Note the picture of Hevelius himself observing a celestial object through the eyepiece of the quadrant at his observatory in order to measure the altitude of the object from the horizon. Courtesy of Reprint-Verlag-Leipzig.

Figure 8.17. Another plate from a 1969 facsimile of Part I (*Pars Prior*) of Hevelius' *Machinae Coelestis*, originally published in 1673. 34.6 × 22.5 cm (page size). Note the picture of Hevelius and his wife Elizabeth, each observing a celestial object through the eyepieces of the sextant at their observatory in order to measure the angular distance between the two objects. Courtesy of Reprint-Verlag-Leipzig.

8.4   THE TELESCOPE

The history of the telescope and its importance to astronomy is rich and extensive but beyond the scope of this book. The interested reader is referred to Henry C. King's 1955 classic *The History of the Telescope*. But a few points need to be made because they have relevance to celestial mapping.

The first telescope appeared during the second half of 1608 and was probably made in Holland. There were several contenders for the honor, two of whom applied for a patent in The Hague, but the patent office denied the applications on the grounds that it was too easy to make such an instrument. News of the basic design traveled quickly, and soon telescopes began to appear in other places. These were of the refractive type, using a combination of convex and concave lenses. But the early versions had a magnification of only a few powers, and the quality of the lenses was poor. When Galileo heard about the instrument, he produced his own version in the summer of 1609, which had a power of about 8 or 9 (Figure 8.18). Over the next few months, he continued to make improvements in his lenses, so that in the fall he produced a 20-power telescope of better quality. It was with this instrument that he made his famous discoveries in late 1609 and early 1610, which were summarized in his famous book *Sidereus Nuncius* (see Section 3.9.3 for details).

During the next several decades, advances were made in telescope design. For example, Kepler discovered that by combining two convex lenses, one could achieve a larger field of view and greater magnification, but the image was upside down, and there were distortions in the shape of the image (spherical aberration) and inaccuracies in the color (chromatic aberration). These aberrations could be reduced by making the lenses flatter, but now the distance between the lenses needed to be extremely long in order to create a proper focus at high magnification, and this resulted in a dimmer image and unwieldy instruments that were difficult to move, sometimes requiring ropes and pulleys. In some cases, to reduce the weight, the telescope tube was dispensed with, and there were simply a series of rings between the two lenses (Figure 8.19). Smaller boutique versions had tubes made out of metal and were quite ornate. Less expensive telescopes were mass-produced and were composed of sliding parts made out of pasteboard that were covered with leather or sharkskin, with bony rings separating each section (Figure 8.20).

In 1668, Newton devised a reflecting telescope that concentrated light by reflecting it from a parabolic metal mirror back to the lens in the eyepiece. Not only did this allow for a shorter tube, but it produced brighter images. Such reflecting telescopes became very popular until the 1730s, when achromatic lenses were developed. This lessened chromatic aberration and resulted in a resurgence of refracting models. Over the next 150 years, improvements in telescope design, lens-making, and mirror materials (from metal to silvered glass) improved both refracting and reflecting telescopes. There were also improvements in the mounts, with the equatorial design gaining prominence over older designs since it allowed for easier tracking of the stars by offsetting the rotation of the Earth on its axis. This especially became important in the 1880s, when dry photographic plates and spectroscopes began to routinely complement the telescope in studying very faint stars and nebulae, since

Figure 8.18. A beautiful frontispiece from volume 4 of a mid-1700s French book. 16.5 × 9.5 cm (page size). Note Galileo describing the telescope (termed here "The Dutch Spying-glass") and his discoveries of 1609 to Lord Sagredo and the Venetian nobility on the tower of St. Mark's in Venice. See also Color Plate Gallery.

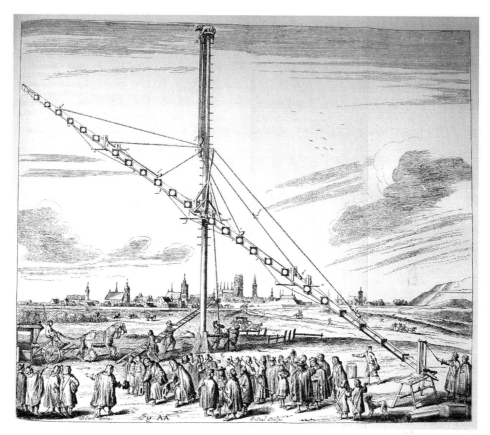

Figure 8.19. A third plate from a 1969 facsimile of Part I (*Pars Prior*) of Hevelius' *Machinae Coelestis*, originally published in 1673. 34.6 × 22.5 cm (page size). Note the picture of Hevelius himself at the lower right pointing to his extremely long refracting telescope, which does not have a tube in order to save weight and is moved by his assistants using a series of ropes and pulleys. Courtesy of Reprint-Verlag-Leipzig.

such observations required long time exposures that could be more easily accomplished if the telescope moved in the same direction and speed as the object being photographed. In the 20th Century, the reflecting telescope completely eclipsed the refracting telescope, since large mirrors that were able to collect light from distant, faint objects exceeded the technological size limits of lenses, which would sag under the increased weight. Along with these giant telescopes, sensitive micrometers, new photographic emulsions and charge-coupled devices, and super-fast computers have recently allowed observations of the heavens to be made with degrees of accuracy that are quantum leaps ahead of those in previous centuries.

More specifically for celestial cartography, telescopes allowed for more accuracy in mapping. In the 18th Century, some of the more traditional instruments for detecting the positions of stars and planets, such as octants, sextants, and quadrants,

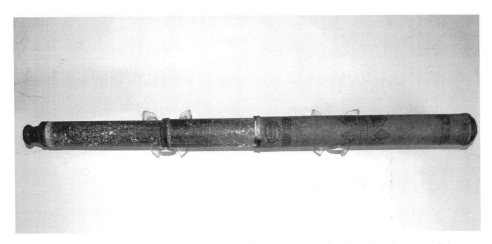

Figure 8.20. Small spyglass-type refracting telescope made by Semitecolo, a well-known Venetian telescope maker, ca. 1800. It is made out of pasteboard, with bony rings separating each section. The main tube is 25.5 cm long. Note the beautiful floral and Greek key design and the screw-on eyepiece cap. *See also* Color Plate Gallery.

were equipped with telescopic eyepieces, making them even more accurate. In addition, regular telescopes were used to observe and record faint stars and nebulae, allowing for the heavens to be better represented on star maps. For example, John Flamsteed used both free-standing telescopes and telescopic eyepieces on his 10-foot mural quadrant and 7-foot sextant at Greenwich in compiling his comprehensive star catalog, which included and accurately plotted many more stars than had previously been recorded. As the equatorial mount came into wider usage in the 19th Century, heavenly objects could be located easier from equatorial coordinate systems (Shubinski, 2011), and there was no need for constellation images to guide the observer to a star or nebulous object. In addition, because so many faint stars could be visualized, they began to clutter up the constellation images on star maps, and it became difficult to separate out the dots being used to represent the stars from the stipples and shadings being used for the constellations. As we shall see in Chapter 10, these changes led to the gradual elimination of constellation images during the 19th Century, a trend that has continued to the present day.

One special kind of telescope that had been successfully used to measure star positions for celestial mapping was the meridian (or transit) circle telescope. This instrument observed the time that stars (or other heavenly objects) passed through the north–south meridian circle line in the sky due to the rotation of the Earth. At the same time that this occurred, the angular distance of the object from the zenith was measured. By knowing the latitude and longitude of the telescope's location, the star's right ascension and declination could then be determined. The idea of such an instrument was mentioned in antiquity, and Tycho Brahe took great advantage of such an instrument when he constructed his large meridian quadrant (see Section 8.3.5.1 and Figure 8.15). By using a telescope with a finely graduated reticle in the

Figure 8.21. An image of a meridian circle telescope from Sir Robert Ball's *The Story of the Heavens*, published in 1897. 23 × 15.3 cm (page size). Note the fixed heavy mount that allows the telescope to swivel along the celestial meridian circle line and observe heavenly bodies that cross into its field of view.

eyepiece, greater precision in positioning could be obtained. Figure 8.21 shows an example of a meridian circle telescope.

The locations of bright stars in the sky have great use for terrestrial navigation, especially when they are located near the meridian. For this reason, specialized atlases have been published that highlight this situation. Figure 8.22 shows Map II from an

MAP II.

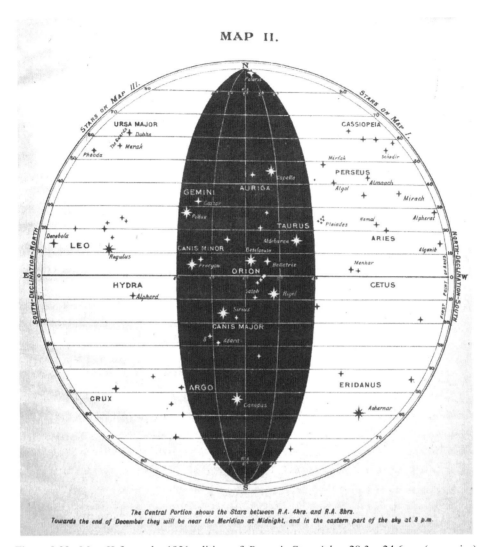

The Central Portion shows the Stars between R.A. 4hrs. and R.A. 8hrs.
Towards the end of December they will be near the Meridian at Midnight, and in the eastern part of the sky at 8 p.m.

Figure 8.22. Map II from the 1921 edition of *Brown's Star Atlas*. 28.3 × 24.6 cm (page size), 20.7 cm dia hemisphere. Note the focus on bright stars and the highlighting of the area near the meridian (here emphasizing the star Sirius and the constellation Orion), which reflects the use of this atlas for navigation at sea. *See also* Color Plate Gallery.

atlas published in 1921 entitled *Brown's Star Atlas, Showing All the Bright Stars, with Full Instructions How to Find and Use Them for Navigational Purposes and Board of Trade Examinations.* The central portion in blue shows stars near the meridian at midnight at the end of December (note the constellation of Orion crossing in the center). The previous Map I shows the meridian stars at midnight for the end of October, and Map III shows them for the end of February. This atlas also contains explanatory text and examples of problems and answers for navigational use at sea.

8.5 NON-STELLAR HEAVENLY BODIES

8.5.1 Sun, planets, and asteroids

The telescope had another benefit for celestial cartography: now features on other worlds could be seen and mapped. Jupiter had bands and a giant reddish spot. Saturn had strange appendages that seemed to change over time (Figure 8.23), which Christiaan Huygens (1629–1695) deduced in the 1650s were actually two sides of a large ring surrounding the planet. Venus went through phases, and some observers thought they saw markings on its surface. Mars typically was shown with surface features. Even the Sun had spots and other irregularities that were thought to be volcanoes and mountains. A leading proponent of this idea was Athenasius Kircher (see Section 7.2), whose depictions were widely reproduced (Figure 8.24).

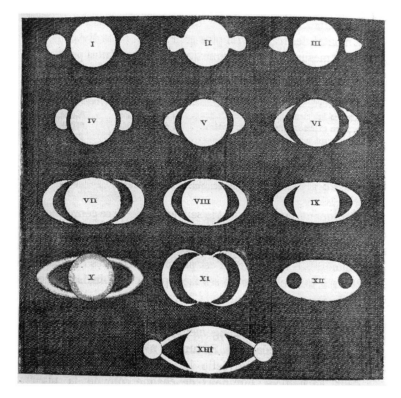

Figure 8.23. Early telescopic representations of Saturn, from Camille Flammarion's *Les Terres du Ciel*, published in 1884. 26.8 × 17.7 cm (page size). Due to imperfect telescope optics, note the rings variously seen as attached lobes (II: Scheiner, 1614); detached lobes (I: Galileo, 1610; III: Riccioli, 1640; IV–VII: Hevelius, 1640–50); a true ring (VIII–IX: Riccioli, 1648–50, X: Eustache de Divinis, 1647); and other odd shapes (XI: Fontana, 1648; XII: Gassendi, 1645; XIII: Riccioli, 1650).

Figure 8.24. Engravings of the Sun and known planets, from Mallet's *Description de l'Univers*, 1683. Each image is approximately 14.5 × 10 cm. Note the Sun as seen by Athanasius Kircher, with mountains and volcanoes; Mercury in a crescent phase; Venus and Mars with surface features; Jupiter with bands and its Red Spot, and Saturn with prominent rings.

The solar system held more surprises that were revealed by the telescope. For millennia, the number of "wandering stars" was limited to what could be seen by the naked eye and was considered to be the five planets (Figure 8.24). But this number was to increase in the 18th and 19th Centuries. In 1781, using a reflecting telescope of his own design, William Herschel (1738–1822) observed an unusual object that he thought was a comet but which was in fact a new planet. Its naming produced a great controversy. Herschel wanted to name it *Georgium Sidus* (George's Star) after his patron, George III of England. Astronomers in continental Europe objected, since the planet was not a star. The name was endorsed even less by the Americans, who had just won their freedom from England and wanted no part in honoring its king in this manner. They wanted to name the planet after its discoverer, Herschel. Thus, for many years this planet appeared as *Georgium Sidus* in English texts and as Herschel in European and American texts (Figures 8.25, 8.26, and 8.27). But in time, the great German astronomer, Johann Bode (1747–1826), proposed the name "Uranus", which in ancient Greek mythology was the name for the Sky who together with Gaea (the Earth) sired Cronus and the other Titans and was the grandfather of Zeus. This name was endorsed by other continental astronomers and later was adopted by the English and the Americans. Incidentally, Bode discovered that the planet had been observed in December, 1690 by John Flamsteed, who did not recognize it as a planet but catalogued it as the star "34 Tauri".

It had long been noted that there was a wide space between the orbits of Mars and Jupiter, and it was surmised that another planet might be lurking in the area. This notion was reinforced by Bode's Law (see Section 6.5.2). After the discovery of Uranus, which supported the validity of this Law, there was a concerted effort by a group of central European astronomers to find a new planet between Mars and Jupiter. A likely culprit was indeed found, but not by a member of this group.

On January 1, 1801, while updating a star catalog by Francis Wollaston (see Section 10.1.2.1), the Sicilian astronomer Giuseppe Piazzi (1746–1826) noted a star-like object that over the course of the next few days appeared to move with respect to nearby stars. Initially he thought it was a comet, but later calculations of its orbit and the absence of a coma led to the conclusion that it was a planet-like object that was located between Mars and Jupiter. He initially named this object Ceres Ferdinandea, after the Roman goddess of harvests and his patron King Ferdinand III of Naples, but the latter name was subsequently dropped. According to Cunningham, Marsden, and Orchiston (2011), Ceres' delayed announcement was accompanied by accusations from other astronomers that Piazzi withheld vital information from them so that he could assume full credit for its discovery and recovery.

Over the next few years, three other bodies were found in the same region: Pallas in 1802, Juno in 1804, and Vesta in 1807. After a 38-year hiatus, a fifth such body, Astraea, was found in 1845. Since these objects did not show a disk in the telescope, it was realized that they were a new class of heavenly body, and William Herschel proposed that they be called asteroids, which means "star-like".

These bodies soon began appearing in celestial maps and books. For example, Figure 8.25 is an astronomy print from the 1823 edition of Delamarche's *Geographe*. On it appear the first four asteroids (which here are still being called *planete*). An enlargement of the print shown in Figure 8.26 shows Ceres to the left, along with information stating that it was discovered by Piazzi in Palermo on January 1, 1801, and that its revolution time is 4 years, 7 months, and 10 hours.

With the use of astrophotography in the late 1800s, the discovery rate of new asteroids escalated. Today, tens of thousands have been catalogued.

There was another planetary controversy, this time surrounding the discovery of the planet Neptune in 1846. For many years, perturbations in the orbit of Uranus had led astronomers to believe that these were being caused by the gravitational pull of a more distant undiscovered planet. Its approximate location was predicted on

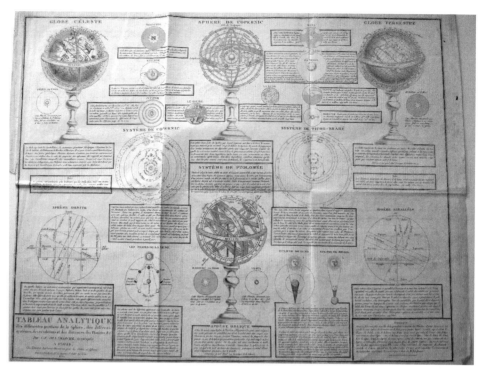

Figure 8.25. An astronomy print labeled "Tableau Analytique", from Delamarche's 1823 edition of *Geographe*. 49.3 × 62.3 cm. Note the depictions and explanatory texts of various cosmological systems and great circles, globes, and members of the solar system.

mathematical grounds in England by John Couch Adams (1819–1892) and in France by Urbain Jean Joseph Leverrier (1811–1877). The latter's prediction was verified first through the telescope, and Leverrier was given credit for the discovery. This set off a firestorm among scientists in England, who claimed that Adams had actually developed his calculations first but that the delay in verifying the new planet's existence through direct observation was due to the slowness of response by England's Astronomer Royal at the time, George Airy (1801–1892). However, Leverrier's claim won the day. But like Uranus, it took a while for the new planet's name to reach the public, especially in America. For example, as seen in Figure 8.27, the 19th edition of a popular American textbook of astronomy written around 1860 by Asa Smith (who we shall meet in Section 9.5.4) continued to display the names "Herschel or Uranus" for the 7th planet and "Leverrier or Neptune" for the 8th planet.

There was a third controversy involving a planet, but this time not related to its discovery. In 1878, Italian astronomer Giovanni Schiaparelli (1835–1910) reported seeing channels, or *canali*, on the Martian surface. His findings resulted in a series of maps suggesting that much of the planet was divided into islands and waterways

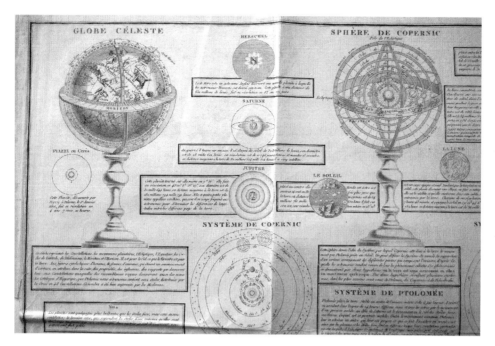

Figure 8.26. An enlargement of the upper left part of the plate shown in Figure 8.25. Note to the left a celestial globe and a depiction of the asteroid Ceres, discovered by Piazzi in 1801. Note in the center the planets Herschel (now called Uranus), Saturn with a true ring, Jupiter orbited by its four Galilean moons, and a Kircher-style depiction of the Sun showing mountains and volcanoes.

(Figure 8.28). This subsequently led to a notion that Mars was occupied by intelligent life and that the *canali* were in fact canals built by Martians to divert water from the poles into population and farming regions. This idea galvanized Mars observers, with some confirming the presence of these canals and others failing to see them (Figure 8.29). A popularizer of the canal theory in America was wealthy amateur astronomer Percival Lowell (1855–1916), who built his own observatory in Arizona and began producing canal-laden drawings and maps of Mars in 1894. He also wrote several books describing the canals and arguing for intelligent life on the Red Planet. However, as Lane (2006) points out, the maps of Schiaparelli and Lowell were highly abstract and represented the synthesis of many observation trials when the seeing conditions through a telescope temporarily cleared to reveal a piece of the surface; none of the maps represented a complete picture that was actually viewed *in toto*. Lane further argues that these artificial maps led to an iconic view of Mars that wasn't supported by direct observations. In time, the building of larger telescopes, new photographic technology, and a more critical eye revealed the canal lines to be optical illusions, and the canal furor died down (Figure 8.30).

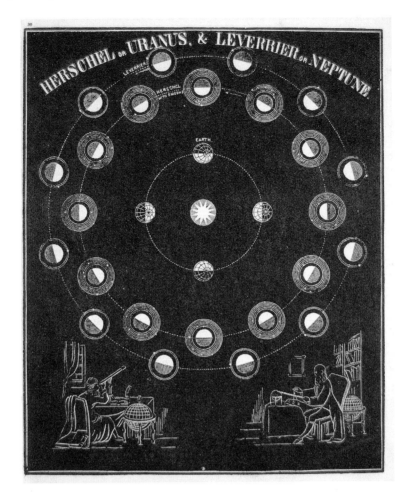

Figure 8.27. This figure is from the 19th edition of Asa Smith's *Illustrated Astronomy*, written around 1860. 25 × 20.6 cm. Note the use of the name "Herschel" for Uranus and "Leverrier" for Neptune in the United States during the early to mid-1800s.

Beginning in the last century, probes that have traveled in space have allowed us to map the rocky planets (Mercury, Venus, and Mars) in great detail and observe the gas giants (Jupiter, Saturn, Uranus, and Neptune) close up (Figure 8.31). Although no life has been seen, other marvels have astounded us, such as evidence that water once flowed on Mars and the impact of a comet hitting Jupiter. We will no doubt learn more about our solar system in the years to come.

8.5.2   Moon

The mapping of the lunar surface has had a long and varied history that has been well reviewed in E.A. Whitaker's *Mapping and Naming the Moon* and Sheehan

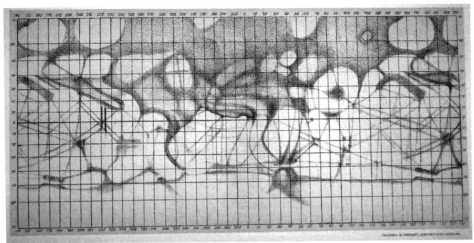

Figure 8.28. A reproduction of Schiaparelli's map drawing of Mars, 1881–1882, taken from Sir Robert Ball's *The Story of the Heavens*, published in 1897. 23 × 15.3 cm (page size). Note the prominent system of canals, some of which are in doubles. *See also* Color Plate Gallery.

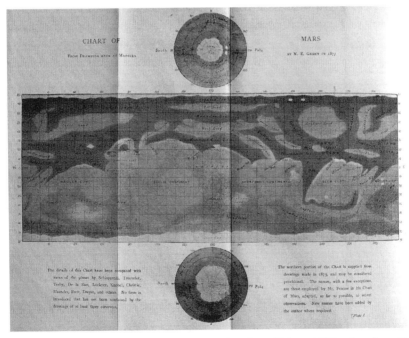

Figure 8.29. A reproduction of N.E. Green's map drawing of Mars, 1877, taken from the 1894 American edition of Flammarion's *Popular Astronomy*. 23.2 × 31.1 cm (pull-out page size). Note the absence of a prominent canal system and the statement that this is a compilation of the drawings of more than ten observers (including Schiaparelli) and that no item is introduced that has not been confirmed by three observers. *See also* Color Plate Gallery.

Figure 8.30. A contemporary view of "Mars and Syrtis Major", *http://grin.hq.nasa.gov/*, image #PR99-27D. Compare with older images shown in Figures 8.28 and 8.29. Courtesy of NASA. *See also* Color Plate Gallery.

and Dobbins' *Epic Moon*. Only a few highlights of this history can be summarized here.

Since prehistoric times, people have looked up at the Moon and seen images in the light and dark markings of its surface, such as a rabbit, a dog, a dragon, a tree, and a "Man in the Moon". Some considered the dark spots to be seas and believed that the Moon was habitable. The first pre-telescopic lunar map was drawn by the English physician and natural philosopher William Gilbert (1544-1603), who also coined the term "selenographia". His map showed 15 features: one sea, two bays, two headlands, a promontory, and nine land masses (Pumfrey, 2011).

When scientists such as Thomas Harriot (1560–1621) in England and Galileo Galilei (1564–1642) in Italy turned their telescopes toward the Moon and sketched the lunar surface, they discerned additional features, such as craters and mountains. It became apparent that rather than being an exotic, smooth ethereal sphere, as believed by Aristotle, the Moon was more Earth-like. This led to a need to develop a map of the Moon to locate and record what was being observed. One of the first attempts was by Michel van Langren (Langrenus) (1600–1675), a member of a prominent Flemish mapmaking family who would become the Royal Cosmographer to King Felipe IV of Spain. In 1645 he published a 34 cm diameter map of the Moon.

Figure 8.31. A contemporary "Solar System Montage" showing the planets from Mercury (top) to Neptune (bottom), *http://grin.hq.nasa.gov/*, image #PIA01341. Compare with older drawings shown in Figures 8.24 and 8.25. Courtesy of NASA. *See also* Color Plate Gallery.

Although produced in a limited print run, it nevertheless managed to stimulate at least one forged copy, which is in the University of Strasbourg Library (see comparison of the two maps in Whitaker, 1999).

8.5.2.1 Hevelius' *Selenographia* and Riccioli's *Lunar Map*

The first true lunar atlas was *Selenographia, sive Lunae descriptio* ..., published in 1647 by Johannes Hevelius (1611–1687), whose instruments we discussed above and whose other accomplishments were mentioned in Section 6.3. This book achieved great acclaim throughout Europe due to its scientific accuracy and visual artistry. Many of his lunar observations were made with a telescope, even though he eschewed this instrument in his stellar positioning work. Hevelius made the engravings himself (which numbered over 130), and the book was produced in his observatory's printing press at Gdansk.

The atlas included three large plates of the full Moon, each measuring about 29 cm in diameter. One plate (labeled "Fig. P" at the bottom) depicted the full Moon as it might appear through the telescope under ideal conditions. Another (labeled "Fig. R") showed the same features but also included topographical characteristics that might appear at other phases. The third (labeled "Fig. Q") was the only one giving the names of features, and it more schematically used conventions found on terrestrial maps, such as rows of "termite hills". All three plates depicted two putti at each corner: the two pairs at the top held labeled banners, and the two pairs at the bottom were playing with astronomical instruments. They also were the first lunar maps to show the regions near the edge that periodically appeared and disappeared due to libration (Wlodarczyk, 2011)—see Figure 8.33 for example).

There was also a series of smaller maps showing the lunar phases day by day. One of the smaller maps of the lunar surface at full Moon from the *Selenographia* is shown in Figure 8.32. Like the lunar images found elsewhere in the atlas, this figure displayed accurate and clear features. But besides being a map, this image was also a volvelle (see Section 8.2). Hevelius and others had observed that as the Moon moved in the sky, the angular position of its axis appeared to vary with reference to the background stars along its path, and this volvelle could be used to measure this variation. First, the movable lunar disk was rotated to match the Moon's orientation in the sky at one time period. The shift in angular position at another time period then could be measured in degrees with reference to the outer circular scale printed on the page with the help of the string pointer. In addition, the angular position of individual lunar features could be determined with respect to the background stars and to other features on the Moon.

Hevelius' system of lunar nomenclature was widely used by others, but it was not without competition. To Hevelius' eye, topographical features on the Moon seemed to resemble those found on Earth, and he named them accordingly. This is illustrated in a well-known "double Moon" plate (Figure 8.33) produced by the Doppelmayr/ Homann team (see Section 7.8). Hevelius' mapping system is shown on the left. By rotating this image 90 degrees counter-clockwise, note that the dark-gray area dominating the lower center of the Moon seems to resemble the Mediterranean Sea. Hevelius labelled this "Mare Mediterraneum". He named the landform in the middle "Sicilia" and the crater in its center "Mt. Aetna". Although influential and creative, Hevelius' system was cumbersome to use due to the length of the Latin names he employed.

In 1651 the Italian Jesuit astronomer Giovanni Riccioli (1598–1671) published a book entitled *Almagestum novum*, wherein he included many of his telescopic observations and argued against the Copernican cosmology (Graney, 2010). Included in this book was a lunar map that named features after scientists and other famous people. Now, as shown in Figure 8.33 on the right, "Mt. Aetna" is named "Copernicus". The names used by Riccioli were not only easier to remember, but his system held promise for astronomers and other scientists that their names might

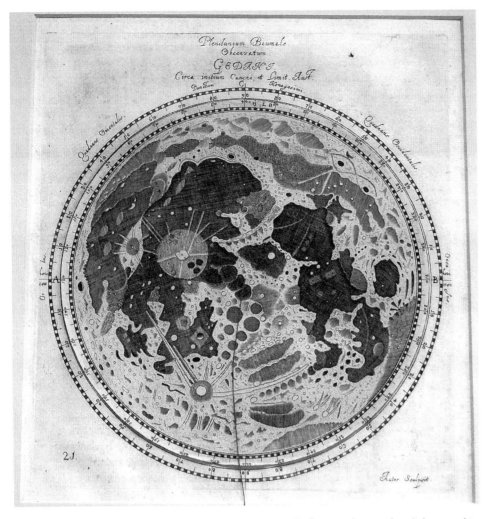

Figure 8.32. Plate showing a map of the full Moon from the first true lunar atlas, *Selenographia*, by Hevelius, which was published in 1647. 16.4 cm dia. inner rotating disk, 17.6 cm dia. outer circular ring. Note that this plate is actually a volvelle (the only one in the atlas), with a revolving Moon and the original measuring string that allowed for the angular measurement of the lunar axis with respect to the background stars. *See also* Color Plate Gallery.

some day be attached to a lunar feature. For some 140 years, the systems of lunar cartography developed by Hevelius and Riccioli competed with each other (and in many cases were depicted side by side in prints from other books and atlases). Gradually, Riccioli's system won out, and today the names of the vast majority of the larger lunar features that we see from the Earth originated with this Jesuit scholar.

Figure 8.33. This image is from a copper engraving by Johann Doppelmayr and published by Homann Publications, ca. 1730. 47.9 × 57.2 cm, each hemisphere 27.7 cm dia. This print depicts the lunar nomenclature system developed by Hevelius on the left and by Riccioli on the right. Hevelius named the features of the Moon after geographical features on Earth (which can be seen by turning the image 90 degrees counter-clockwise), whereas Riccioli named them for famous people and scientists. Note in the Hevelius map *Sicilia* with *Mt. Aetna* in the center, which Riccioli named *Copernicus*, a name that persists to the present day. *See also* Color Plate Gallery.

8.5.2.2   Later maps of the Moon

Despite the popularity of the maps by Hevelius and Riccioli, other lunar maps were published in the late 17th and 18th Centuries that deserve mention. In 1679, Jean Dominique Cassini (Cassini I), who we met as the patriarch of the famous French Cassini family in Section 7.14, published a 54 cm diameter lunar map that was made at the Paris Observatory but which was produced in a limited printing run. This was followed in 1692 by a smaller version in preparation for an eclipse of the Moon later in the year. Forty prominent lunar features were listed in the margins in the order in which they would be eclipsed, and the map was widely distributed. Hence, it became quite popular and was widely copied in the 1700s and early 1800s in a variety of French dictionaries, astronomy books, and almanacs (Figure 8.34). Besides the listing of the 40 features, this map and its derivatives can be recognized by the prominent Greek letter "phi" image that is found in *Mare Serenitatis*. Another interesting map of the 18th Century was completed by Tobias Mayer (1723–1762) around 1750 but was published posthumously by Georg Lichtenberg in 1775. Its importance stems

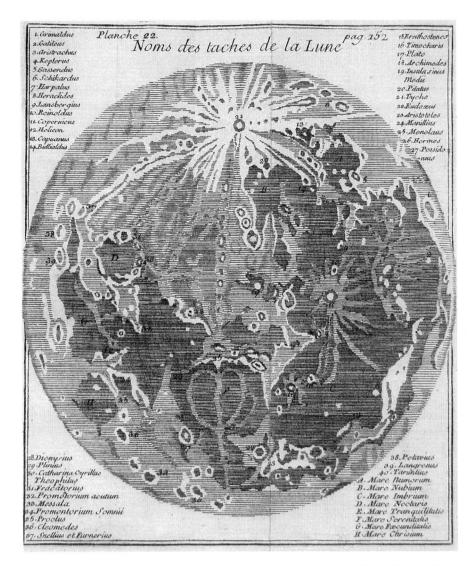

Figure 8.34. Map of the Moon, probably from Bion's *L'Usage des Globes Célestes et Terrestres ...*, 1728. 14.8 × 12.8 cm, 13 cm dia. lunar disk. Note the feature resembling the Greek letter "phi" in *Mare Serenitatis* (*bottom center*) and the 40 lunar features written in the margins. This is a copy of Cassini's 1692 lunar map drawn in anticipation of the total eclipse of the Moon on July 28 of that year.

from the fact that many of the positions of the lunar features were carefully measured from telescopic observations then plotted on the map using trigonometrical formulae. Lines of latitude and longitude were later added to this 20 cm diameter map by Lichtenberg, making it the most accurate lunar map produced up to that time.

By the early 1800s, advances in telescopic work made it possible to construct large-scale maps of the Moon. The first such map was begun by the German cartographer William Lohrmann (1796–1840) and was to be constructed in 25 square sections showing the lunar surface in great and meticulous detail. However, Lohrmann died before completing his work. But between 1834 and 1836, German astronomer Johann Mädler (1794–1874) and his friend Wilhelm Beer (1797–1850), who owned his own observatory, published a detailed map of the Moon in four sections that became the gold standard for some 30 years. In 1837, Mädler published a one-third smaller version of this map in one image that measured nearly 32 cm in diameter. Others copied this image during the rest of the 19th Century, particularly in Germany (Figure 8.35).

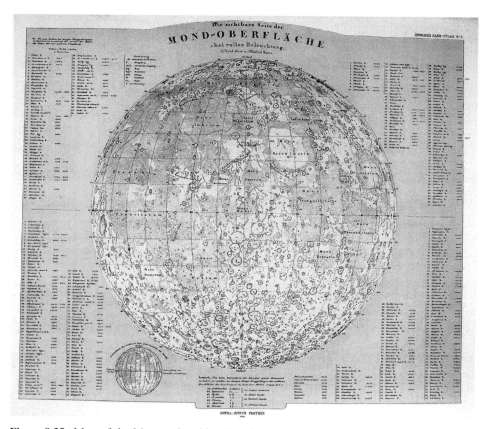

Figure 8.35. Map of the Moon printed in 1876, from the popular *Stieler's Hand-Atlas*, which was begun in 1816 and went through many editions until well into the mid-20th Century. 33.6 × 41.1 cm, 28.3 cm dia. lunar disk. Note the careful attention to detail and the almost photographic appearance of the lunar surface. Note also that the heading states that this image was taken from the famous lunar map produced by Beer and Mädler (in 1837). *See also* Color Plate Gallery.

The next advance came with the development of photography. In fact, one of the first photographs ever taken was a picture of the Moon by Louis Daguerre in 1838. Later photographs revealed unprecedented details on the lunar surface that were recorded and published in atlases with greater precision than any drawing (Figure 8.36). Now that we have sent satellites and men with cameras to the Moon, and have even imaged its far side, the topography of the Moon has become about as familiar to us as that of the Earth.

Figure 8.36. Photograph of the Moon at day 10 of its cycle, from Camille Flammarion's *Les Terres du Ciel*, published in 1884. 26.8 × 17.7 cm (page size). Note the fine detail and contrast from this early lunar photograph.

8.5.3   Eclipses

Solar eclipses, where the Moon comes between the Sun and the Earth, have long inspired fear and awe in people, particularly when they are total. The observation that in broad daylight the Sun can disappear behind the disk of the Moon is quite spectacular, particularly when the sky blackens and the Sun's corona suddenly appears. Records of solar eclipses have gone back some 3,000 years to ancient Mesopotamia and China. Lunar eclipses, where the Earth comes between the Sun and the Moon, are less dramatic but still have been remarkable events that draw great interest.

In ancient China, it was thought that a solar eclipse was caused by the Sun being devoured by a dragon, which had to be fought off by banging drums and shooting fireworks. However, the classical Greeks understood the causes of eclipses as being related to the geometry of the Sun, Moon, and Earth. In fact, by the time of the Renaissance, manuscripts and printed books showed diagrams of eclipses, from either a pre-Copernican geocentric perspective (Figure 8.37) or a post-Copernican heliocentric perspective. In addition, by noting historic patterns and applying principles of geometry, astronomers began to be able to predict eclipses in advance, contributing somewhat to their demystification. Booksellers would even print and sell information sheets (called broadsides) in advance to inform the public and illustrate how the eclipse would appear from a local area.

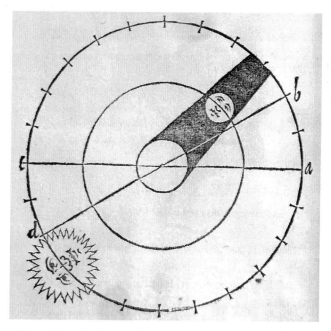

Figure 8.37. An illustration of a lunar eclipse, from the 1647 Leiden edition of Sacrobosco's *De Sphaera*. 15.1 × 9.7 cm, 6.8 cm dia. solar orbit. Note that the perspective is geocentric, with the Earth in the center and the Moon and Sun in orbit around it.

During the 1700s, a particular type of map came into being for solar eclipses. This map showed the path that the Moon's shadow made on the Earth's surface during the eclipse. One of the first of these maps was produced by Edmond Halley (1656–1742) to show the shadow path of the solar eclipse of 1715. In some cases, eclipse maps were produced in advance (such as Halley's, who made it so that people would not be surprised or worried by the event). In other cases, they were made after the fact. Some maps showed the path of one event; others, multiple events (see Figure 8.38). They were sometimes accompanied by diagrams of how the Sun would appear to the observer at different stages of the eclipse. Often, the region being affected was

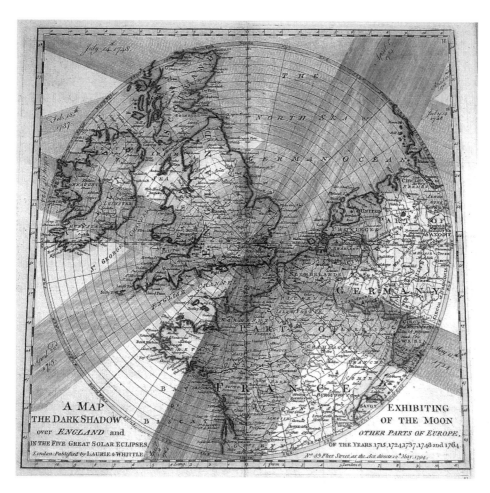

Figure 8.38. Copper engraving entitled "A Map Exhibiting the Dark Shadow of the Moon . . .", produced by Laurie and Whittle in 1794. 29 × 28.7 cm. Note the shadow paths from the five total solar eclipses seen from Britain in the 18th Century (in 1715, 1724, 1737, 1748, and 1764), which are superimposed on a high-quality terrestrial map of the U.K. and northern Europe. *See also* Color Plate Gallery.

printed in great detail, equaling other terrestrial maps in terms of quality. Such maps combined the best of terrestrial cartography and astronomy.

8.5.4   Comets

Another event inspiring fear and awe was the appearance of a comet in the sky. Since ancient times, comets were seen as harbingers of major (usually negative) events, such as a poor harvest or the loss of a war. Their appearances have been recorded in many historical records. For example, what was later called Halley's Comet (see below) was mentioned in a 3rd-Century-BC Chinese document and was pictured in the 11th-Century Bayeux Tapestry, which commemorated the Norman Conquest of England in 1066. Several comets in history were also recorded in the 1493 *Nuremberg Chronicle* (see Section 3.8.3). Cometary paths in the sky have been mapped with respect to the fixed stars, their shapes were diagrammed and categorized, and broadsides alerted people to their occurrence and significance. Even today, now that we can predict the arrival of many periodic comets, people anticipate their appearance with great interest (although with usually not as much trepidation as in earlier times). In a similar manner, meteor showers—most of which are due to the Earth's yearly passage through the debris left by a previous comet crossing our orbit—are anticipated by many, and their dates of occurrence can be predicted with great accuracy.

Because of their great impact on our psyche, comets and their paths have appeared in a number of celestial maps, prints, and books. For example, Hevelius was interested in comets and published a number of maps detailing their paths through the heavens. Many of these appeared later in his book *Machinea Coelestis Pars Posterior*, which was discussed in Section 8.3.5.2 (see Figure 8.39). Other astronomers and cartographers followed suit, and some of the more famous will now be discussed.

8.5.4.1   Stanislaw Lubieniecki's *Theatrum Cometicum*

From 1666 to 1668, three volumes of a book appeared in Amsterdam entitled *Theatrum Cometicum* ... It was written by Stanislaw Lubieniecki (or Stanislas Lubieniczki) (1623–1675), a Polish theologian, astronomer, and historian. This encyclopedic work contained a history of over 400 comets and their impact on terrestrial events, and it focused especially on reports from a number of astronomers on the bright comet of 1664–1665 (Figure 8.40). The book was liberally illustrated with celestial maps of cometary paths and figures showing the appearances of cometary nuclei and tails at different times in their orbits. Since the sources of the observations varied, a number of artistic styles and cartographic traditions popular in the 17th Century were represented.

8.5.4.2   *Edmond Halley's comet*

The life and achievements of this polymath were discussed in Section 4.3.2.4, especially his catalog and map of the southern hemisphere stars. However, Halley

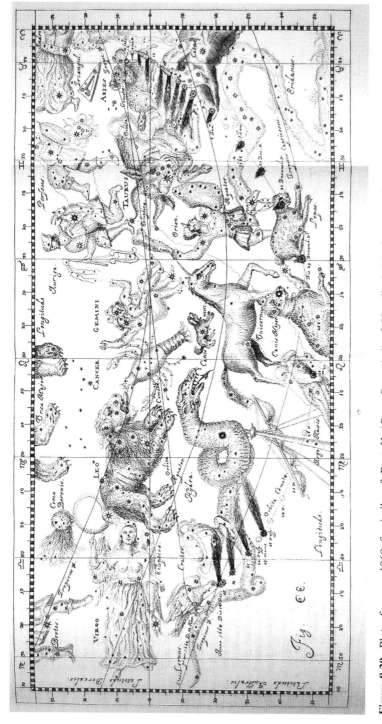

Figure 8.39. Plate from a 1969 facsimile of Part II (*Pars Posterior*) of Hevelius' *Machinae Coelestis*, originally published in 1679. 34.6 × 22.5 cm (page size). Note the path of the bright comet of 1664–1665 shown against the background constellations, which he recorded from his observatory in Gdansk from December 14, 1664 (*left*) to February 18, 1665 (*right*). Courtesy of Reprint-Verlag-Leipzig.

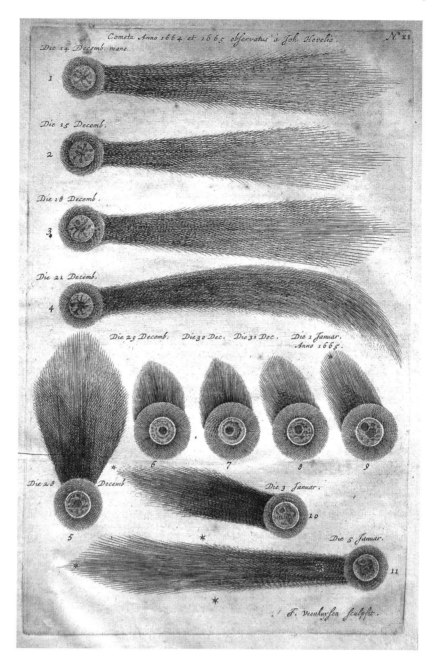

Figure 8.40. Plate from Stanislaw Lubieniecki's *Theatrum Cometicum* (*Pars Posterior*), published in 1667. The images show the bright comet of 1664–1665 according to the observations of Hevelius from December 14, 1664 (*top*) to January 5, 1665 (*bottom*). 27.5 × 17.9 cm. Note the changing features in the nucleus and the variations in the shape and length of the tail (due in part to foreshortening) as the comet orbited around the Sun.

perhaps is best known to the average person for his work on comets, an interest he
first developed by observing the Great Comet of 1680. He studied the records of a
number of early comets and their paths across the sky, and he was influenced by
Newton's work on calculating cometary orbits. In 1705 he published a book in which
he described the parabolic orbits of 24 comets that had been observed from 1337 to
1698. Halley noticed that the cometary appearances in 1531, 1607, and 1682 defined
the same general path, and he concluded that these were one and the same comet. He
predicted that this comet would reappear in December 1758. Although he died before
this time, his prediction was accurate, and from then on the comet was called Halley's
Comet. In future years this most famous of comets continued to appear on a regular
basis around every 76 years, and its time of appearance and orbital path could be
announced in advance in broadsides and almanacs (Figure 9.2).

8.5.4.3 Charles Messier's catalog

Charles Messier was born on June 26, 1730 in Badonviller, which was in the Lorraine
region of France. He was the 10th of 12 children. His father, a civil servant who
arrested people who were in debt, died when he was 11. Charles became interested in
astronomy after observing a prominent six-tailed comet in 1744 and an annular solar
eclipse on July 25, 1748. In 1751 he went to Paris and began working with the
Astronomer of the Navy, Joseph Delisle (1688–1768), who encouraged him to pursue
observations of the heavens and to carefully record what he saw. While looking for
Halley's Comet in 1758, he spotted another comet, which he tracked, and a comet-
like patch that turned out to be a nebula. He recorded it, and it later became the
first entry in his famous catalog of deep-sky objects. After Delisle retired, Messier
continued working at the observatory. Although he made other astronomical observations (e.g., transits of Mercury and Venus; Saturn's rings), his interest in comets
continued, and he began to achieve international fame for his cometary discoveries.
He subsequently was named to membership in several prestigious academies, such
as the Royal Society in London, the Berlin Academy of Sciences, and the Paris
Académie Royale des Sciences. He married in 1770, but both his wife and son died
two years later. In 1771 he was appointed Astronomer of the Navy. He continued his
comet hunting, even through some lean years during the French Revolution, when he
lost his salary and pension. In 1806 Messier was awarded the Cross of the Legion of
Honor for his astronomical work by Napoleon. He suffered a stroke in 1815 and died
in Paris on April 12, 1817.

   According to Warner (1979), Messier found a total of 21 comets. The paths
through the heavens for most of these were published in the *Mémoires, Académie
Royale des Sciences* and elsewhere. Warner lists and describes over 28 publication
entries for him in her book, and the reader is referred there for details. Although the
projections of the background constellations varied, all but one of the listed orientations were geocentric. Messier borrowed heavily from Fortin's atlas for his charts (see
Section 6.4.4.2).

   However, Messier's fame is due more to the catalog he developed that bears his
name than to the comets he discovered. During his comet hunts, he observed and

recorded a number of nebulae or star clusters that could easily be confused with being distant comets. To help other comet hunters and astronomical observers distinguish between such permanent objects in the sky and the more transient comets, he decided to catalog the former. In 1771 he presented the first version of this catalog, which consisted of 45 deep-sky objects, to the Académie Royale des Sciences. His catalog subsequently became quite famous, and the objects listed began to acquire the matching Messier number as their scientific name. Over time, additional deep-sky objects were added to the list, some found by him and some by his assistant and fellow comet hunter, Pierre Mechain. In 1781 Messier compiled his last version of the catalog, which had grown to 103 entries. It was published in the 1784 edition of the French almanac *Connaissance des Temps*.

Much later, in the 20th Century, another seven objects that had been observed by Messier or Mechain were tabulated that had not been included in the 1781 catalog. These seven objects have been added to the list, and today most astronomers accept a total of 110 Messier objects. This current catalog contains examples of a variety of deep-sky objects, such as galaxies, planetary nebulae, open clusters, and globular clusters. Because they were accessible to Messier's relatively small 4-inch telescope, these can easily be seen with the larger telescopes used today by serious amateur astronomers. In fact, a favorite pastime for many of them is the "Messier Marathon", whereby they try to view all 110 Messier objects in a single night, which is possible during a few favorable days of the year, usually in March.

8.5.5 Deep-sky objects

8.5.5.1 *Current classification*

Deep-sky objects refer to faint naked eye and telescopic heavenly bodies that are outside of our solar system and do not include comets or individual stars. Today, they refer to star clusters, nebulae, and galaxies. Star clusters are groupings of stars. Some are tightly bound by gravity into roughly spherical shapes and consist of tens or hundreds of thousands of very old stars; these are called globular clusters (Figure 9.8). In our Milky Way galaxy, they tend to be distributed spherically around the center in the galactic halo. Other star groupings are more loosely bound and consist of less than a few thousand young stars; these are called open clusters. They generally are found in the galactic plane, typically in the spiral arms, and may be associated with H II regions of ionized hydrogen gas, which are associated with new star formation.

Nebulae are interstellar clouds of dust and gas (ionized or non-ionized) that consist of several kinds. Diffuse nebulae are thin and widespread without clear boundaries. Many are associated with star formation, and they can either emit their own light or reflect light from nearby stars. Planetary nebulae are expanding gaseous shells that are ejected by an existing but dying hot central star, which excites the shell and causes it to emit light. Supernova remnants are nebulous ejecta of stars that have violently exploded, sometimes with a luminosity of several billion times that of our

Sun. Finally, dark nebulae are dust clouds that are only visible because they are illuminated by nearby stars.

Galaxies are massive, gravitationally bound systems of stars, clusters, and nebulae that range in size from millions to trillions of stars. Some consist of a rotating disk with a central bulge of older stars and spiral arms that contain major gas and dust centers of new star formation. Some spirals, like our own Milky Way, have a bar-shaped band of stars that extend to either side of the core and merge into the spiral arm structure; others have no prominent bar, only spiral arms emanating from the core. Another type of galaxy are the ellipticals, which have relatively more older stars, less structure, and fewer areas of new star formation than the spirals. There are also intermediate galaxies with properties of both spirals and ellipticals; irregular types with features of neither (like the Magellanic Clouds); and dwarf galaxies with less than a few billion stars that often orbit a single larger galaxy (and which themselves may be classified as spiral, elliptical, or irregular).

Although a few deep-sky objects (like the Pleiades star cluster, the Orion nebula, the Magellanic Clouds, and the Andromeda galaxy) can be seen under dark skies with the naked eye, most were not identified until the advent of the telescope. It is important to note that initially most deep-sky objects were lumped together as nebulae because they looked like indistinct cloudy objects (nebula means "cloud" in Latin). Furthermore, the Milky Way was not seen as being part of a discrete galaxy but as an assemblage of stars in the vast universe. Other deep-sky objects were similarly conceived, with no sense that some were in our galaxy and others were their own "island universes". How did we come to appreciate the different deep-sky objects, and what were some of the major celestial images that contributed to our modern understanding?

8.5.5.2 *From 1600 to 1900*

Let's begin with the early telescopic sightings. In 1610, Galileo (1564–1642) reported in his epic monograph *Sidereus Nuncius* that tiny new stars had popped into view at the higher magnification of his eyepiece as he gazed at the Milky Way, and he counted 21 stars in the nebula in Orion's Head. Later observers, like the Sicilian astronomer Giovan Battista Hodierna (1597–1660) and the Dutch scientist Christian Huygens (1629–1695), also spotted stars in various nebulae that they observed with their telescopes (Hoskin, 2008). This led to the question of whether the nebulae could all be resolved into component stars with proper magnification and lens quality, or whether true unresolved nebulae existed which were composed of some sort of luminous fluid.

Theologians and philosophers also were getting into the act. In 1750, the largely self-taught natural scientist and philosopher Thomas Wright (1711–1786) proposed a model of a universe that was composed of numerous star systems, each of which contained a central "Abode of God" (Hoskin, 1997). Although Wright preferred a spherical shape to these star systems, one alternative model conceived of a flattened ring of stars surrounding the central area. Influenced by this notion, the philosopher Immanuel Kant (1724–1804) likewise conceived of several star systems in the universe, but instead of a ring he saw each of them being shaped like a disk, without a supernatural center. Both Wright and Kant operated more from speculation than from observation, and in a sense their prescient notions were lucky guesses.

After his discovery of Uranus in 1781 (see Section 8.5.1), the great English astronomer William Herschel (1738–1822) began systematically studying nebulae. He was aware of the catalog of the French astronomer and comet-hunter Charles Messier (1730–1817), which had been published in 1771 and had listed a number of nebulae to differentiate them from any new comet that might appear in the area. In the process of surveying the nebulae, Herschel found that many typically revealed masses of stars under the influence of his powerful telescopes, which included a 20 foot reflector of 47 cm aperture. Together with his sister and collaborator, Caroline Herschel (1750–1848), he produced several catalogs listing nebulous objects, and Caroline produced her own as well (Hoskin, 2006).

William developed a theory that stars naturally clustered together over time due to gravitational attraction, and these nebulae were nothing more than dense masses of stars that could be discerned with stronger and better telescopes (Smith, 2011a, 2011b). But this theory was jolted in 1790 when he discovered an object that consisted of a central star surrounded by a luminous area that could not be further resolved. He hypothesized that the central star was condensing out of the surrounding cloud of true nebulosity under the forces of gravity. He speculated that such nebulous material formed from the cataclysmic collapse of dense clusters of stars or from light that became trapped and pooled. Thus, gravity was involved in both star formation and star destruction.

In terms of characterizing the totality of stars in the universe (sometimes termed "Galaxy", to distinguish it from the separate "galaxies" we now conceptualize), Herschel made two assumptions: that his telescope could penetrate to its border in every direction, and that despite the presence of a few groupings here and there, the stars generally were evenly distributed throughout (Smith, 2006). In this way, he could determine the extent and shape of the Galaxy in a given direction by counting the number of stars in that direction. In 1785, he had published his famous diagram for the cross section of the Galaxy, shown in Figure 8.41. Note that he placed our Sun near the center. However, he received a shock when he began observing though his new 40 foot, 1.26 meter reflector in 1789. He realized that even more stars now could be resolved, suggesting that they extended out indefinitely, and that the stars were not as uniformly distributed as he had earlier assumed. This caused him to reject the accuracy of the image shown in Figure 8.41, although it continued to be reproduced until well into the second half of the 19th Century (Hoskin, 2008).

Further work on nebulae continued during the 1800s as telescopes became larger and better in terms of optics and mounts. In the early 1840s, William Parsons (1800–1867), the 3rd Earl of Rosse, built a 54 foot, 1.83 meter reflecting telescope at his estate at Birr Castle, Parsonstown, Ireland to continue with his interests in astronomy. Termed the "Leviathan", it was the largest telescope in the world until the early 20th Century (Figure 8.53). Many discoveries were made with this telescope, including the observation in 1845 of the spiral structure of the nebula M51 (the 51st entry in Messier's catalogue), which we now call the Whirlpool galaxy. A drawing of

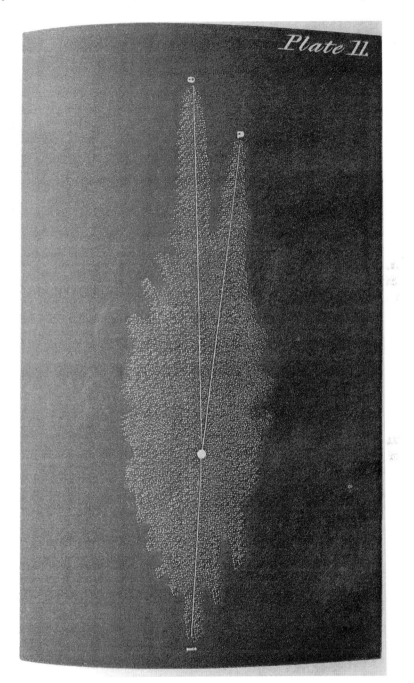

Figure 8.41. Engraving from the 1848 edition of Mitchel's *The Planetary and Stellar Worlds*. 8 × 13.1 cm. It depicts Herschel's famous cross section of the Milky Way. Note our Sun's location near the center and the lines to the edges representing the "streams" of the galaxy.

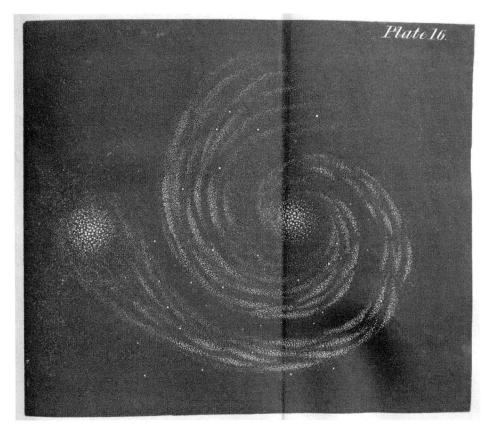

Figure 8.42. Another engraving from the 1848 edition of Mitchel's *The Planetary and Stellar Worlds*. 13.2 × 15.6 cm. It depicts Lord Rosse's famous image of the Whirlpool Galaxy (M51/NGC5194), the first to be seen as a spiral nebula. Note the clear spiral form and the companion galaxy NGC5195 to the left.

this object was widely reproduced (Figure 8.42). In 1888, British amateur astronomer and celestial photographer Isaac Roberts (1829–1904) showed photographic evidence that the Andromeda nebula (which we now know to be a galaxy) also had a spiral structure, thus demonstrating the advantages of this medium for nebular research over drawings at the telescope (Smith, 2008).

In the 1800s, many astronomers (including William Herschel) were in favor of the "nebular hypothesis", which said that stars were formed by gravity acting on gaseous nebulae. This presupposed that true nebulae existed, a notion supported by the growing application of spectroscopy to the heavens. For example, in the 1860s British amateur astronomer and spectroscopist William Huggins (1824–1910) found that many of the nebulae he examined gave spectroscopic evidence of being gaseous in nature; others showed continuous spectra consistent with the presence of unresolved stars. But other astronomers, like Rosse, did not believe that nebulae were truly gaseous but were composed of fine stars that could be resolved only if the telescope was large enough and of good quality. He claimed to be able to resolve even the Orion nebula into individual stars with the Leviathan. In this, he was challenged by John Herschel (1792–1871), William's astronomer son, who countered that his observations concluded that the Orion nebula was a true nebula. Both astronomers declared that each other's instruments were flawed, and there the debate remained until the turn of the Century.

8.5.5.3   *The 20th Century*

By the beginning of the 1900s, most astronomers continued to believe that everything that could be seen was part of one large Galaxy, although a few people speculated that other "island universes" existed beyond telescopic range. Furthermore, our Sun was seen as being relatively close to the center of the Galaxy. The nebular hypothesis continued to be discussed. But as the century progressed, clarity began to emerge, partly due to the increasing sophistication of photography and spectroscopy, as well as to the building of better and larger telescopes. Gradually, our Milky Way galaxy was seen as simply being one of many in the universe, and nebulae were sorted out into the taxonomy mentioned at the beginning of this section. The story of these developments is a long and complicated one, and we can only touch on a few highlights.

In 1914, Lowell Observatory astronomer Vesto M. Slipher (1875–1969) reported spectroscopic evidence that showed many spiral nebulae to exhibit large redshifts, suggesting that they were moving away from the Milky Way at great speeds. This led some astronomers to consider these spirals as far-off star clusters, if not discrete galaxies. Also in 1914, astronomer Harlow Shapely (1885–1972) arrived at Mount Wilson and began studying globular clusters. He discovered that they contained Cepheid variable stars, which allowed him to construct period–luminosity relationships that could be used to measure their distances. He found them to be symmetrically located on either side of the Milky Way galaxy, which he concluded to be immense in size, although his estimate of 300,000 light-years in diameter was later found to be too large since he had dismissed the dimming effects of interstellar dust absorption (Smith, 2006). He also concluded that our Sun was eccentrically located relative to the galactic center. In 1917, Lick astronomer Heber D. Curtis (1872–1942) found that novae in the spiral nebulae were much fainter than those in our galaxy, and he became a proponent of the "island universe" hypothesis, saying that Andromeda was an example of an independent galaxy roughly the same size as our own, which also could have a spiral shape (Curtis, 1920). Shapley and Curtis discussed some of these issues involving one "Big Galaxy" versus many island universes at a meeting at the National Academy of Sciences on April 26, 1920, which has been termed the "Great Debate".

Although no firm conclusion resulted from this meeting, additional insight was gained in late 1923 and 1924 by Mount Wilson astronomer Edwin Hubble (1889–1953). Using the brightness of Cepheid variables in the Andromeda nebula, Hubble confirmed that it was so far away that it must be a separate galaxy, much beyond even

Shapley's size estimate for the Big Galaxy. In 1926, he published a classification scheme that divided the nebulae into galactic nebulae (planetary and diffuse) and extragalactic nebulae (elliptical; spiral, both normal and barred; and irregular) (Smith, 2009). Note the similarities of this scheme to that presented at the start of this section. Hubble's term "extragalactic nebulae" were called "galaxies" by Shapley (who in 1921 had become the Director of the Harvard Observatory), and this name stuck.

In the 1920s, Shapley and his colleagues conducted surveys of galaxy distributions that found many of them to aggregate into clusters. In 1929 and into the 1930s, Hubble followed up on Slipher's redshift findings involving spiral nebulae and compared them with his own distance measures to several extragalactic nebulae. He found a linear relationship between the redshift and the distance. This velocity–distance relationship is now called Hubble's Law, and it suggests that the universe is expanding, thus providing support for the Big Bang theory. Also during the 1920s and 1930s, Hubble and his colleagues found increasing evidence suggesting that the Milky Way was spiral in shape. This notion was confirmed in the early 1950s by Yerkes astronomer William W. Morgan (1906-1994), who used the distribution of bright young blue stars and H II regions to plot the spiral arms in our galaxy.

Thus, the basic classification of deep-sky objects and the characterization of our Milky Way galaxy have been established. Modern use of radio and other wavelength techniques, as well as advances in telescope design such as charge-coupled devises, adaptive optics, and the Hubble Space Telescope (Figure 8.43), has confirmed these findings and has led to the discovery of even more exotic structures in deep space, including quasars, pulsars, black holes, and dark matter. Who knows what will be discovered in the future.

8.6 PLAYING CARDS

Among the more distinctive formats for displaying celestial material are playing cards. The notion of decorating playing cards beyond the basic suit signs is not in itself unusual, however. According to Beresiner (2010) this activity was employed by the Visconti Sforza family in 15th Century Italy. The decoration of playing cards using a variety of themes became popular in Germany in the 16th Century. Starting in the 17th Century, it became a great commercial success in England, and a number of themes were represented, including politics, history, music, and heraldry. The earliest playing cards to employ a cartographic theme are English and go back to 1590. They show maps of the 52 counties in England and Wales, which coincidentally matched the number of cards in a standard deck.

Celestial-themed playing cards are rarer. Figure 8.44 shows an example of such a card from a deck published in France by Hegrad in 1785 entitled *Jeu des Cartes Geographiques*. It is an ace and depicts an armillary sphere (another ace from this deck shows a diagram of the Copernican model of the universe). The remaining non-ace cards in this deck display terrestrial maps. Note that instead of a suit mark, there

Figure 8.43. A contemporary image labeled "Magnificent Details in a Dusty Spiral Galaxy", which is NGC4414 as shown by the Hubble Space Telescope, *http://grin.hq.nasa.gov/*, image #PR99-25. Compare with the older engraving of a spiral galaxy shown in Figure 8.42. Courtesy of NASA. *See also* Color Plate Gallery.

is a red colored square. Other colors are used to indicate the three other suits in this unusual playing card set.

8.7 FRONTISPIECES AND TITLE PAGES

As we have seen, celestial books and atlases have historically conveyed a great deal of information about the heavens. Two types of image have dominated: constellation maps and cosmological maps. In many cases, the images that accompanied the text were both informative as well as artistic, and they were a good barometer of the times. During the Golden Age (17th and 18th Centuries), Baroque was in vogue, and celestial images were replete with acanthus leaves, shells, scrolls, ribbons, banners, and angelic putti (for example, see Figures 7.1 and 7.5). Classical elements also were

Figure 8.44. A celestial-themed card from a set of rare playing card maps by Hegrad entitled *Jeu des Cartes Geographiques*, published in 1785. 11.3 × 6.4 cm (card size). Note the image of an armillary sphere on an ace (the non-aces had geographical maps), and instead of a suit mark there is a red colored square. *See also* Color Plate Gallery.

present as a result of the Enlightenment. As we shall see, this changed during the 19th and 20th Centuries, with images in celestial books and atlases showing more straightforward views with less artistic embellishment. In addition, the impact of the telescope necessitated less need for constellation images and more need for accuracy of stellar placements. Faint stars and deep-sky objects could now be visualized, and the surface of planetary and other solar system bodies could be seen in more detail.

However, textual images were not the only ones present in celestial books and atlases. Less well-known were those embedded in the title pages or appearing in the often decorative plate preceding and usually facing the title page, the so-called frontispiece. Swedish museum curator and art historian Inga Soderlund (2010) has likened the frontispiece to a beautiful façade at the entrance of a building, promising wonders to be found within. Frontispieces were very popular in the Renaissance, especially adorning Bibles and scholarly works, and their use continued well into the 20th Century. In terms of quality, wealth of detail, complexity of ideas, and allegories, celestial frontispieces reached their zenith in the 1600s and early 1700s. Thereafter, they became simpler and more narrowly related to the subject matter, perhaps reflecting the growing empiricism of the text as astronomical science advanced. They have become less common in recent decades.

During the early days of printing in Europe, title pages likewise were more elaborate than today's versions, including not only information about the name of the author and publisher, year of printing, and place of publication, but also extensions of the title that gave information about the content of the book. They sometimes included images that were either highly decorative in nature (especially where there was no frontispiece) or were symbolic of the printer. These printer's marks (or logotypes) have sometimes been called colophons, but this term now is reserved for publication information or production notes located on the reverse side of the title page or at the end of the book. Printer's marks were popular in the Renaissance, varying from elaborate allegorical designs, to alchemical symbols of elements used in the printing process, to simple and identifiable pictures (such as the famous dolphin and anchor of the Aldine Press).

Although the focus of this section will be on images that adorned frontispieces and illustrated title pages, it should be mentioned that not all astronomy books had images at the front. Two prominent examples are Johann Bode's *Uranographia* and Jean Fortins' *Atlas Celeste de Flamsteed*, neither of which had a frontispiece. The title page from Bode's atlas simply reflected written information about the atlas and had no images. The title page from Fortin's 1795 edition likewise had no images, but that from the 1776 edition had a beautiful printer's mark.

In her survey of 291 illustrations adorning the front of 17th Century books on astronomy, Soderlund (2010) found that 65% comprised an entire illustrated page, either a frontispiece or illustrated title page. Besides serving as a textual introduction, she concluded that these images also had a commercial purpose, encouraging the reader to buy the book by engaging him in the beauty and mystery of astronomy and helping him to identify with an intellectual elite. In the remaining 35%, the images were less prominent and were subordinate to the text. In general, the larger and more expensive the book or atlas, the more likely it was to have an elaborate frontispiece that used fiction, myth, and allegory to invoke antiquity and identification with an erudite topic.

Based on my review of European and American astronomical books and atlases going back from the present to the beginning years of printing, one may classify the pre-text images into four types. These are: printer's marks; allegorical images using classical mythology and important astronomers from antiquity; pictures of instruments and people that are contemporary with the time the book was published; and images and schematics that are extensions of the book's content. In addition, one sometimes finds combinations of these categories appearing in both the frontispiece and title page. Let's examine these categories.

8.7.1 Printer's marks

A beautiful Renaissance example of a printer's mark can be found in Alessandro Piccolomini's *La Sfera del Mondo* (*The Sphere of the World*), which was originally published in 1540 (see Section 5.5). The title page from the 1579 edition of the book is shown in Figure 8.45. The title appears at the top, followed by Piccolomini's name, a brief description of the book, the printer's mark, the publisher's name and location ("Vinegia", or Venice), and the date of publication. The printer's mark is elaborate, with an image of a sea god in the middle with two tails. This nautical theme likely reflects the fact that the printing was done in Venice.

But printer's marks were not only a Renaissance phenomenon. Figure 8.46 shows a 20th Century example from the *Atlas Céleste* by Dien and Flammarion (see Section 10.1.3.1). This title page is from the eleventh edition, published in 1904. The usual information is listed, along with a simple printer's mark that shows a shield depicting a crossed anchor and pick behind a vertical sword. Below this shield are the publisher's superimposed initials "G/V" (for Gauthier–Villars), surrounded by a banner with a phrase in Latin attributed to St. Benedict: "Behold, work on, and don't be disheartened". This is a fitting motto for a busy family-run publishing house specializing in work from scientific societies and well acquainted with the Latin language.

8.7.2 Allegorical images from the past

Many of the great atlases of the Golden Age were introduced by remarkable frontispieces that were themselves works of art rivaling the beautiful plates that followed. Allegorical images proliferated, with Baroque and Classical trappings. One of the most common elements, that of a beautiful woman who sometimes held an astronomical instrument, represented the discipline of astronomy itself and could be identified as Urania, one of the muses in Apollo's circle, or Astronomia, one of the seven liberal arts. Deities representing the Sun (e.g., Apollo), the Moon (e.g., Diana), or the planets also were common. Astronomers were usually represented by older scholarly men holding a telescope, armillary sphere, or astrolabe. Pleasant dream-like images abounded, such as gardens, temples, monuments, and other references to the past. Both pagan and Christian elements were present as well, suggesting a link between astronomy and religion in terms of eternal ideas and interest in the heavens. Child-like putti were also a common feature, sometimes pictured holding up banners or astronomical instruments (as if they were toys). Their presence gave the total image a playful quality and suggested that reading the following text would be fun as well as educational. Not until later would realistic images of a more scientific astronomy appear: large telescopes, observatories, detailed planetary surfaces, etc.

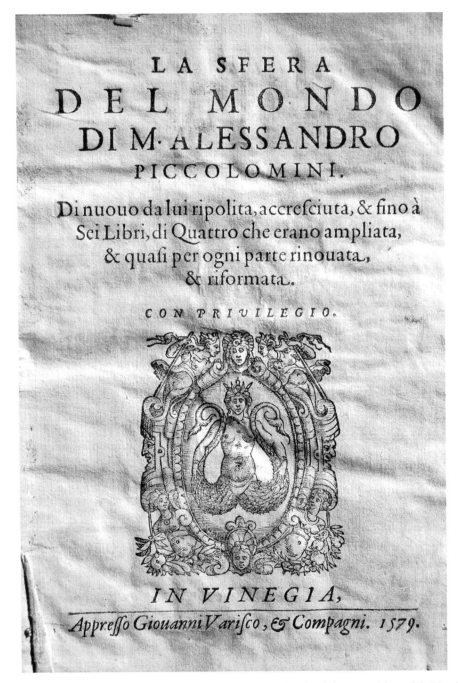

Figure 8.45. Title page from the 1579 edition of Piccolomini's *La Sfera del Mondo*. 20.2 × 14.3 cm (page size). Note the nautical theme of the printer's mark, reflecting its publication in Venice.

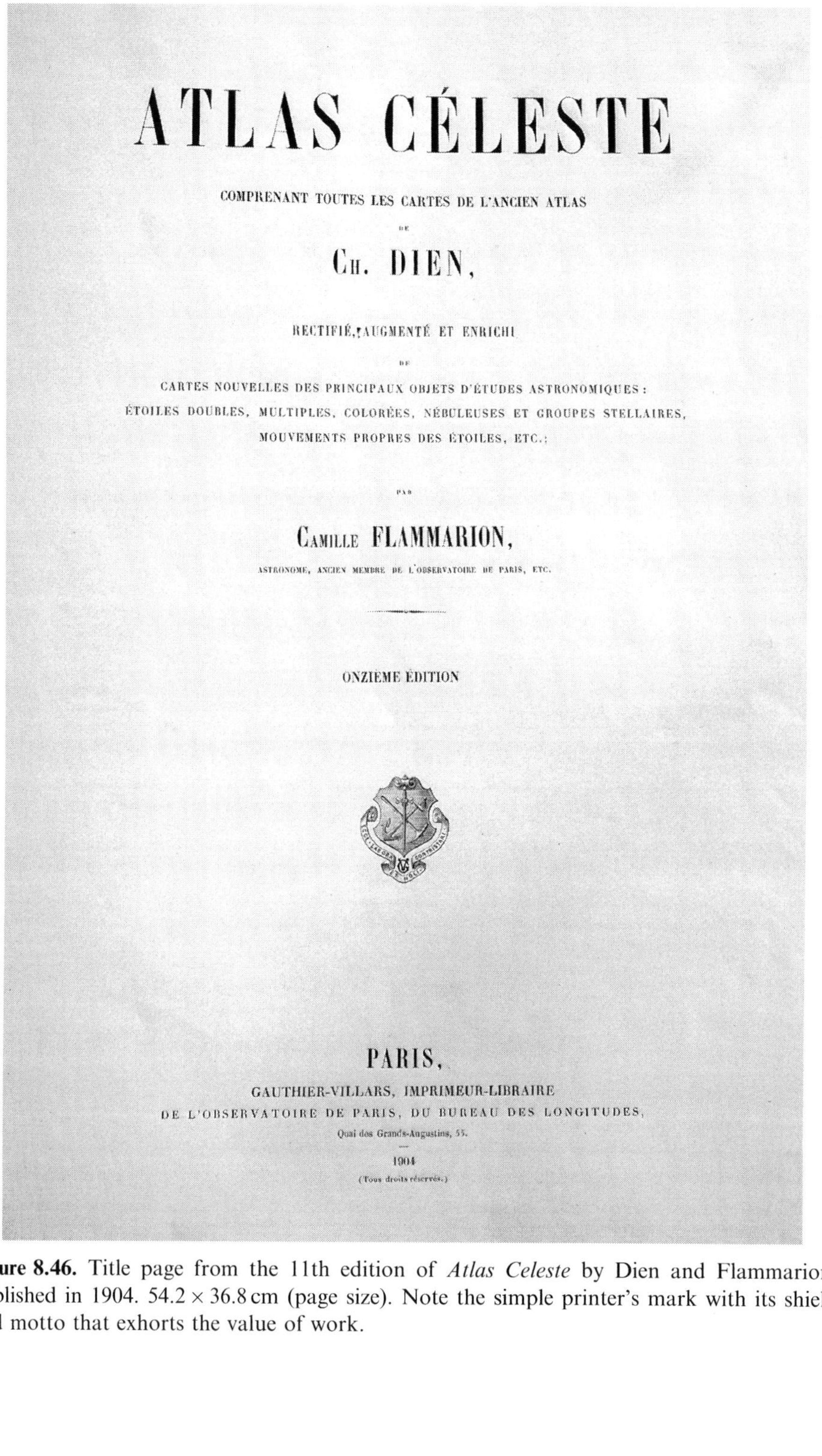

ATLAS CÉLESTE

COMPRENANT TOUTES LES CARTES DE L'ANCIEN ATLAS

DE

Ch. DIEN,

RECTIFIÉ, AUGMENTÉ ET ENRICHI

DE

CARTES NOUVELLES DES PRINCIPAUX OBJETS D'ÉTUDES ASTRONOMIQUES :
ÉTOILES DOUBLES, MULTIPLES, COLORÉES, NÉBULEUSES ET GROUPES STELLAIRES,
MOUVEMENTS PROPRES DES ÉTOILES, ETC.;

PAR

Camille FLAMMARION,

ASTRONOME, ANCIEN MEMBRE DE L'OBSERVATOIRE DE PARIS, ETC.

ONZIÈME ÉDITION

PARIS,

GAUTHIER-VILLARS, IMPRIMEUR-LIBRAIRE
DE L'OBSERVATOIRE DE PARIS, DU BUREAU DES LONGITUDES,
Quai des Grands-Augustins, 55.

1904

(Tous droits réservés.)

Figure 8.46. Title page from the 11th edition of *Atlas Celeste* by Dien and Flammarion, published in 1904. 54.2 × 36.8 cm (page size). Note the simple printer's mark with its shield and motto that exhorts the value of work.

In books of the period that contained such elaborate frontispieces, the associated title page was often drab by comparison, containing written material that included mini-abstracts of the book's text. They typically were devoid of images, in a sense surrendering to the beauty and complexity of the frontispiece. A case in point is from Hevelius' *Selenographia*, the first true lunar atlas (see Section 8.5.2.1). Figure 8.47 shows the title page and Figure 8.48 shows the frontispiece from this famous atlas. Note that the title page has the usual identifying information but no printer's mark. It also includes text stating that in addition to the Moon, the atlas depicts planets as seen through the telescope. But it is the frontispiece that is a true wonder, full of imagery and allegory. At the top, between the Moon to the left and the Sun (complete with sunspots) to the right, is the allegorical figure Contemplation, covered with eyes, ascending aloft on the back of an eagle. She is carrying a telescope with her, with which to view and contemplate the heavens (true to her name). Below her, floating in the Baroque sky, are two putti holding a banner referencing Isaiah 40, verse 26: "Lift up your eyes on high, and behold who hath created these things . . .". At the bottom are two scientists from the past: on the left, Ibn al-Haytham, holding a geometric diagram and standing on the pedestal of reason, and on the right Galileo, holding a telescope and standing on the pedestal of the senses. Together, they display a brief version of the atlas' title page, as if to indicate that the book represents the uniting of reason and sensation. Below them is a picture of the skyline of Gdansk, home of Hevelius' observatory and the place where the book was published. Thus, through a mixture of Baroque and heavenly elements, as well as deceased scientists from the Islamic and Christian lands, wisdom is united in *Selenographia* through the union of the mind and the senses, brought together observationally by the telescope. Heady stuff!

Not to be outdone is the frontispiece from Cellarius' stunning *Atlas Coelestis, seu Harmonia Macrocosmica*, first published in 1660 and perhaps the most beautiful celestial atlas ever produced (see Section 7.1). The title page and frontispiece of the 1708 edition produced by Valk and Schenk are shown in Figure 8.49 and 8.50, respectively. Note on the title page the relative simplicity of the format, but the elaborate and descriptive title that contains much information about the contents of the atlas. The frontispiece is identical to the original 1660 edition, except for the phrase "Apud G. Valk, et P. Schenk" engraved below the title in the banner at the top. Flanking the banner are four putti, two of whom are holding it as well as a zodiacal ring, which is illuminated by a central Sun and around which rolls a sphere, probably representing the solar system. At the periphery, two other putti are sighting the heavens with cross-staffs, zeroing in on the constellations Libra and Virgo and the Sun and Moon, separated by a comet. Below are some clouds with three angels peeking out. At the bottom center is Urania, the muse of astronomy, holding an armillary sphere and pointing up at the heavens. She is surrounded on a balcony overlooking a Classical garden by six famous astronomers of the past, dressed in period costumes. According to Robert van Gent (2006), these are (from left to right): Tycho Brahe, holding a pair of dividers in his right hand, which rest on a celestial globe; Claudius Ptolemy, pointing to a passage from his Earth-centered book, the *Almagest*; a mystery figure wearing a turban who van Gent takes to be

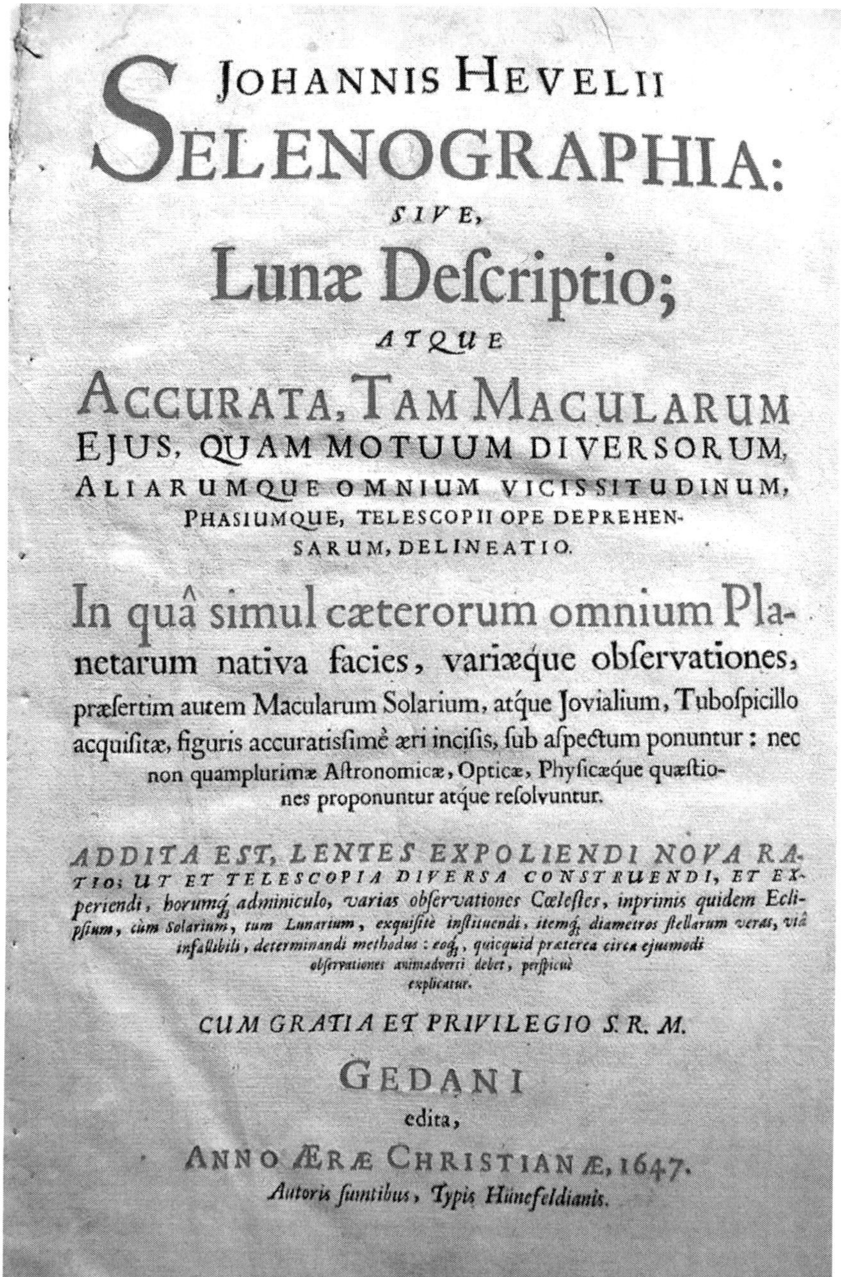

SELENOGRAPHIA:
JOHANNIS HEVELII
SIVE,
Lunæ Deſcriptio;
ATQUE
ACCURATA, TAM MACULARUM
EJUS, QUAM MOTUUM DIVERSORUM,
ALIARUMQUE OMNIUM VICISSITUDINUM,
PHASIUMQUE, TELESCOPII OPE DEPREHEN-
SARUM, DELINEATIO.
In quâ simul cæterorum omnium Pla-
netarum nativa facies, variæq́ue obſervationes,
præſertim autem Macularum Solarium, atq́ue Jovialium, Tuboſpicillo
acquiſitæ, figuris accuratisſimè æri inciſis, ſub aſpectum ponuntur: nec
non quamplurimæ Aſtronomicæ, Opticæ, Phyſicæq́ue quæſtio-
nes proponuntur atq́ue reſolvuntur.
ADDITA EST, LENTES EXPOLIENDI NOVA RA-
TIO; UT ET TELESCOPIA DIVERSA CONSTRUENDI, ET EX-
periendi, horumq; adminiculo, varias obſervationes Cœleſtes, inprimis quidem Ecli-
pſium, cùm Solarium, tum Lunarium, exquiſitè inſtituendi, itemq; diametros ſtellarum veras, viâ
infallibili, determinandi methodus: eoq;, quicquid præterea circa ejusmodi
obſervationes animadverti debet, perſpicuè
explicatur.
CUM GRATIA ET PRIVILEGIO S. R. M.
GEDANI
edita,
ANNO ÆRÆ CHRISTIANÆ, 1647.
Autoris ſumtibus, Typis Hünefeldianis.

Figure 8.47. Title page from the first true lunar atlas, *Selenographia*, by Hevelius, which was published in 1647. 34.9 × 22.5 cm (page size). Note the amount of information given, including comments that not only the Moon but also the planets will be shown as they appear naturally through the telescope. *See also* Color Plate Gallery.

Figure 8.48. Frontispiece from the first true lunar atlas, *Selenographia*, by Hevelius, which was published in 1647. 30.8 × 18.5 cm. Note at the top that Contemplation is ascending into the heavens carrying her telescope, with which to see the Moon on the left and the Sun on the right, and that below al-Haytham (representing reason) and Galileo (representing the senses) are united holding a banner showing the title page of the atlas hanging over the city of Gdansk.

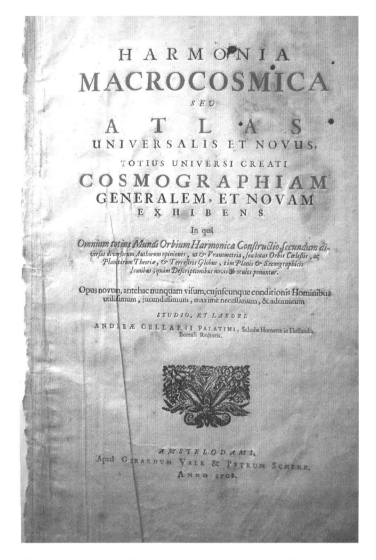

Figure 8.49. Title page from the Valk and Schenk edition of Cellarius' *Harmonia Macro-cosmica*, 1708. 51.2 × 30.7 cm (page size). Note that like the example shown in Figure 8.47, there is a lot of information given in the title page about the contents of the atlas. *See also* Color Plate Gallery.

the great Islamic astronomer al-Battani; the Castilian king Alfonso the Wise, for whom are named the Alfonsine tables, holding a model of the Copernican solar system; Philip Lansbergen, a writer of a popular astronomy text of the time who is using a long pointer to acknowledge the heliocentric world system above (but who Soderlund suggests might instead be the atlas' author, Cellarius); and seated,

Figure 8.50. Frontispiece from the Valk and Schenk edition of Cellarius' *Harmonia Macro-cosmica*, 1708. 43 × 26.2 cm. Note the six famous astronomers at the bottom surrounding Urania, the muse of astronomy. They are pointing at the heavens or at important astronomical instruments, models, and books of their times. *See also* Color Plate Gallery.

Nicolas Copernicus, pointing to an armillary sphere with his left hand and writing with his right. A number of astronomical instruments are at his feet, along with a large book which might represent his famous *De revolutionibus orbium coelestium*. Altogether, this is a fitting frontispiece to set the tone for a beautiful and complete celestial atlas.

Famous astronomers are also featured on the frontispiece of Doppelmayr's *Atlas Coelestis* (see Section 7.8), which is shown in Figure 8.51. At the top, two putti hold up a diagram of our solar system, complete with the known planets and their moons and a comet orbiting around the central Sun. Surrounding our solar system are examples of other star systems, suggesting that we are not alone in God's universe. Below this diagram is a scene with palm trees and two sphinxes, suggesting the wisdom of the ancients, and a central celestial globe on a pedestal that is being unveiled (possibly as a result of the information in Doppelmayr's atlas). In the foreground are four famous astronomers of the past with descriptive labels hanging from each palm tree. From left to right they are: Ptolemy, holding a model depicting his geocentric universe; Copernicus, pointing up to his heliocentric world view; Kepler, pointing down at the navigational and celestial scientific instruments that surround a decorative cartouche for the atlas; and Tycho Brahe, holding a book labeled *History of the Heavens*. This stunning frontispiece is full of imagery and allegory that places this atlas among the writings of some of the historical giants of astronomy.

8.7.3   Contemporary instruments and people

An early example of an illustration in a title page depicting a contemporary astronomical instrument comes from a Renaissance edition of Johannes de Sacrobosco's famous *Tractatus de Sphaera*, more commonly known as *De Sphaera* (see Section 3.5.2). The title page shown in Figure 8.52 is from a 1647 edition published in Leiden. The image used as a printer's mark is an armillary sphere, which we discussed in Section 1.4.2. Although telescopes were in use by the time of this edition, armillary spheres were still valuable instruments in positional astronomy. This image also was symbolic of the major theme of this book, which was to discuss the geocentric universe composed of a central Earth surrounded by the sphere of the heavens containing the stars and planets.

The telescope is depicted in the frontispiece of *Steele's Popular Astronomy*. This book was written by Joel Dorman Steele (1836–1886), a New York state educator, who together with his wife wrote a number of important and popular science and history textbooks in the post–Civil War period. This book covers a variety of topics in astronomy, such as the solar system, the stars and constellations, and spectrum analysis. At the end is a large section that teachers can use as a guide for class work. The frontispiece is from an 1899 edition of this book revised and updated by Mabel Loomis Todd. The beautifully engraved frontispiece shown in Figure 8.53 depicts the giant 6-foot diameter reflecting telescope belonging to Lord Rosse, a suitable instrument to introduce this fine American textbook.

A contemporary astronomer is depicted in the frontispiece of the last book written by O.M. Mitchel entitled *The Astronomy of the Bible* (see Section 9.6). This frontispiece is shown in Figure 8.54. Note the engraving of the handsome and distinguished American astronomer, O.M. Mitchel himself, from a photograph taken in 1855. The choice to show Mitchel here might have been prompted by the book's posthumous publication and the fact that this book also contains an extensive

Figure 8.51. Frontispiece from the 1742 edition of Doppelmayr's *Atlas Coelestis*. 47 × 27.2 cm. Note at the top a large depiction of our solar system surrounded by the solar systems of other stars, and below a central celestial globe and the images of four famous astronomers of the past (Ptolemy, Copernicus, Kepler, and Tycho Brahe).

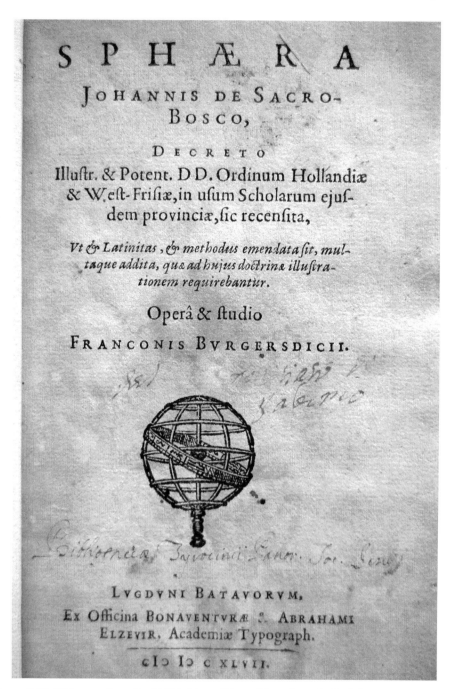

Figure 8.52. Title page from the 1647 Leiden edition of Sacrobosco's *De Sphaera*. 15.2 × 9.7 cm
(page size). Note the image of an armillary sphere, an important astronomical instrument of the
time, that also served as a printer's mark for this book.

Figure 8.53. Frontispiece from Todd's 1899 revised edition of *Steele's Popular Astronomy*. 12.5 × 18.5 cm (page size). Note the beautiful engraving of Lord Rosse's giant telescope in Ireland, one of the largest in the world.

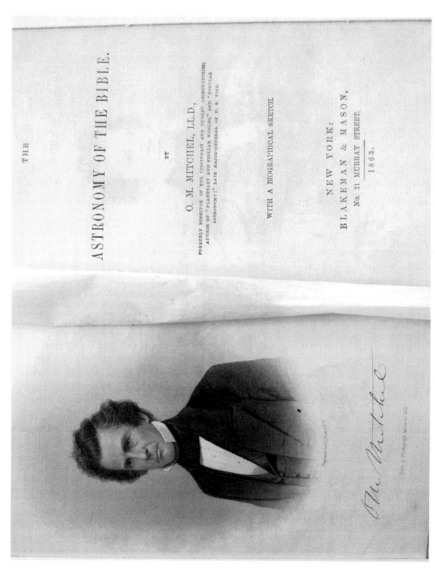

Figure 8.54. Frontispiece and title page from Mitchel's *The Astronomy of the Bible*, published in 1863. 18.6 × 12.2 cm (page size). Note the life-like engraving of Mitchel on the frontispiece and the tissue paper (here rolled up) that was meant to prevent the ink from the frontispiece from staining the title page.

biography of its author. But showing an image of the author in the frontispiece was not unprecedented; the portrait of John Flamsteed adorns the frontispiece of his famous *Atlas Coelestis*, which also was published posthumously (see Section 6.4).

8.7.4   Extension of the book's content

In some books, the image on the frontispiece is not meant to be an allegory or a decorative element, but rather it is an extension of the material in the text to which the reader may refer. An example appears in a book by Denison Olmsted (see Section 9.5.3) entitled *An Introduction to Astronomy: Designed as a Text-book for the Use of Students in College*, which was published in 1839. The 1876 edition was revised by E.S. Snell, a professor at Amherst College, and its frontispiece is shown in Figure 8.55. Note that this is labeled "Fig. 1" at the top. It shows drawings made at the telescope of the full Moon, the Moon near quadrature, Saturn and its rings, and Jupiter with its four large moons. The frontispiece was not meant to be simply a decorative element. It was meant to be used by the reader, since there were no other comparable planetary images in the text.

Another example of a text extension into the frontispiece is to be found in an 1857 book by Hannah Mary Bouvier entitled *Familiar Astronomy, or an Introduction to the Study of the Heavens* (see Section 9.7). The frontispiece of this book consisted of two celestial hemispheres, shown in Figure 8.56. One was centered on the north celestial pole and the other on the south celestial pole. The stars and faint constellation images were drawn in white against a blue background. Prominent stars in each constellation were connected by lines, and stellar magnitudes were indicated by the size of the star's image. A number of constellations were depicted, including many that are now obsolete. These two hemispheres were meant to be consulted by the reader as he or she read the book, particularly the very complete section on the constellations. This devise of showing a celestial hemisphere in the frontispiece that can be used by the reader has been employed by others, most notably by Arthur Norton in his *Norton's Star Atlas* (see Section 10.2.3.1)

8.7.5   Combinations

An engraving that combines several of the above elements into one is found in the *Uranometria* by Johann Bayer (see Section 6.2), which shows a beautiful allegorical frontispiece image that includes the atlas' title page complete with a small printer's mark at the bottom. The illustrated title page from the 1661 edition is shown in Figure 8.57. I am grateful to Professor Christopher Ocker at the San Francisco Theological Seminary in San Anselmo, California, for his translation assistance.

The title reminds us that the atlas contains charts of all the constellations using a "new" method and that the images are engraved on copper plates. At the top are three figures, likely representing (from left to right) the Sun (Apollo); Eternity, holding back two lion-like beasts of ignorance and representing the eternity of the heavens; and the Moon (Diana). At their feet is written the Greek and Latin for "Let no one unlearned in geometry enter eternity!" Below are two figures who frame a

Figure 8.55. Frontispiece from Snell's 1876 revised edition of Olmsted's *An Introduction to Astronomy: Designed as a Text-book for the Use of Students in College.* 22.9 × 14.6 cm (page size). Note the lunar and planetary images, which were the only comparable images of these solar system bodies in the book, and were thus meant to be consulted by the reader.

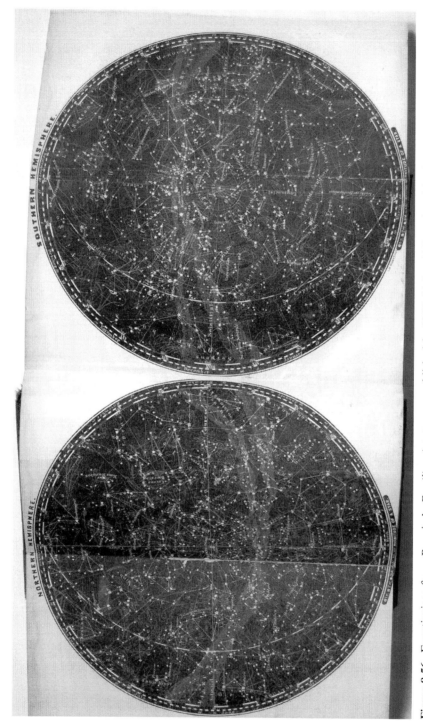

Figure 8.56. Frontispiece from Bouvier's *Familiar Astronomy*, published in 1857. 21.3 × 44.7 cm (combined page size), 20.6 cm dia. hemispheres. Note the very faint images but the prominent stick figures of the constellations in these northern and southern celestial hemispheres, which were meant to be consulted by the reader as he or she perused the text. *See also* Color Plate Gallery.

Figure 8.57. Combined frontispiece and title page from the 1661 edition of Bayer's *Uranometria*. 28.7×19 cm. Note the addition of a small printer's mark at the bottom and the names of various owners of the atlas handwritten in the margins, including a monastery entry at the bottom. *See also* Color Plate Gallery.

banner serving as the atlas' title page. To the left is Atlas, pointing to an astrolabe, standing on a pedestal that reads: "To Atlas, master of the most ancient astronomers". To the right is Hercules, holding a celestial globe, standing on a pedestal that reads: "To Hercules, disciple of the most ancient astronomers." To explain this, one may refer to a variant of the myth of Atlas stating that as a legendary king of Mauretania he was knowledgeable in astronomy and discovered the nature of the sphere, and that he taught this knowledge to Hercules. Between the figures is an image for the sign of the zodiac, Capricorn. Below that is the skyline of Augsburg (where the 1st edition of the atlas was published in 1603). Differences between the 1603 and the 1661 frontispieces included, in the latter, the naming of Ulm as the city of publication, the publisher's name (Iohannis Gorlini), and the year "MDCLXI" in the title information, and at the bottom, the replacement of a medallion of a child's head with a printer's mark having the initials "I/G", standing for the publisher.

Another kind of combination consists of books having images on both the frontispiece and the title page. An example comes from Wilkins' *Elements of Astronomy* (see Section 9.5.1). The first edition of this small textbook for school children was published in 1822. Figure 8.58 shows the frontispiece and title page from the 1829 edition. Note the printer's mark on the title page, with its flowery design. On the preceding page (conveniently labeled "Frontispiece") is a diagram of our solar system showing the orbits of the planets and their moons, plus several asteroids and an intersecting comet. Rimming the last planet, Uranus (Neptune had not yet been discovered) is the ecliptic, complete with symbols for the zodiac constellations, representing the stars beyond. Along the left margin is a diagram showing the "Proportional Distances of the Planets from the Sun", suggesting that this frontispiece was not simply decorative but was meant to be used by the students as an extension of the text.

Another example is from O.M. Mitchel, whom we have met above. The book is entitled *The Orbs of Heaven, or, The Planetary and Stellar Worlds*, and it is the British edition of his 1848 book *The Planetary and Stellar Worlds*. Both versions were written expositions of ten of his lectures on astronomy, although in the 1853 British edition his last name was incorrectly spelled with two "l's", and there are more illustrations than in the American edition. In addition, there is a long 76 page appendix of sections taken from a book by Yale professor Denison Olmsted (whom we also met earlier) entitled *Mechanism of the Heavens* that deals, among other things, with telescopes, such as Lord Rosse's behemoth, and historical issues, such as the work of Galileo.

Images supporting these appendix topics appear on both the title page and frontispiece of this book. As shown in Figure 8.59, the frontispiece is a tribute to Galileo, shown working at his telescope and making entries to his notebook, symbolizing the birth of scientific observational astronomy. In the title page, where the printer's mark would normally be, is an engraving celebrating Lord Rosse's behemoth telescope, a great leap forward from the time of Galileo. Curiously, Mitchel's name does not appear here but is given on a subsequent page.

An unusual example of a frontispiece that combines allegorical material with material pertinent to the content of the book and also includes title and publisher information is shown in Figure 8.60. This book is entitled *Mythographi Latini*, and it

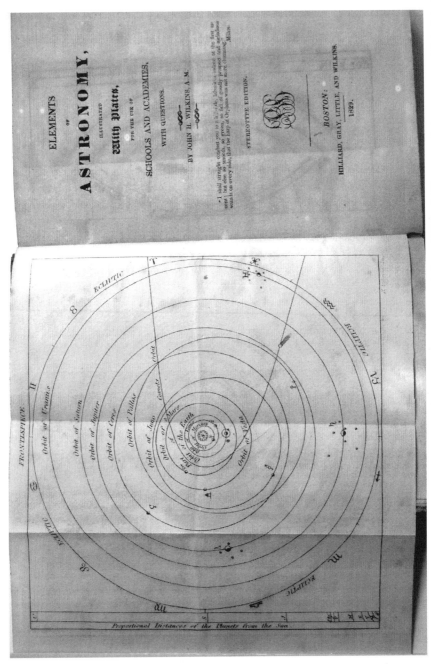

Figure 8.58. Pull-out frontispiece and title page with decorative printer's mark from Wilkins' *Elements of Astronomy*; published in 1829. 17.2 × 19.5 cm (frontispiece page size). Note the plethora of solar system information conveyed in the frontispiece, which was meant to be used by the reader.

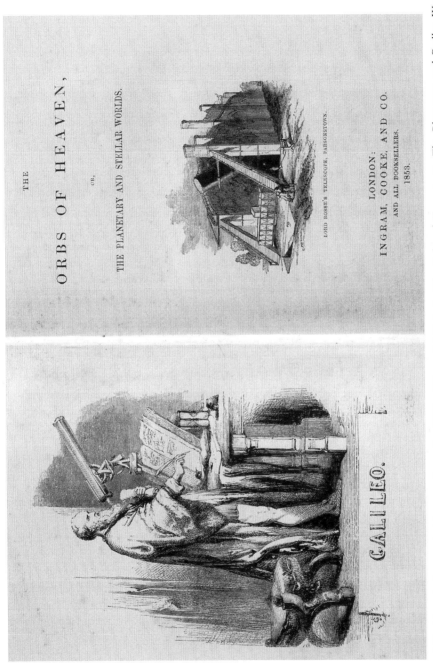

Figure 8.59. Frontispiece and title page from the 1853 edition of Mitchel's *The Orbs of Heaven, or, The Planetary and Stellar Worlds*. 18.7 × 12.3 cm (page size). Note that both images feature a telescope, one from the past being used by Galileo in the frontispiece and the other contemporary to the time the book was written: Lord Rosse's "Leviathan", the largest telescope in the world at that time.

Figure 8.60. Frontispiece from the 1681 edition of Munckerus' *Mythographi Latini*. 17.6 × 10 cm. Note that it includes the book's title, some publishing information, and an image that depicts a number of mythological characters engaged in activities related to their histories.

is composed of several separate books bound together that discuss Greco-Roman mythology according to ancient Latin sources. It was compiled in 1681 by Thomas Munckerus, who was the Rector at the Gymnasium in Delft. Included in this compilation was the text of the 1482 Ratdolt edition of Hyginus' *Poeticon Astronomicon*, a book written in the 2nd Century AD that summarized the mythology of the constellations and the location of some of their principal stars. True to the subject matter, the frontispiece shows a number of mythological characters engaged in activities related to their histories, thus giving an appropriate introduction to later discussions of constellation mythology. Note that Zeus and his retinue are pictured at the top, while below them Helios crosses the heavens in a chariot. On the water, Poseidon stands holding a trident. On land, a Minotaur joins others in pouring wine or water from a vase, while an eagle picks at the liver of a reclining Prometheus. In the underworld below, the three-headed dog Cerberus howls as Charon ferries an unfortunate couple across the river Styx.

8.7.6   Conclusions

Thus, we have seen that a number of images have appeared in frontispieces and title pages of celestial books and atlases that serve a number of purposes, from publisher's advertisements, to allegorical references, to contemporary acknowledgments of people and instruments, to text extensions. Many of the images depicted were quite beautiful, being both informative and artistically constructed using Baroque and Classical elements. Not only did they set the tone for the subject matter of the book, but they used allegory and vivid pagan, Christian, and Islamic imagery to link the past with the present, suggesting the universality of the thematic content. The time-lessness and vastness of the cosmos were emphasized, along with the modernity that was promised in the pages to follow. Famous mythological characters or real ancient and contemporary astronomers were prominently featured, along with various astronomical tools and instruments, to highlight the scholarship of the book. The result was a beautiful and informative work of science and art that rivaled or even surpassed the spectacular images that followed in the text.

According to Shirley (2009), by the 1930s the images depicted in frontispieces and title pages became more minimalistic and functional. Today, in most cases, the images have disappeared, and only the basic printing information is included at the beginning of most books and atlases. Although this is congruent with modern sensibilities, something has been lost in terms of beauty and artistry. The great images of the past deserve a moment of our time for reflection and appreciation, as the author and publisher intended.

8.8   BIBLIOGRAPHY

Ashbrook, J. (1984a) Chapter 25. Herschel's "large 20-foot" telescope. *The Astronomical Scrapbook*. Cambridge, U.K.: Cambridge University Press, pp. 127–132.

Ashbrook, J. (1984b) Chapter 49. The first true mapper of the Moon. *The Astronomical Scrapbook*. Cambridge, U.K.: Cambridge University Press, pp. 247–251.

Ashbrook, J. (1984c) Chapter 58. Naming some early minor planets. *The Astronomical Scrapbook*. Cambridge, U.K.: Cambridge University Press, pp. 297–302.

Ashbrook, J. (1984d) Chapter 76. The discovery of spiral structure in galaxies. *The Astronomical Scrapbook*. Cambridge, U.K.: Cambridge University Press, pp. 397–404.

Ashbrook, J. (1984e) Chapter 78. Johann Bayer and his star nomenclature. *The Astronomical Scrapbook*. Cambridge, U.K.: Cambridge University Press, pp. 411–418.

Aughton, P. (2004) *The Transit of Venus*. London: Phoenix & Orion Books.

Beresiner, Y. (2010) A fine hand: Cartographic and map playing cards 1590–1798. *Journal of the International Map Collectors' Society*, **122**, 15–23.

Brashear, R. and Lewis, D. (2001) *Star Struck: One Thousand Years of the Art and Science of Astronomy*. San Marino, CA: Huntington Library.

Chapman, A. (2009) A new perceived reality: Thomas Harriot's Moon maps. *A&G*, **50**, 1.27–1.34.

Cunningham, C.J., Marsden, B.G., and Orchiston, W. (2011) Giuseppe Piazzi: The controversial discovery and loss of Ceres in 1801. *Journal for the History of Astronomy*, **42**, 283–306.

Curtis, H.D. (1920) Modern theories of the spiral nebulae. *Journal of the Royal Astronomical Society of Canada*, **14**, 317–327.

Dahl, E.H. and Gauvin, J-F. (2000) *Sphaere Mundi: Early Globes at the Stewart Museum*. Montreal: McGill-Queen's University Press.

Duke, D.W. (2006) Analysis of the Farnese globe. *Journal for the History of Astronomy*, **37**, 87–100.

Edmonds, M.G. (2011) An initial assessment of the accuracy of the gear trains in the Antikythera Mechanism. *Journal for the History of Astronomy*, **42**, 307–320.

Evans, J. (1998) *The History and Practice of Ancient Astronomy*. Oxford, U.K.: Oxford University Press.

Evans, J., Carman, C.C., and Thorndike, A.S. (2010) Solar anomaly and planetary displays in the Antikythera Mechanism. *Journal for the History of Astronomy*, **41**, 1–39.

Field, J.V. (1996) European astronomy in the first millennium: The archaeological record. In: C. Walker (ed.), *Astronomy before the Telescope*. New York: St. Martin's Press.

Freeth, T.; Bitsakis, Y.; Moussas, X. *et al.* (2006) Decoding the ancient Greek astronomical calculator known as the Antikythera Mechanism. *Nature*, **444**, 587–591.

Gingerich, O. (1993) Astronomical paper instruments with moving parts. In: R.G.W. Anderson; J.A. Bennett; and W. F. Ryan (eds.), *Making Instruments Count*. Cambridge, U.K.: Variorum/Ashgate Publ., pp. 63–74.

Goss, J. (1993) *The Mapmaker's Art: An Illustrated History of Cartography*. New York: Rand McNally.

Graney, C.M. (2010) The telescope against Copernicus: Star observations by Riccioli supporting a geocentric universe. *Journal for the History of Astronomy*, **41**, 453–467.

Hoskin, M. (1997) *The Cambridge Illustrated History of Astronomy*. Cambridge, U.K.: Cambridge University Press.

Hoskin, M. (2006) Caroline Herschel's catalogue of nebulae. *Journal for the History of Astronomy*, **37**, 251–255.

Hoskin, M. (2008) Nebulae, star clusters and the Milky Way: From Galileo to William Herschel. *Journal for the History of Astronomy*, **39**, 363–396.

Hoskin, M. (2011a) William Herschel and the nebulae, part 1: 1774–1784. *Journal for the History of Astronomy*, **42**, 177–192.

Hoskin, M. (2011b) William Herschel and the nebulae, part 2: 1785–1818. *Journal for the History of Astronomy*, **42**, 321–338.

House, D. (1975) *Francis Place and the Early History of the Greenwich Observatory*. New York: Science History Publications.

Johnston, P.A. (1985) *Celestial Images—Astronomical Charts from 1500 to 1900*. Boston, MA: Boston University Art Gallery.

Kanas, N. (2002) Mapping the solar system: Depictions from antiquarian star atlases. *Mercator's World*, **7**, 40–46.

Kanas, N. (2005) Are celestial maps really maps? *Journal of the International Map Collectors' Society*, **101**, 19–29.

Kanas, N. (2005) Volvelles! Early paper astronomical computers. *Mercury*, March/April, 33–39.

Kidwell, P.A. (2009) The astrolabe for latitude 41°N of Simon de Witt: An early American celestial planisphere. *Imago Mundi*, **61**, 91–96.

King, D.A. (1996) Islamic astronomy. In: C. Walker (ed.), *Astronomy before the Telescope*. New York: St. Martin's Press.

King, H.C. (1955) *The History of the Telescope*. Cambridge, MA: Sky Publishing.

Kremer, R.L. (2011) Experimenting with paper instruments in Fifteenth- and Sixteenth-Century astronomy: Computing syzygies with isotemporal lines and salt dishes. *Journal for the History of Astronomy*, **42**, 223–258.

Lachieze-Rey, M. and Luminet, J-P. (2001) *Celestial Treasury: From the Music of the Spheres to the Conquest of Space*. Cambridge, U.K.: Cambridge University Press.

Lane, K.M.D. (2006) Mapping the Mars canal mania: Cartographic projection and the creation of a popular icon. *Imago Mundi*, **58**, 198–211.

Meller, H. (2004) Sky disk of Nebra. *National Geographic*, January, 76–87.

Mueller, K. (2010) How to craft telescopic observation in a book: Hevelius's *Selenographia* (1647) and its images. *Journal for the History of Astronomy*, **41**, 355–379.

North, J. (1995) *The Norton History of Astronomy and Cosmology*. New York: W.W. Norton.

OUP (2002) *The Shorter Oxford English Dictionary*, 5th edn., Vol. 2, N–Z. Oxford, U.K.: Oxford University Press.

Pannekoek, A. (1989) *A History of Astronomy*. New York: Dover Publications.

Pumfrey, S. (2011) The *Selenographia* of William Gilbert: His pre-telescopic map of the moon and his discovery of lunar libration. *Journal for the History of Astronomy*, **42**, 193–203.

Raeder, H.; Stromgren, E.; and Stromgren, B. (transl. and eds.) (1946) *Tycho Brahe's Description of His Instruments and Scientific Work as Given in "Astronomiae Instauratae Mechanica"*. Copenhagen: I Kommission Hos Ejnar Munksgaard.

Ronan, C.A. (1967) *Their Majesties' Astronomers*. London: Bodley Head.

Schaefer, B.E. (2005) The epoch of the constellations on the Farnese Atlas and their origin in Hipparchus's lost catalogue. *Journal for the History of Astronomy*, **36**, 167–196.

Sheehan, W.P and Dobbins, T.A. (2001) *Epic Moon: A History of Lunar Exploration in the Age of the Telescope*. Richmond, VA: Willmann-Bell.

Shirley, R. (2009) *Courtiers and Cannibals, Angels and Amazons*. Houten, The Netherlands: Hes & De Graaf.

Shubinski, R. (2011) How the equatorial mount changed astronomy. *Sky & Telescope*, February, 58–61.

Smith, R.W. (2006) Beyond the big galaxy: The structure of the stellar system 1900–1952. *Journal for the History of Astronomy*, **37**, 307–342.

Smith, R.W. (2008) Beyond the galaxy: The development of extragalactic astronomy 1885–1965, part 1. *Journal for the History of Astronomy*, **39**, 307–342.

Smith, R.W. (2009) Beyond the galaxy: The development of extragalactic astronomy 1885–1965, part 2. *Journal for the History of Astronomy*, **40**, 71–107.

Snyder, G. (1984) *Maps of the Heavens*. New York: Abbeville Press.

Soderlund, I.E. (2010) *Taking Possession of Astronomy: Frontispieces and Illustrated Title Pages in 17th Century Books on Astronomy*. Stockholm: Center for History of Science at the Royal Swedish Academy of Sciences, *http://www.center.kva.se/bilder/Avhandling_5.pdf*

Stephenson, B.; Bolt, M.; and Friedman, A.F. (2000) *The Universe Unveiled: Instruments and Images through History*. Chicago: Adler Planetarium & Astronomy Museum.

Turner, G.L.E. (1996) Later Medieval and Renaissance instruments. In: C. Walker (ed.), *Astronomy before the Telescope*. New York: St. Martin's Press.

Van Gent, R.H. (2006) *Andreas Cellarius, Harmonia Macrocosmica of 1660: The Finest Atlas of the Heavens*. Cologne: Taschen.

Vertesi, J. (2007) Picturing the moon: Hevelius's and Riccioli's visual debate. *Studies in History and Philosophy of Science*, **28**, 401–421.

Volkoff, I.; Franzgrote, E.; and Larsen, A.D. (1971) *Johannes Hevelius and His Catalog of Stars*. Provo, UT: Brigham Young University Press.

Warner, D.J. (1979) *The Sky Explored: Celestial Cartography 1500–1800*. Amsterdam: Theatrum Orbis Terrarum.

Whitaker, E.A. (1999) *Mapping and Naming the Moon: A History of Lunar Cartography and Nomenclature*. Cambridge, U.K.: Cambridge University Press.

Wlodarczyk, W. (2011) Libration of the Moon, Hevelius's theory, and its early reception in England. *Journal for the History of Astronomy*, **42**, 495–519.

9

Mapping the stars in early America

Since the earliest days of the Republic, people in the United States have been fascinated by events in the sky and have followed the paths of the Moon, planets, and stars. Although Europeans set the standard for celestial cartography, many of its elements took root in America during the period between the Revolutionary and Civil Wars. A number of books and atlases were printed that oriented people to the heavens, which contributed to the early popularity of astronomy in the United States and its later emergence as a world leader in astronomical observation and science (see Section 8.5.5).

There were many antecedents in colonial America that led to this development. Although Ptolemaic ideas began to be taught shortly after the founding of Harvard College in 1636, Copernican principles were in the curriculum by 1659. Connecticut governor John Winthrop Jr. (1606–1676) brought the first major telescope to America in 1663. A practicing physician and perhaps the colony's best scientist, Winthrop made many observations of the sky, including the planets, and he communicated his observations to colleagues in the newly formed Royal Society in England. Later in life, he bequeathed his $3\frac{1}{2}$ foot refractor to Harvard in 1672. There it was put to good use by Thomas Brattle, who in 1680 reported some of his cometary observations to Astronomer Royal John Flamsteed and Sir Isaac Newton in England. Observations of the sky continued at Harvard and at other universities, such as Yale and the College of William and Mary. Colonists such as Ezra Stiles skillfully applied known astronomical knowledge to current problems, such as calculating cometary orbits from known Newtonian principles. Craftsmen such as David Rittenhouse became prolific manufacturers of optical instruments such as orreries and presaged the great American telescope makers of the 1800s (e.g., Alvin Clark & Sons, Warner & Swasey). The pinnacle of colonial astronomy was the 1769 transit of Venus, which fortuitously happened to be best viewed from North America and where the observations of several colonists contributed to a value of the Sun's distance approximating today's value (Mendillo *et al.*, 1976).

Although excitement from this event led to calls for a national observatory, any momentum for this occurrence was squelched by the Revolutionary War, and it was not until the 1800s when national pride, the Industrial Revolution, and private philanthropy led to the building of over 200 telescopes and vaulted America into international astronomical prominence. But compared with Europe, the relative lack of telescopes in pre-Revolutionary America kept observational astronomy at an amateur level. In addition, a focus on day to day issues affecting the struggling colony related to agriculture, manufacturing, navigation, and surveying encouraged astronomy to remain at a practical, rather than at a theoretical "text book", level. However, this also piqued the interest of the average person, leading to broad support for issues related to the sky. Nowhere is this better exemplified than in the widespread use of almanacs.

9.1   THE ALMANACS

Stimulated by the need to predict weather events and to know the best dates to plant and harvest crops, as well as to entertain people at a time when books and newspapers were scarce, almanacs were very popular in early American society, going back to the mid-1600s. One of the best known was *Poor Richard's Almanack*, written by the famous polymath and politician, Benjamin Franklin (1706–1790). It sold thousands of copies over successive years beginning in 1732. Many topics were covered, such as weather forecasts, ideal planting dates, hygenic information, notices of Quaker meetings, and ways of protecting one's house from lightning. Astronomical information also was included, such as ephemerides (i.e., tables showing the location of a heavenly body over a sequence of dates), information on eclipses and lunations, and astronomical and astrological material related to constellations. Another well-known almanac is the *Old Farmer's Almanac*, which began in 1792. After 215 years of continuous publication, it can still be found today, both in print and in an Internet version!

Another famous almanac was written by Benjamin Banneker (1731–1806), who was a self-taught descendant of freed African slaves. Raised to be a farmer, this remarkable man later conducted independent astronomical observations of the sky, and he theorized about the existence of planets around stars other than the Sun. He also developed mathematical and surveying skills, and as a result of these he was hired to assist the Surveyor General in laying out the city plan for Washington, D.C. The first issue of his almanac was for the year 1792, and several yearly issues were printed throughout the rest of the decade. Like Franklin's almanac, Banneker's dealt with a number of topics, including planting information, weather forecasts, humorous anecdotes, and fine prose and poetry. It also contained information on lunations, conjunctions, and eclipses, as well as essays on his theories of extra-solar planets and the possibility of life on them. The ephemerides were so accurate in plotting the location of heavenly bodies in the sky that sailors used them for navigation. Banneker's almanac was widely read and respected, and its various issues were great commercial successes.

Almanacs continued to be popular well into the 19th Century in great variety, and they often reflected regional items of interest. For example, the 1847 issue of *The New-England Almanac, and Farmers' Friend*, which was published in Connecticut and written by Nathan Daboll, contained many features found in other almanacs, such as ephemerides giving the predicted location of heavenly bodies and weather forecasts for each day of the year, planting and agricultural advice, and anecdotes and homilies (see Figure 9.1). But also reflecting the whaling interests of the area, there was a table giving information about ships in the "whale fishery" at a number of northeastern seaports, including information on tonnage, masters and agents, and sailing dates. Also included was information on the average number of voyages that were made by "Sperm and right whalers" in the years 1842 to 1845.

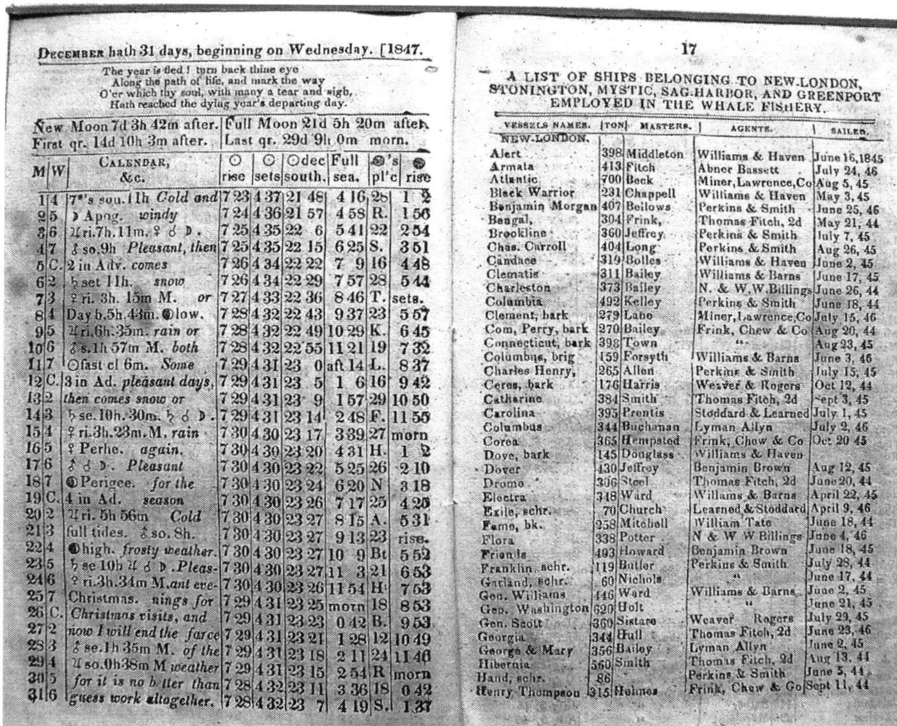

DECEMBER hath 31 days, beginning on Wednesday. [1847.

The year is fled! turn back thine eye
Along the path of life, and mark the way
O'er which thy soul, with many a tear and sigh,
Hath reached the dying year's departing day.

New Moon 7d 3h 42m after.	Full Moon 21d 5h 20m after.
First qr. 14d 10h 3m after.	Last qr. 29d 9h 0m morn.

M	W	CALENDAR, &c.	☉ rise	☉ sets	☉ dec south.	Full sea.	☽'s pl'c	☽ rise
1	4	7*'s sou. 11h *Cold and*	7 23	4 37	21 48	4 16	28	1 2
2	5	☽ Apog. *windy*	7 24	4 36	21 57	4 58	R.	1 56
3	6	♃ ri. 7h. 11m. ♀ ☌ ☽.	7 25	4 35	22 6	5 41	22	2 54
4	7	♂ so. 9h *Pleasant, then*	7 25	4 35	22 15	6 25	S.	3 51
5	C.	2 in Adv. *comes*	7 26	4 34	22 22	7 9	16	4 48
6	2	♄ set 11h. *snow*	7 26	4 34	22 29	7 57	28	5 44
7	3	♀ ri. 3h. 15m M. *or*	7 27	4 33	22 36	8 46	T.	sets.
8	4	Day b. 5h. 43m. ☽ low.	7 28	4 32	22 43	9 37	23	5 57
9	5	♃ ri. 6h. 35m. *rain or*	7 28	4 32	22 49	10 29	K.	6 45
10	6	♂ s. 1h 57m M. *both*	7 28	4 32	22 55	11 21	19	7 32
11	7	☉ fast cl 6m. *Some*	7 29	4 31	23 0	aft 14	L.	8 37
12	C.	3 in Ad. *pleasant days,*	7 29	4 31	23 5	1 6	16	9 42
13	2	*then comes snow or*	7 29	4 31	23 9	1 57	29	10 50
14	3	♄ se. 10h. 30m. ♄ ☌ ☽.	7 29	4 31	23 14	2 48	F.	11 55
15	4	♀ ri. 3h. 23m. M. *rain*	7 30	4 30	23 17	3 39	27	morn
16	5	♀ Perhe. *again.*	7 30	4 39	23 20	4 31	H.	1 2
17	6	♂ ☌ ☽. *Pleasant*	7 30	4 30	23 22	5 25	26	2 10
18	7	☽ Perigee. *for the*	7 30	4 30	23 24	6 20	N	3 18
19	C.	4 in Ad. *season*	7 30	4 30	23 26	7 17	25	4 25
20	2	♃ ri. 5h 56m *Cold*	7 30	4 30	23 27	8 15	A.	5 31
21	3	full tides. ♂ so. 8h.	7 30	4 30	23 27	9 13	23	rise.
22	4	☽ high. *frosty weather.*	7 30	4 30	23 27	10 9	Bt	5 52
23	5	♄ se 10h ♃ ☌ ☽. *Pleas-*	7 30	4 30	23 27	11 3	21	6 53
24	6	♀ ri. 3h. 34m M. *ant eve-*	7 30	4 30	23 26	11 54	H	7 53
25	7	Christmas. *nings for*	7 29	4 31	23 25	morn	18	8 53
26	C.	*Christmas visits, and*	7 29	4 31	23 23	0 42	B.	9 53
27	2	*now I will end the farce*	7 29	4 31	23 21	1 28	12	10 49
28	3	♂ se. 1h 35m M. *of the*	7 29	4 31	23 18	2 11	24	11 46
29	4	♃ so. 0h 38m M *weather*	7 29	4 31	23 15	2 54	R	morn
30	5	*for it is no better than*	7 28	4 32	23 11	3 36	18	0 42
31	6	*guess work altogether.*	7 28	4 32	23 7	4 19	S.	1 37

17

A LIST OF SHIPS BELONGING TO NEW-LONDON, STONINGTON, MYSTIC, SAG-HARBOR, AND GREENPORT EMPLOYED IN THE WHALE FISHERY.

VESSELS NAMES.	TON	MASTERS.	AGENTS.	SAILED.
NEW-LONDON.				
Alert	398	Middleton	Williams & Haven	June 16, 1845
Armata	413	Fitch	Abner Bassett	July 24, 46
Atlantic	700	Beck	Miner, Lawrence, Co	Aug 5, 45
Black Warrior	231	Chappell	Williams & Haven	May 3, 45
Benjamin Morgan	407	Bellows	Perkins & Smith	June 25, 46
Bengal,	304	Frink,	Thomas Fitch, 2d	May 21, 44
Brookline	360	Jeffrey	Perkins & Smith	July 7, 45
Chas. Carroll	404	Long	Perkins & Smith	Aug 26, 45
Candace	319	Bolles	Williams & Haven	June 2, 45
Clematis	311	Bailey	Williams & Barns	June 17, 45
Charleston	373	Bailey	N. & W. W. Billings	June 26, 44
Columbia	492	Kelley	Perkins & Smith	June 18, 44
Clement, bark	279	Lane	Miner, Lawrence, Co	July 15, 46
Com, Perry, bark	270	Bailey	Frink, Chew & Co	Aug 20, 44
Connecticut, bark	398	Town	"	Aug 23, 45
Columbus, brig	159	Forsyth	Williams & Barns	June 3, 46
Charles Henry,	265	Allen	Perkins & Smith	July 15, 45
Ceres, bark	176	Harris	Weaver & Rogers	Oct 12, 44
Catharine	384	Smith	Thomas Fitch, 2d	Sept 3, 45
Carolina	395	Prentis	Stoddard & Learned	July 1, 45
Columbus	344	Buchanan	Lyman Allyn	July 2, 46
Corea	365	Hempsted	Frink, Chew & Co	Oct 20 45
Dove, bark	145	Douglass	Williams & Haven	
Dover	430	Jeffrey	Benjamin Brown	Aug 12, 45
Dromo	306	Steel	Thomas Fitch, 2d	June 20, 44
Electra	348	Ward	Williams & Barns	April 22, 45
Exile, schr.	70	Church	Learned & Stoddard	April 9, 46
Fame, bk.	258	Mitchell	William Tate	June 18, 44
Flora	338	Potter	N & W W Billings	June 4, 46
Friends	403	Howard	Benjamin Brown	June 18, 45
Franklin, schr.	119	Butler	Perkins & Smith	July 28, 44
Garland, schr.	60	Nichols	"	June 17, 44
Gen. Williams	446	Ward	Williams & Barns	June 2, 45
Geo. Washington	620	Holt	"	June 21, 45
Gen. Scott	360	Sistare	Weaver Rogers	July 29, 45
Georgia	344	Hull	Thomas Fitch, 2d	June 23, 46
George & Mary	356	Bailey	Lyman Allyn	June 2, 45
Hibernia	560	Smith	Thomas Fitch, 2d	Aug 13, 44
Hand, schr.	86		Perkins & Smith	June 5, 44
Henry Thompson	315	Holmes	Frink, Chew & Co	Sept 11, 44

Figure 9.1. This figure shows two pages from Nathan Daboll's 1847 edition of *The New-England Almanac, and Farmer's Friend.* 15.8 × 10.7 cm (page size). Note on the left an ephemeris for December showing the daily rising and setting of the planets, the Sun, and the Moon, and the predicted weather for the month (but with the ending caveat: "now I will end the farce of the weather for it is no better than guess work altogether"). On the right is a table containing information about whaling ships belonging to several northeastern seaports. *See also* Color Plate Gallery.

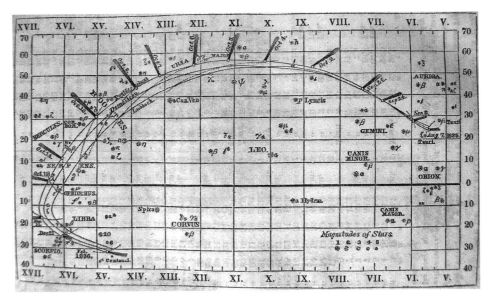

Figure 9.2. A map of the path of Halley's Comet, from *The American Almanac and Repository of Useful Knowledge for the Year 1835*. 7.3 × 13.4 cm. Note its projected path from August 7, 1835 (*upper right*) well into February 1836 (*lower left*), with reference to the background constellations and key stars.

Some almanacs were more national in scope and served as mini-encyclopedias. An example was *The American Almanac and Repository of Useful Knowledge*. This gave much useful information, such as the branches of government and governmental officials currently active in the United States and in each individual state; census and actuarial statistics; tables listing prominent American colleges, churches, newspapers, and financial institutions; information on the governments and histories of European countries; and even steamboat explosions! It also contained important calendar and celestial information. For example, the 1835 issue included tide tables, ephemerides dealing with the location of heavenly bodies in the sky; lunar and solar risings and settings in key American cities for each day of the year; eclipses and occultations; and notable events for the year: in this case, a transit of Mercury and the reappearance of Halley's comet (complete with a diagram of its projected path with respect to the background constellations)—see Figure 9.2.

9.2 JEDIDIAH MORSE

Another source of celestial information in the United States were general texts on geography. For example, several books written by Jedidiah Morse (1761–1826), the "Father of American Geography", contained such information. Born in Connecticut and educated at Yale during the American Revolution, Morse studied theology and founded a girls' school and a theological seminary. Later in life, he conducted a major study of American Indians, and one of his children, Samuel, developed the telegraph. But most of Jedidiah's life was involved with the writing of comprehensive, scholarly geographical texts, first for his students and later for the general public. Many of these were popular and went through several editions.

In *The American Universal Geography*, which was first published in 1793, Morse included an extensive section on astronomy, in which he presented different historical models of the solar system; added a table that gave the diameters, periods of revolution, and other information (e.g., conjunctions, oppositions) on the planets; discussed the nature of comets; and described several aspects of the fixed stars and their constellations. He also defined the great circles that were projected in the sky (e.g., celestial equator, meridian, ecliptic), and he illustrated these in a figure that depicted an armillary sphere, as well as a diagram of the Copernican model of the solar system. Morse also discussed the problems of determining terrestrial latitude and longitude from celestial observations, and he gave concrete examples of how to do the calculations in the text.

9.3   EARLY AMERICAN STAR MAPS

9.3.1   Bartholomew Burges

Bartholomew Burges was a Bostonian lecturer in navigation and astronomy who had fought in the Revolutionary War. In 1789 he published the first chart of the solar system produced in the new United States entitled *The Solar System Displayed*. According to Johnston (1985), it measured 50.8×61 cm in total page size and included the path of a comet that was scheduled to appear in the same year as well as the planet Herschel (now Uranus), which was discovered only eight years earlier. It was also accompanied by explanatory text.

9.3.2   Enoch Gridley

Another early celestial plate printed in the United States was made by Enoch G. Gridley, an engraver and print maker who was active in the late 1700s and early 1800s. His work appeared in several geography books that were published at that time, and this plate has been said to be included in some of Jedidiah Morse's geography books, although I have yet to find an example. Gridley also engraved a portrait of Aaron Burr ca. 1801–1802 that is in the National Portrait Gallery in Washington, D.C.

The right part of this plate (see Figure 9.3) that is labeled *The Solar System* shows the relative diameters of the seven known planets (with Uranus called "Herschel"), the apparent size of the Sun as viewed from each planet, and the proportional size and location of the planetary orbits around the Sun, all located in a circle measuring 15.5 cm in diameter. In the middle of the plate is a 6 cm diameter compass rose (top), useful for describing the directions of the Earth's winds, and a 6.3 cm diameter

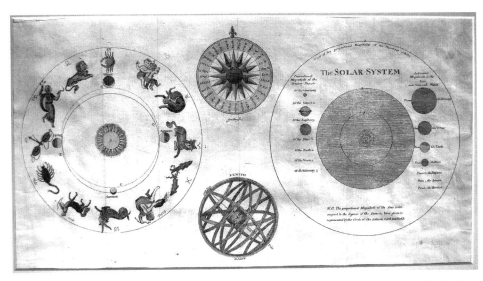

Figure 9.3. Print engraved by Enoch G. Gridley, ca. 1800. 18.4 × 35.4 cm (image size). Note from left to right images depicting the Earth's orbit surrounded by the figures and symbols of the constellations of the zodiac (14.5 cm dia.); a compass rose (*top*) and an armillary sphere (*bottom*); and information concerning the diameters and orbits of the known planets of the solar system, giving the name Herschel for what later was called the planet Uranus (15.5 cm dia.). *See also* Color Plate Gallery.

armillary sphere (bottom), which shows some of the great circles in the sky. Finally, on the left is a depiction of the Earth's orbit surrounded by the constellation figures that comprise the zodiac, all located in a circle of 14.5 cm diameter.

This plate was probably printed around the turn of the 1800s, and Gridley's name is written just below the compass rose. The area containing the images measures approximately 18.2 × 35.2 cm. Although the plate had limited scientific information, it nevertheless gave early Americans a sense of the location, size, and appearance of the planets. It also exposed them to a number of topics related to astronomy in a decorative and easy-to-digest manner.

9.3.3   William Croswell

William Croswell was an American cartographer and teacher of navigation in Boston who was born in 1760 and died in 1834. In 1810 he published a celestial print entitled *A Mercator Map of the Starry Heavens* ..., which according to both Warner (1979) and Johnston (1985) was the earliest known American-produced map of the stars and constellations. It covered the area of the sky from 66.5 degrees N dec. to 66.5 degrees S dec., used a Mercator projection with an external orientation, and according to Warner measured 41.5 × 83.5 cm. The brightest stars were identified by Bayer letters and, in some cases, names, and the path of the comet of 1807 was shown, along with the position of several planets. Two new constellations were introduced that were never again copied: Sciurus Volans (the flying squirrel) and Marmor Sculptile (the bust of Columbus) (in Figure 9.4, located at top and bottom of IV-hour line, respectively).

9.4   ELIJAH H. BURRITT

9.4.1   Life and times

In addition to geography books and free-standing prints, there were also texts devoted exclusively to astronomy, many of which included celestial maps and plates. One of the most popular of these celestial books was *The Geography of the Heavens*, which together with an accompanying *Atlas* was written in 1833 by Elijah H. Burritt. Born into a poor Connecticut farm family in 1794, Burritt showed early skill in mathematics and astronomy. With the assistance of some friends, he attended Williams College, where he supported himself as a teacher, writer, and journalist. He soon realized that there was a need for a comprehensive but inexpensive text on astronomy in the United States, which he wrote. Burritt died five years later in 1838 of yellow fever.

9.4.2   *The Geography of the Heavens* and its *Atlas*

Burritt's book was comprehensive and covered a number of topics. For example, the 5th (1845) edition included a detailed description of the stars and the mythology of the constellations; sections on variable and double stars, clusters, nebulae, meteors and comets; chapters on the known planets (including Herschel) and illustrations of their appearances in a telescope; diagrams explaining planetary phases, conjunctions, oppositions, and retrograde motions; illustrations of solar and lunar eclipses and the phases of the Moon; and discussions of complex topics, such as gravitation and the precession of the equinoxes. The book concluded with a set of mathematical problems related to celestial matters for the reader to solve (complete with solutions), and a number of tables and ephemerides concerning the Sun, the planets, the fixed stars, and other heavenly bodies in the sky. All in all, it summarized much of what was known at the time in astronomy.

The atlas (entitled *Atlas Designed to Illustrate the Geography of the Heavens*) was a complete representation of the heavens, but in pictorial form. For example, the 1856 edition included illustrations of a number of heavenly objects, such as double stars, star clusters, nebulae, and comets, and a picture of the Harvard College Observatory's 15″ refractor, which for many years was the largest telescope in the United States. There also was a double-page "plan" of the solar system, which illustrated the relative sizes and distances of the planets from the Sun, the inclination of their orbits to the plane of the ecliptic, and information concerning their satellites.

A major feature of Burritt's atlas was a set of six geocentric celestial maps that were printed on pages measuring 41.9 × 36.8 cm using a sinusoidal projection in a grid system that was oriented to the celestial equator, complete with right ascension

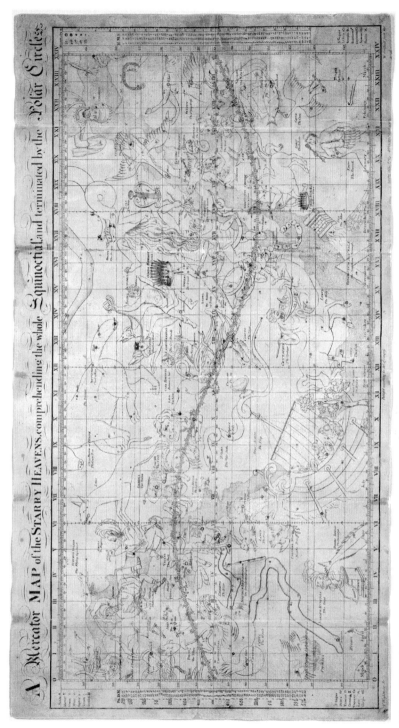

Figure 9.4. The first American star chart, by William Croswell, produced in 1810. 51.5 × 94.5 cm (page size). Note the simply drawn images and the presence of some unique American constellations (see text). Courtesy of the Norman B. Leventhal Map Center at the Boston Public Library.

and declination lines. Two maps were hemispheres centered on the north and south equatorial poles. The other four covered the area from 50 degrees N dec. to 50 degrees S dec. and were centered on the location of the Sun in the zodiac during the time of the vernal or autumnal equinoxes, or the winter or summer solstices. The maps showed all the stars that were visible to the naked eye. Accompanying the stars were 99 constellation figures, most of which depicted gods or animals from Greek mythology or contemporary instruments of science. Following a practice that commonly occurred in celestial atlases of the times, new constellations were included to please patrons or to fill in gaps in the sky. For example, in Figure 9.5 note the constellation

Figure 9.5. This figure is from a colored version of the 1835 edition of Elijah Burritt's *Atlas Designed to Illustrate the Geography of the Heavens*, showing the constellations in the Virgo/Leo region of the sky. 33.3 × 31.7 cm. Note the now-extinct constellation of Noctua the Owl, which is perched at the tip of Hydra's tail at the bottom left. *See also* Color Plate Gallery.

of Noctua the owl perched at the end of the tail of Hydra the water snake. Variously depicted in other atlases as an extinct flightless bird similar to the dodo, a rock thrush, or a mocking bird (see Section 4.4), this area of the sky subsequently lost its avian connection and was incorporated into the constellation Hydra.

In 1848 an edition of Burritt's book and atlas was published that included as an author the prominent Director of the Cincinnati Observatory, O.M. Mitchel. The atlas was quite different from other editions of Burritt, and it will be described in more detail in Section 9.6 on Mitchel.

Burritt's book and atlas borrowed heavily from English sources (especially the atlas of Alexander Jamieson—see Section 6.5.5.2), and the atlas contained errors in the positional accuracy of the stars. But, despite these imperfections, Burritt's work brought the European celestial cartographic tradition to the United States. The book went through several editions from 1833 to 1856. By 1876, there were some 300,000 copies of the atlas in circulation. The book and atlas were relatively inexpensive, and they encouraged readers to perform their own observations. They thus played an important role in popularizing astronomy in America. For example, one professional astronomer, S.W. Burnham, was drawn into the field as a result of reading them. Also, Richard H. Allen, the well-known author of *Star Names: Their Lore and Meaning*, acknowledged Burritt's work for "stimulating a boyhood interest in the skies" (Allen, 1963, p. 15).

9.5   ASTRONOMY EDUCATION IN THE SCHOOLS

9.5.1   John H. Wilkins

Burritt was not the only person educating Americans in astronomy in the period before the Civil War. Popular books on the topic were part of the general school curriculum as well. One such textbook was written by "John H. Wilkins, A.M." and was descriptively entitled *Elements of Astronomy, Illustrated with Plates, for the Use of Schools and Academies, with Questions.* Originally published in Boston in 1822, it went through several editions.

In addition to the general text, (which followed a number of written endorsements from people who liked earlier editions), the 1829 "stereotype edition" included a pull-out frontispiece showing the solar system, images of the planets as seen through a telescope, questions and problems for the student to answer, and a set of plates at the end that diagrammed eclipses, orbits, and other geometrical celestial events. This is the earliest American school textbook I have seen that gives the name Uranus rather than Herschel as the preferred name for the 7th planet.

9.5.2   John Vose

Another student textbook was *A Compendium of Astronomy*, which had the following lengthy but descriptive subtitle: *Intended to Simplify and Illustrate the Principles of the Science, and Give a Concise View of the Motions and Aspects of the Great Heavenly*

Luminaries. Adapted to the Use of Common Schools, as well as Higher Seminaries. It was written in 1836 by John Vose (1766–1840), who had been the principal of Pembroke Academy and had published other works on astronomy. The book contained text and tables dealing with a number of topics, such as a history of the planets (including Herschel); moons and asteroids of the solar system; meteors and comets; a catalog of future eclipses; information on parallax and refraction; and a discussion of the fixed stars and constellations.

At the end of Vose's book was a set of plates (which interestingly were identical to those in another book written about the same time: Tobias Ostrander's 1834 edition of *The Planetarium and Astronomical Calculator*). These plates illustrated a number of astronomical themes, such as a depiction of three historical models of the solar system; a comparison of planetary diameters and the size of the Sun as seen from each planet; the appearance of the planets and the Moon through a telescope; diagrams illustrating parallax, the tidal action of the Moon, the Earth's tilt, and eclipses; and pictures of selected constellations, star clusters, and nebulae. This little book justified its subtitle and served as a useful guide to the heavens.

9.5.3   Denison Olmsted

An important book on astronomy written for older students was entitled *An Introduction to Astronomy: Designed as a Text-book for the Use of Students in College*. This popular book was written by Denison Olmsted, Professor of Natural Philosophy and Astronomy at Yale. Born in 1791, Olmsted graduated with a degree in physics from Yale in 1813. He became the Chairman of the Chemistry Department at the University of North Carolina in 1817, but he returned to Yale in 1825, where he remained until his death in 1859. He is most noted for his work involving Leonid meteor showers, which he began in 1833, and his subsequent demonstration that meteors were cosmic, not atmospheric, in origin.

Olmsted wrote many books on the natural sciences. His *An Introduction to Astronomy* was based on an earlier book with the same main title but with the subtitle: *Designed as a Text-book for the Use of Students of Yale College*, which was published in 1839. The first stereotype version of his book with the more general "... *Students in College*" subtitle was published in 1844, and it went through many editions over the next 40+ years. As befitting its intended audience, this book was more advanced than many written for students. For example, the 1876 "third stereotype edition" (revised by E.S. Snell, Professor of Natural Philosophy at Amherst) contained mathematical formulae and information on orbital mechanics in addition to the more descriptive material and the beautiful plates at the front and back illustrating the lunar surface, comets, nebulae, and double stars.

9.5.4   Asa Smith

Perhaps the best known of the early American student textbooks on astronomy was Asa Smith's *Illustrated Astronomy* (subtitled: *Designed for the Use of the Public or*

Common Schools in the United States). Smith was the principal of Public School No. 12 in New York City. He wrote the first edition of his book in 1848, and it became popular, going through many editions. The format consisted of a series of lessons in a question-and-answer format (including some tables) on one page and a plate (white or color images on a black background) illustrating the material on the facing page. For example, the 19th (ca. 1860) edition contained 54 lessons that dealt with a variety of astronomical topics, such as astronomical history; the zodiac; the Sun and the planets; telescopic views of the Moon and depictions of its phases; diagrams of eclipses, parallax, and orbital mechanics; and illustrations of comets, double stars, and nebulae. At the end of the book was a series of sidereal maps that showed the stars and constellations that were visible at different times of the year, along with an accompanying description of their location and mythology (Figure 9.6). The stars were plotted using a stereographic projection in a geocentric coordinate system that was centered on the celestial equator. The maps were interspersed with essays on a number of special topics, such as leap years, the equation of time, zodiacal light, and mathematical problems for the reader to solve.

9.6   O.M. MITCHEL

Professional astronomers also contributed to the popularization of astronomy in early America, especially in the area of telescopic observation. A case in point was the distinguished astronomer Ormbsy MacKnight (O.M.) Mitchel, a true "Renaissance man" who was the founder and director of the Cincinnati Observatory and who could make a claim as the "Father of American Astronomy". He led a distinguished and fascinating life that bears re-telling in some detail.

9.6.1   Life and times

Mitchel was born in an area that is now called Union County, Kentucky on August 28, 1810 (although some sources list his birth year as 1809). His father died when he was about three years old, and the family subsequently moved to Lebanon, Ohio (just north of Cincinnati) to be with other relatives. As an early adolescent, he worked as a clerk in a country store, but his intelligence and motivation led to his being appointed to the Military Academy at West Point on June 23, 1825 by a special waiver, since he was below the usual age of admission. He graduated in 1829, 15th in a class of 46 that included Robert E. Lee and Joseph E. Johnston, who later became famous Confederate generals. As a 2nd Lieutenant, he was assigned to duty as an Assistant Professor of Mathematics at the Academy. In 1831 he married a military widow, Louisa Trask, and the next year he resigned his military commission to study and then practice law in the rapidly growing frontier city of Cincinnati, Ohio. He also held an appointment as Chief Engineer for the Little Miami Railroad from 1836 to 1837.

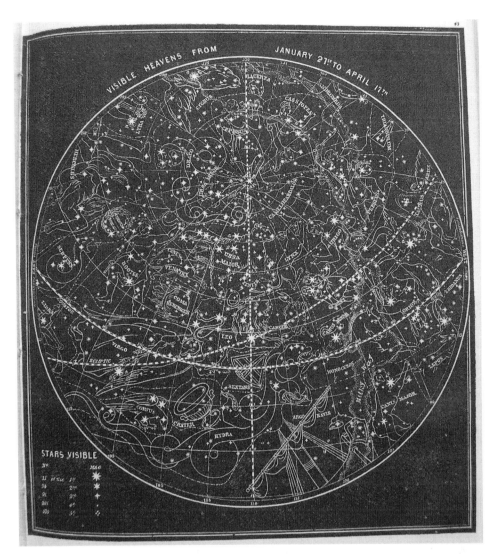

Figure 9.6. This constellation map is from the 19th edition of Asa Smith's *Illustrated Astronomy*, written around 1860. It shows the stars and constellations visible from January 21 to April 17. 24.2 × 22 cm, 21.3-cm dia. hemisphere. Note that Noctua the owl is no longer depicted at the tip of Hydra's tail at the bottom left, as in Figure 9.5.

In 1834, Mitchel was appointed Professor of Mathematics, Philosophy, and Astronomy at the new Cincinnati College. He was an active teacher, but he became frustrated at the lack of observing facilities and began advocating for the construction of a major observatory at the college. To ascertain the local public interest, he gave a series of lectures on astronomy in the spring of 1842. They were a great hit, and he announced his plans to fund his observatory privately through membership in a new

Cincinnati Astronomical Society. On June 16 of the same year, he sailed for Europe to find a suitable telescope, visit established observatories, and learn observational techniques from experienced astronomers. In Munich he located a lens of excellent quality with an effective diameter of over 11 inches, which he purchased. A four-acre site overlooking Cincinnati was subsequently donated for the observatory. The cornerstone for the pier of the new telescope was laid on November 9, 1843 by former President John Quincy Adams, who had long supported American astronomy. The completed telescope arrived in February 1845, and the observatory became operational on April 14. At the time, the observatory's 11-inch refracting telescope was the largest in the Western Hemisphere, until the installation in 1847 of the 15-inch refractor at Harvard College.

After a fire burned Cincinnati College to the ground, there were no funds left to support Mitchel as director of the observatory, and he had to depend on his diverse background in mathematics, engineering, and law to earn income. For example, during the 1840s and 1850s, he worked as a surveyor and then as a consulting engineer for the Ohio and Mississippi Railroad; was Adjutant-General of Ohio in 1847–1848; gave a series of popular public lectures on astronomy in several American cities, including Cincinnati, Boston, New York, and New Orleans; and returned to Europe to sell railroad bonds, which resulted in a substantial commission.

Mitchel also conducted an active program at his observatory. His first serious astronomical observation with the new telescope was a transit of Mercury on the Sun's disk that took place on May 8, 1845. He subsequently discovered that the bright star Antares was in fact a double star, and he initiated a program of testing Professor Otto Struve's findings on the orbital characteristics of double stars. From 1846 to 1848, he published the *Sidereal Messenger*, the nation's first astronomical journal. During 1854 to 1859, he made nearly 50,000 observations of faint stars. His program included observations of sunspots, planets, comets, and nebulae. He also used his engineering background to invent two instruments: a revolving disk chronograph that allowed an observer to make and send accurate timings of celestial events, and a fast and accurate apparatus he called a "declinometer" that consisted of a small telescope mounted on a transit instrument that could be used for measuring differences in the declination of heavenly bodies. Mitchel received much recognition for his work, including several honorary college degrees, membership in the American Philosophical Society, and fellowship in the Royal Astronomical Society.

Over time, conditions at the Cincinnati Observatory declined, in part due to the scarcity of funds and in part to the growing air pollution over Cincinnati. In addition, Mitchel wanted more time to conduct scientific observations, which his fund-raising obligations prevented him from doing. In 1852 he was asked to provide plans for the building of the new Dudley Observatory in Albany, New York. In 1860 he was appointed as Director of this facility and paid a regular salary for his work. He continued in this role until the next year, when circumstances intervened that dramatically changed his life.

This event was the Civil War. For patriotic reasons, Mitchel re-enlisted in the Union Army, putting his astronomical work on hold. On August 9, 1861 he was appointed to the rank of Brigadier General. Due to the advocacy of the citizens of

Cincinnati, he was assigned to fortify and protect the city (which was strategically important for its river harbor and manufacturing facilities). He commanded the 3rd Division of the Army of the Ohio during the campaigns of Tennessee and Northern Alabama during the winter of 1861–1862, and he was involved with the occupations of Bowling Green, Kentucky, and Nashville, Tennessee, in February 1862.

In April of that year, Mitchel became a part of the famous early Civil War raid known as the "Great Locomotive Chase", an event that in recent years has stimulated several books and a popular movie of the same name. James Andrews, a Virginian and a Union spy, proposed the mission, which involved a group of men infiltrating into the South, boarding a train as passengers near Atlanta, taking over the train, and burning bridges as it moved north in order to disrupt Confederate troop movements. Mitchel authorized Andrews' plan and allegedly helped him to enlist volunteers for the raid. Although the raid failed, with Andrews and seven of his men being hanged as spies, 19 of the 22 raiders ultimately received Medals of Honor, among the first Americans to be so decorated.

As this raid was being undertaken, Mitchel commanded a force that seized Huntsville, Alabama, and he made several forays directly into enemy territory. After being promoted to the rank of Major General, he was assigned to command the 10th Army Corps at Hilton Head, South Carolina. He arrived there on September 16 and infused new life into the men. Unfortunately, yellow fever was rampant in this area. Mitchel was stricken with this disease and subsequently died on October 30, 1862.

9.6.2   Mitchel's edition of Burritt's book and atlas

During his early years at the Cincinnati Observatory, Mitchel undertook a revision of Burritt's book and star atlas. The book listed both Burritt (who had died) and Mitchel as authors, and it was entitled *The Geography of the Heavens, and Class Book of Astronomy, Accompanied by a Celestial Atlas*. It was published in 1848, with other issuances in 1849 and 1850. It was similar in format to earlier editions, although it had no pictures of the planets as viewed through the telescope. However, each constellation was introduced with a summary of its telescopic objects (e.g., stars, nebulae, clusters), and there was an interesting description of the discovery of Neptune, including a comparison of its orbital characteristics as predicted by both Adams and Leverrier before the planet was discovered in 1846 (see Section 8.5.1). Like other editions, the constellations were keyed to the star maps in the corresponding atlas.

The atlas also was published in 1848 (with another issuance in 1849), but it only listed Mitchel's name. It was entitled: *Atlas Designed to Illustrate Mitchel's Edition of the Geography of the Heavens*. As may be inferred by the single authorship and title, it definitely had Mitchel's unique influence, and it was quite different from other editions of Burritt's atlas. There were now 24 star charts (instead of 6) showing the positions of stars down to 6th magnitude. There were also images of deep-sky objects and planets taken from observations at the Cincinnati Observatory. Unlike other Burritt editions, the stars in Mitchel's atlas were white against a black background, and the constellation boundaries were very faint (which was part of a trend in

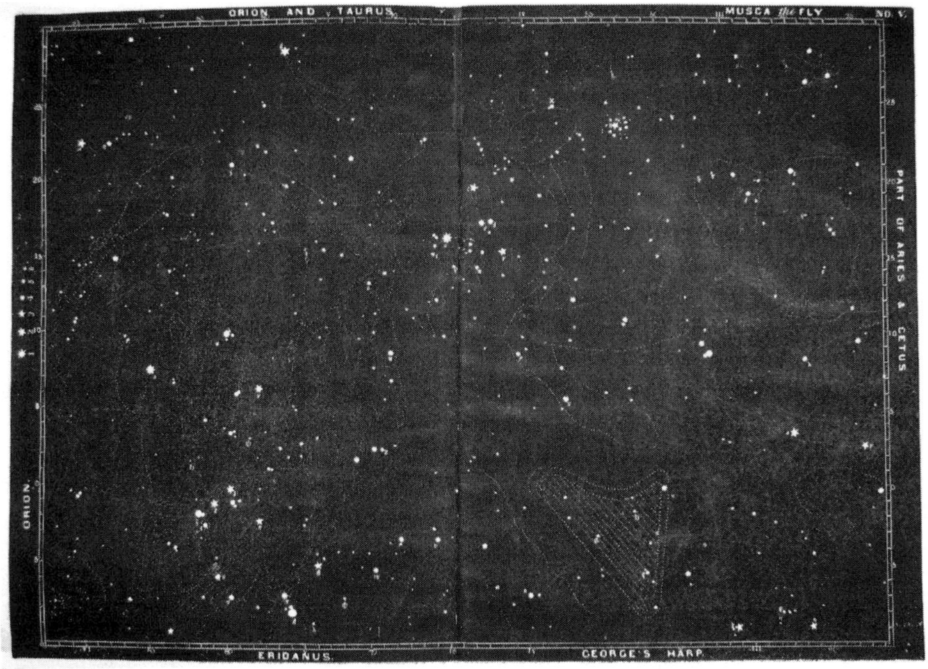

Figure 9.7. A double-page star chart showing the region involving Orion, Taurus, Musca, Eridanus, and George's Harp, from Mitchel's *Atlas Designed to Illustrate Mitchel's Edition of the Geography of the Heavens* ... 23.3 × 32.7 cm. Note the differences from Burritt's atlas (Figure 9.5), including the white stars on black background and the faint constellation outlines. *See also* Color Plate Gallery.

the 19th Century—see Chapter 10). There were also some differences in the plotting of the stars in the two atlases, and Mitchel showed Felis the cat, which was not found in Burritt. A comparison of Figure 9.7 with Figure 9.5 clearly shows some of these differences. Despite the name association with Burritt, Mitchel's publication was very much its own atlas.

9.6.3   *The Planetary and Stellar Worlds*

Mitchel wrote several other books on astronomy that were well received and went through many editions, even into the 1900s. A review of these books gives an indication of the scope of Mitchel's work and the state of American astronomy in the mid-1800s. The first book was published in 1848, and its complete title was: *The Planetary and Stellar Worlds: A Popular Exposition of the Great Discoveries and Theories of Modern Astronomy. In a Series of Ten Lectures.* As indicated by the sub-title, this book was a written exposition of ten of his lectures, preceded by a preface in which he reviewed the history of the construction of the Cincinnati Observatory. Topics considered in this book included the apparent motions of celestial bodies in the heavenly sphere; solar and lunar eclipses; a historical review of early discoveries in astronomy; the laws of motion and gravitation and how they were used to discover Uranus, Neptune, and the asteroids; the motions of comets; the scale of the universe (discussing parallax and the velocity of light); and the motions and life spans of the fixed stars. Although this book presents a heavy dose of Newtonian theory and the astrophysics that was in use at the time, Mitchel's descriptions are clear, straightforward, and oriented toward the public, with no diagrams or mathematical equations to interrupt the prose.

But perhaps the most appealing feature of this book occurs at the end, with the appearance of 17 plates and descriptions of drawings made from telescopic views of the heavens. Included are comets, star clusters, and nebulae as seen through the Cincinnati Observatory telescope by Mitchel as well as through the telescopes of Herschel, Struve, and Lord Rosse. Besides their beauty, these plates indicate the state of knowledge of nebulous objects in the mid-19th Century. For example, Plate V depicts an engraving of the Globular Cluster in Hercules, which is taken from one of Mitchel's drawings (see Figure 9.8). In his description Mitchel states: "This is doubtless one of the 'Island Universes' found in space, and of an extent not inferior to that of the Milky Way" (Mitchel, 1848, p. 330). He did not realize that this cluster of stars was in fact a part of our galaxy, thus showing the limits of understanding at that time of intra-galactic and extra-galactic deep-sky objects. But in his description of a later plate showing a drawing of the Whirlpool Nebula in Canes Venatici (Plate XVI), Mitchel correctly infers the galactic nature of the two components with this concluding statement: "The curious spiral form is exhibited with great beauty, and seems to indicate the action of some powerful and controlling law, in this remote body or cluster of universes; for such indeed it seems to be" (Mitchel, 1848, p. 336).

Editions of this book appeared in Great Britain under the title *The Orbs of Heaven, or, The Planetary and Stellar Worlds. A Popular Exposition of the Great Discoveries and Theories of Modern Astronomy*. Although the lecture/chapter sequence and text were identical in the two versions, Mitchel's last name was spelled with two "l's" in the British editions. There were other changes as well. For example, the 1853 edition published in London is fully illustrated throughout with pictures of people, places, and celestial objects, and it contains plates not found in the American version. There also is a long appendix that is abstracted from Yale astronomy professor Denison Olmsted's book *Mechanism of the Heavens* that includes sections on the telescope, observatories, the calendar, Copernicus, Galileo, Kepler, and Newton.

9.6.4 *Popular Astronomy*

Another book written by Mitchel was *Popular Astronomy: A Concise Elementary Treatise of the Sun, Planets, Satellites, and Comets*, first published in 1860. In contrast to his first book, this offering is mainly oriented towards the solar system. The text format is more traditional, with chapters that are organized by subjects, diagrams that are used to illustrate the text, and tables at the end that summarize the

Figure 9.8. Engraving from the 1848 edition of Mitchel's *The Planetary and Stellar Worlds*. 13.2 × 7.8 cm. Note the drawing of the Globular Cluster in Hercules as seen through the Cincinnati Observatory refractor.

"elements" of various members of the solar system (e.g., their size, distances from the Earth, orbital periodicities). The frontispiece shows an engraving of the Dudley Observatory. There are separate chapters on the Sun, each of the planets, the Moon, the asteroids, comets, the laws of motion and gravitation, instrumental astronomy, and the nebular hypothesis.

Engraved drawings are interspersed throughout to show the appearance through the telescope of various members of the solar system. For example, there are two depictions of Mars as seen through the Cincinnati Observatory telescope in August 1845 (see Figure 9.9). Notably, in the lower Martian image dated "Aug. 5th", there is a nearly circular white extension protruding from the left side of the south polar cap (which is located at the top of this inverted telescopic image). This likely represents the area announced by Mitchel in 1846 as remaining discrete while the rest of the polar cap receded during the Martian spring. He speculated that this bright area was a mountainous region where snow persisted longer than elsewhere owing to the higher elevation. However, recent evidence from orbiting robotic spacecraft have not imaged prominent mountains here but instead have found evidence for a south-facing scarp that retains frost longer because the area is protected from sunlight. Nevertheless, this feature has retained the name "Mountains of Mitchel" in honor of its discoverer.

9.6.5   *The Astronomy of the Bible*

The final book related to Mitchel's work was actually published posthumously in 1863. This is *The Astronomy of the Bible*, which consists of seven lectures he gave on the relationship of astronomical findings to biblical teachings. The book also contains an engraving of Mitchel and a very complete biographical section. The lectures cover a number of heady topics: hypotheses concerning the formation of the universe; ways in which astronomical findings explain the omnipotence and wisdom of God, creation, and miracles (e.g., the halting of the motion of the Sun for a day at the command of Joshua); and how the language of the Bible impacts on our thinking. Although his arguments suggest harmonious relationships between astronomical findings and biblical teachings, he stops short of making dogmatic assertions, and he seems to put his faith in the power of science to establish more congruity in the future. Certainly, Mitchel was not afraid to tackle important scientific and philosophical topics in the same confident manner that he dealt with challenging issues in his own life.

9.7   HANNAH MARY BOUVIER

Hannah Mary Bouvier was born in Philadelphia in 1811. When she was three, her family moved to western Pennsylvania, where her father pursued a newspaper and printing career. She had an active mind and received most of her education from her father. When he decided to pursue a law career, the family moved back to Philadelphia in 1823. There, she attended private school and was a very good student.

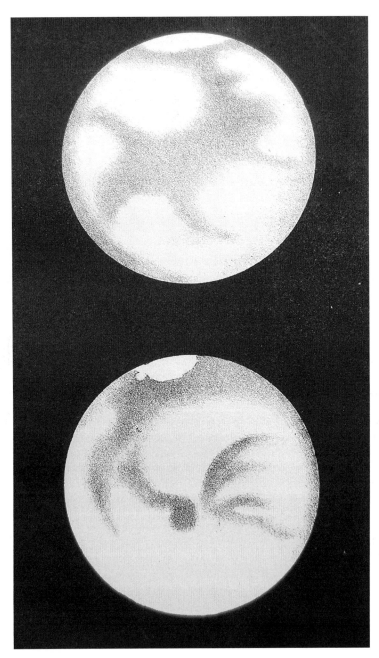

Figure 9.9. Engraving from the 1860 edition of Mitchel's *Popular Astronomy*. It depicts two drawings of Mars as seen through the Cincinnati Observatory refractor in August 1845. 14.3 × 8.1 cm. Note in the lower image the nearly circular white extension protruding from the left side of the polar cap at the top of the planet. This likely represents the "Mountains of Mitchel", named for its discoverer.

She was educated in linguistics, painting, and music, and according to Malpas (2008) she helped her lawyer father draw up legal documents. She married Robert Evans Peterson, a lawyer and publisher who was also interested in the sciences, and the couple had two children. She worked in the family publishing business. Initially stimulated by her husband, Hannah developed her own interests in the sciences, especially astronomy. In 1850 she wrote her first book, *Familiar Science*, which was published under her husband's name. It was very popular, selling some 250,000 copies (Malpas, 2008). She wrote other books as well, including popular cookbooks that are still available today. She died on September 4, 1870.

Bouvier wrote a popular astronomy book under her own name: *Familiar Astronomy, or an Introduction to the Study of the Heavens*. Published by Childs & Peterson in 1857, it was 499 pages long, had over 200 engravings, and was written in a question-and-answer format that was typical at the time (Figure 9.10). Its frontispiece has already been discussed (see Section 8.74 and Figure 8.56). A number of topics were included in this complete book, such as descriptions of the planets, information on the constellations and images of deep-sky objects within them, a review of various cosmological systems and the history of astronomy, and information on astronomical

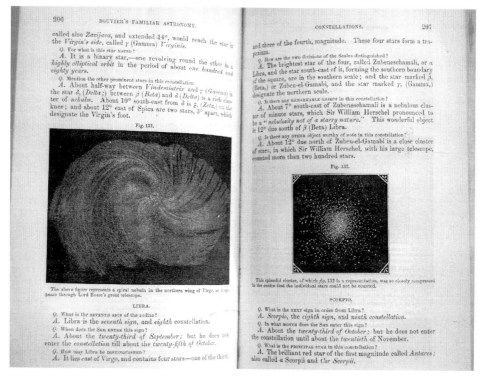

Figure 9.10. Two pages from Bouvier's *Familiar Astronomy*, published in 1857. 21.3 × 12.8 cm (page size). Note the question-and-answer format and the fine engravings of a spiral nebula (really a galaxy) in Virgo on the left and a star cluster in Libra on the right.

instruments and globes. The book was very well received, and according to Malpas it resulted in letters of endorsement from prominent American and English astronomers, such as George Airy (England's Astronomer Royal), John Herschel, and Denison Olmsted.

9.8   BIBLIOGRAPHY

Allen, R.H. (1963) *Star Names: Their Lore and Meaning*. New York: Dover Publications.

Ashbrook, J. (1984) Chapter 30. The Clarks and some of their refractors. *The Astronomical Scrapbook*. Cambridge, U.K.: Cambridge University Press, pp. 155–160.

Bell, T.E. (2011) The great telescope race. *Sky & Telescope*, June, 28–33.

Johnston, P.A. (1985) *Celestial Images—Astronomical Charts from 1500 to 1900*. Boston: Boston University Art Gallery.

Kanas, N. (2002) Mapping the solar system: Depictions from antiquarian star atlases. *Mercator's World*, **7**, 40–46.

Kanas, N. (2004) Early American contributions to celestial cartography. *Journal of the International Map Collectors' Society*, **96** (Spring), 15–26.

Kanas, N. (2005) Are celestial maps really maps? *Journal of the International Map Collectors' Society*, **101**, 19–29.

Kanas, N. (2005) Astronomer-general of early America (O. M. Mitchel). *Mercury*, November/December, 23–31.

King, H.C. (1955) *The History of the Telescope*. Cambridge, MA: Sky Publishing.

Malpas, B.D. (2008) Ambassadors to the heavens. *GardenStateLegacy.com*, Issue I, September, 1–4.

Mendillo, M., DeVorkin, D., and Berendzen, R. (1976) History of American astronomy: A chronological perspective. *Astronomy*, July, 20–63.

Mitchel, O.M. (1848) *The Planetary and Stellar Worlds: A Popular Exposition of the Great Discoveries and Theories of Modern Astronomy. In a Series of Ten Lectures*. New York: Baker & Schribner.

Mitchel, O.M. (1860) *Popular Astronomy. A Concise Elementary Treatise on the Sun, Planets, Satellites, and Comets*. New York: Phinney, Blakeman & Mason.

Mitchel, O.M. (1863) *The Astronomy of the Bible*. New York: Blakeman & Mason.

Mitchell, O.M. (1853) *The Orbs of Heaven, or, The Planetary and Stellar Worlds. A Popular Exposition of the Great Discoveries and Theories of Modern Astronomy*. London: Ingram, Cooke, & Co.

Pittenger, W. and Bogle, J.G. (1999) *Daring and Suffering: A History of the Andrews Railroad Raid*, 3rd edn (1887). Nashville, TN: Cumberland House Publishing.

Warner, D.J. (1979) *The Sky Explored: Celestial Cartography 1500–1800*. Amsterdam: Theatrum Orbis Terrarum.

Wesemael, F., Del Duchetto, K., and Racine, R. (2011) From J. Winthrop, Jr, to E.E. Barnard: The arduous path to the first sighting of Amalthea. *Journal for the History of Astronomy*, **42**, 125–139.

10

The transition to non-pictorial star maps

It is rare to find constellation images in modern-day star maps. Sometimes lines are drawn connecting major stars to produce stick-figure representations of the constellations. In other cases, there are no representations, just boundary lines drawn around the area of the sky assigned to the constellations. These maps typically show stars and deep-sky objects beyond naked eye visibility whose positions may have been computer-generated and which are located in a coordinate system that is current and very precise. What made previous maps appealing were their anthropomorphic and historic qualities (which in past times were relevant to their mapping function), but such features are now seen as so much unnecessary superficial clutter.

In this chapter we will trace the demise of the pictorial star atlas and its replacement by its modern non-pictorial counterpart. The transition occurred during the 1800s, and in the 1900s advances in technology created even more accurate representations of the heavens. Today's star maps are the result of this evolution.

10.1 THE 1800s: A CENTURY OF TRANSITION

10.1.1 Factors pushing for change

During the 19th Century a number of factors contributed to the transition from pictorial to non-pictorial star maps. First, the construction of increasingly larger and more accurate telescopes and mountings allowed astronomers to see increasingly fainter and more distant stars and deep-sky objects. This created a two-edged sword in terms of celestial cartography: star maps could be produced that included objects well beyond naked eye resolution, but this increased the need for even more accurate star maps to allow astronomers to correctly show these newly discovered objects in the heavens. Second, other technological refinements, such as more precisely calibrated micrometers and astronomical instrument circles, allowed astronomers to position the stars in the heavens with unprecedented accuracy. However, the lines and colors of constellation figures sometimes cluttered up the details and obscured the positioning of faint stars. Third, the development of astrophotography in the mid-19th Century greatly extended the reach of the observer into space, allowing for a quantum jump in ability to see and position faint objects. In fact, some atlases were simply catalogs of photographs taken with powerful telescopes, which seemed to speak for themselves in terms of locating deep-sky objects in relationship to the background stars. Finally, as astronomy became ever more precise and scientific, the presence of mythological figures on star maps became embarrassing to some astronomers, reflecting pre-scientific notions and astrological symbolism. Such images were just not in keeping with the astronomical *Weltanschauung* of the times.

The impact of these factors was slow in coming, and at the beginning of the 1800s pictorial star atlases were still being produced. We have already discussed Bode's famous 1801 *Uranographia* in Section 6.5.4 as the epitome of classical pictorial star atlases, containing more constellations and stars than any of its predecessors. Elsewhere in Chapter 6, we discussed other pictorial celestial atlases produced in Europe in the early 1800s, including those by Alexander Jamieson, Kornelius Reissig, the Society for the Diffusion of Useful Knowledge, M.C.G. Riedig, and the "lady" of *Urania's Mirror*. In addition, in Chapter 9 we considered some early American star maps with constellation figures by William Croswell and Elijah H. Burritt. However, early in the 19th Century the seeds were being planted to reduce the impact of constellation images in star maps, as we shall now examine.

10.1.2 Star maps with subdued constellation images

10.1.2.1 *Wollaston's Portraiture of the Heavens*

Francis Wollaston was born on November 23, 1731 in London. Like many in his educated family, he studied at Sussex College, Cambridge, entering in 1748 and graduating in 1754. He was ordained a priest in 1755, and the next year he began preaching at St. Anne's in Soho. He married in 1758, and he and his wife had 19 children. He subsequently held a number of other religious positions, finally settling in 1769 at the rectory in Chislehurst in Kent, where he remained for the rest of his life. He died on October 31, 1815. During his life, he raised issues concerning the philosophy of religion, some of which were controversial.

Wollaston also had an interest in observational astronomy, for which he was elected a Fellow of the Royal Society in 1769. He also published widely in this area. According to Warner (1979), in an article for the 1784 *Philosophical Transactions*, he included a *Map of 107 stars, besides those marked by Bayer, in the constellation of Corona Borealis* ... that illustrated his premise that astronomers should use telescopes with eyepiece crosswires to more accurately re-chart all of the stars. In 1800 he produced seven celestial maps that could be joined together to form a large planisphere 77 cm in diameter centered on the north equatorial pole down to 65 degrees N dec. using a polar equidistant projection with external orientation; it was labeled *The Northern Circumpolar Region Laid Down from Observation*, ... This was accompanied by a catalog of 260 circumpolar stars that was published in his *Fasciculus Astronomicus*. He later produced two celestial hemispheres labeled *Northern* (or *Southern*) *Celestial Hemisphere on the Plane of the Equator* that were published in 1809.

In 1811, Wollaston published an atlas of 10 celestial maps of the sky entitled *A Portraiture of the Heavens, as they Appear to the Naked Eye*. According to Ashworth (1997), this was the first major star atlas to appear after Bode's *Uranographia*. In contrast to Bode's great work, Wollaston's maps were much simpler and cleaner. The number of constellations was reduced, and the figures were faint and produced only in outline form. Instead, the emphasis was more on the stars and their positions, and there was additional information on variable stars and changes in the fixed stars.

10.1.2.2 Stieler's Hand-Atlas

This popular book was the leading German world atlas from 1871 through the middle of the 1900s. It was named for Adolf Stieler (1775–1836), a German cartographer who worked at the Justus Perthes Geographical Institute in Gotha, where it was published. Altogether, *Stieler's Hand-Atlas* went through 10 editions from 1816 to 1944. Each edition was issued in parts; for example, the 1st edition, which contained 50 maps, was begun in 1816 and completed in 1823. Due to Stieler's death, other editors were involved with subsequent editions. According to *Wikipedia.org*, the dates for subsequent editions were: 2nd (1845–1847), 3rd (1852–1854), 4th (1862–1864), 5th (1866–1868), 6th (1871–1875), 7th (1879–1882), 8th (1888–1891), 9th (1901–1905), and 10th (1920–1925). Beginning with the 6th edition of 90 maps, the fame of the series was assured due to its high scientific level and the quality of its maps. The 9th edition of 100 maps was the first to be completed on cylindrical printing machines using lithographic techniques, which dropped its price and made it more accessible to the public. Editions of this popular atlas also were published in English, French, Italian, and Spanish.

Because of the high standards set by this atlas, its long life, and the fact that it included celestial as well as terrestrial maps, it is instructive to compare different versions to look for changes in constellation presentation. Figure 10.1 shows the map of the northern celestial region (listing 1875 as its printing date) from the 6th edition. The internal image dimensions are 27.1 × 33.6 cm. Note the presence of a number of constellations that are faintly depicted as a background to the numerous stars. Dotted lines demarcate the constellation boundaries, but they are also faint. Stellar magnitudes are indicated by slight variations in the size and shape of the star images. The overall look is cluttered.

In contrast, compare this map with that in Figure 10.2, which is from the 9th edition (listing 1905 as its printing date). This plate is larger, measuring 32.3 × 39.4 cm. Here, only the zodiac constellations are shown, appearing more obvious than in the earlier example. However, the dotted lines showing the constellation boundaries are more pronounced. Similarly, stellar magnitude differences are

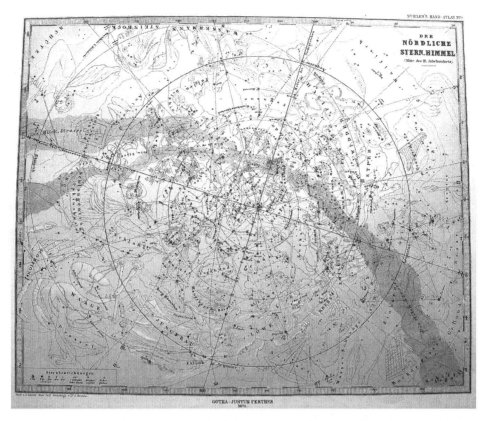

Figure 10.1. A map of the northern celestial region printed in 1875 from the 6th edition of *Stieler's Hand-Atlas.* 27.1 × 33.6 cm. Note the presence of a number of faint constellation images, constellation boundaries that are not very pronounced, stellar magnitudes indicated by slight variations in the size and shape of the star images, and an overall cluttered look. *See also* Color Plate Gallery.

more differentiated, with much larger circles indicating the brighter stars (e.g., compare the large black circle of Sirius near the right edge of this plate with its less obvious counterpart in Figure 10.1). The overall look in Figure 10.2 is less cluttered and more obviously demarcates the extent of the constellations and the magnitudes of the stars. Also, nebulae and novae are more clearly indicated (note the large hexagonal symbol for a nova near the left margin just above the Scorpion). Although constellation images have not completely disappeared, indicating their presence is clearly not the intent of this plate.

10.1.2.3   *Von Littrow's Atlas des Gestirnten Himmels*

Austrian astronomer Joseph Johann von Littrow was born in Horsovsky Tyn (now Bischofteinitz), Bohemia on March 13, 1781. He entered Charles University

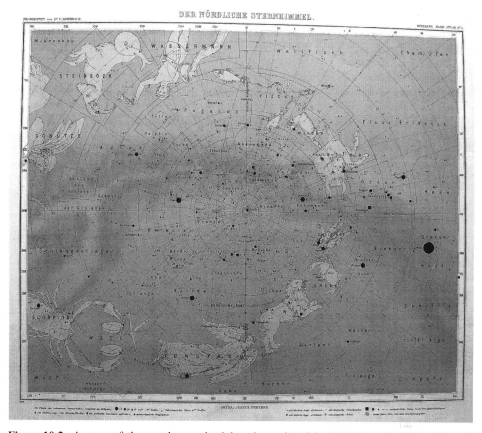

Figure 10.2. A map of the northern celestial region printed in 1905 from the 9th edition of *Stieler's Hand-Atlas*. 32.3 × 39.4 cm. Note, in contrast to Figure 10.1, only the zodiac constellations are imaged, the constellation boundaries are more pronounced, stellar magnitudes are more differentiated, and nebulae and novae are indicated by hexagonal symbols. *See also* Color Plate Gallery.

in 1799 and graduated with degrees in law and theology in 1802, although he also studied astronomy. For a time, he worked as a private tutor for a count in Silesia. In 1807 he was appointed Professor of Astronomy at Cracow University. He established an observatory at Kazan University in 1810, and in 1816 he became co-director of the observatory at Ofen (Buda). In 1819 he was appointed Professor of Astronomy at the University of Vienna and became the director of its observatory, positions he held until his death. He is known for his Littrow map projection and for several professional books and papers, including *Die Wunder des Himmels* in 1834, which was popular and went through some 10 editions well into the 1900s. He died in Vienna on November 30, 1840. One of his children, Karl Ludwig, also became an astronomer and succeeded him as the director of the Vienna Observatory.

In 1839, von Littrow published a celestial atlas entitled *Atlas des Gestirnten Himmels: für Freunde der Astronomie* (*Atlas of the Starry Heavens: for Friends of Astronomy*). It consisted of two double-page hemispheres centered on the north and south equatorial poles, 12 double-page geocentric constellation plates of the sky oriented to the celestial equator, and additional images at the end of star alignments, double stars and clusters, and deep-sky objects. In the constellation plates (Figure 10.3), the naked eye stars were emphasized, and the constellation images were faint and simply drawn. There were no lines connecting the major stars in these plates, although the star alignment plate (XXIX/XXX) had such lines between stars, both within and between constellations. This atlas proved to be quite popular and went through several editions.

10.1.3   Star maps with connecting-line constellation images

10.1.3.1   The Atlas Céleste of Dien and Flammarion

Charles Dien was a French astronomer and globe maker who was born in 1809 and died in 1870. He also produced celestial atlases that were precise and popular. In 1831 he published his *Uranographia* in Paris, which was intended for mounting on the wall. According to Ashworth (1997), it was composed of two polar maps at the top and a wide equatorial map at the bottom. Notably, there were no traditional constellation figures shown on any of the maps. Rather, the principal stars of a constellation were joined by lines to produce geometrical shapes. In a sense, these shapes served to identify the constellation without all the clutter, and they were an alternative to showing only the stars. This connecting-line format was to became a popular compromise between elaborate and no constellation images, as we shall see in this section, and it exists today on many "star-finders" and celestial maps for amateur astronomers.

Nicolas Camille Flammarion was born on February 26, 1842 at Montigny-le-Roi, France. He became interested in astronomy as a student, and at the early age of 16 he wrote a book on cosmology. He also assisted Leverrier (credited with discovering Neptune) at the Paris Observatory. After working at the Bureau of Longitudes from 1862 to 1867, Flammarion returned to the Paris Observatory and became involved with its double-star program. He also observed the Moon and Mars. Some of his ideas were eccentric. For example, he believed that there was life on the Earth's Moon and on many planets throughout the universe, and that there were canals and vegetation on Mars. He was a prolific writer and became famous through his books, which included *La Pluralité des Mondes Habités* in 1862, *Les Terres du Ciel* in 1884, and *La Planète Mars* in two volumes in 1892/1909. His book *L'astronomie Populaire* (*Popular Astronomy*), published in 1880, sold over 100,000 copies and was translated into English in 1894 by J. Ellard Gore (Figures 1.3, 2.3, 2.9, 3.16, and 8.29). In 1877, Flammarion founded the Astronomical Society of France, and he published an illustrated edition of Messier's famous catalog. He was also a supporter of amateur astronomy. With funds given to him by an admirer, he set up a private observatory in 1882 at Juvisy, from where he

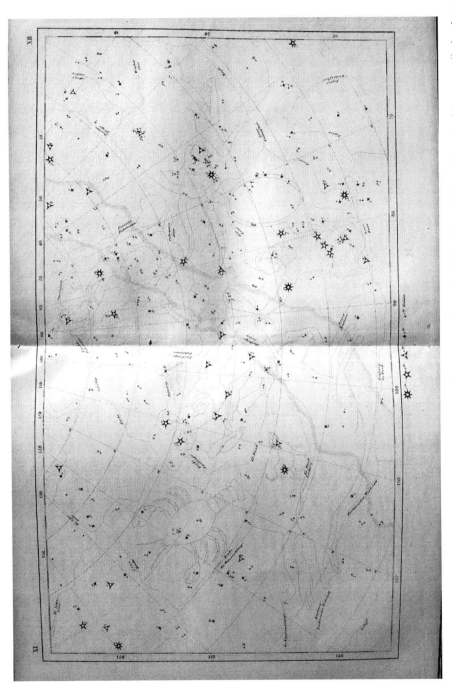

Figure 10.3. Maps XI/XII from Littrow's *Atlas des Gestirnten Himmels*, published in 1839. 18.6 × 32.9 cm. Note the faint constellation images and the prominent stars in the belt of Orion at the bottom to the right of center.

continued his observations of double stars, the Moon, and Mars. He also studied psychic phenomena and wrote science fiction. In 1922 he received the Legion of Honor for his work in astronomy. He died on June 3, 1925.

In 1864 Dien produced a comprehensive, high-quality celestial atlas entitled *Atlas Céleste*, showing the location of over 100,000 stars and nebulae. He asked Flammarion to collaborate with him, and both names were listed on the 1865 and subsequent editions of this successful book, even after Dien's death. The atlas continued to be produced well into the 1900s and had a number of plates of constellations (all credited to Dien) and other items of astronomical interest. Both traditional and contemporary features were included. For example, in the 1904 11th edition, the first two plates (A and B) were of the northern and southern skies and measured 34.9 cm and 34.6 cm in dia, respectively (Figure 10.4). They were centered on the equatorial poles using a geocentric orientation. Simple constellation figures were drawn in, and there was no coordinate system, just hours of right ascension on the rim of the hemispheres. Interestingly, some key asterisms were indicated by connecting lines and were superimposed on the figures (e.g., the Big Dipper, the Little Dipper, the Square of Pegasus extending into Andromeda).

Plates 1–23 were of specific geocentric constellations and were more contemporary. They measured about 23.8 × 33.9 cm. There were no constellation figures, although there were connecting lines for the featured constellations (Figure 10.5). In addition, there were loose and curvy constellation boundaries, which astronomer George Lovi (1987) has pointed out varied from atlas to atlas until they were standardized in 1930 (see Section 10.2.2). The stars for the principal constellations were shown down to the 9th magnitude, and many were given Bayer letters. Fuzzy patches indicated nebulae. The coordinate system was detailed, demarcating each degree of declination and every four minutes of right ascension.

Plate 24 was a south celestial hemisphere showing stars from the catalog of Brisbane, with constellation boundaries indicated but no images or connecting lines. The remaining five plates were attributed to Flammarion and showed celestial hemispheres indicating the proper motions of stars (using arrow vectors), the distribution of multiple stars in the sky, the proper motions of multiple stars, the orbits of double and multiple stars, and white-on-black images of important nebulae. All in all, this was a precise atlas for serious users.

10.1.3.2   Proctor's Half-Hours with the Stars

Richard Anthony Proctor was born in Chelsea, England on March 23, 1837. He was the youngest of four children whose father, a solicitor, died when he was 13. Young Richard was schooled at home and was a great reader. He attended King's College in London and then received a scholarship to attend St. John's College, Cambridge, where he studied mathematics. His mother died during his second year, and he fell in love and married an Irish girl before he graduated in 1860. He decided to become a writer and focus on astronomy, which was of great interest to him.

In 1865 he published his first magazine article (on double stars) and his first book (on Saturn and its ring system). Astronomers received his book favorably, since he

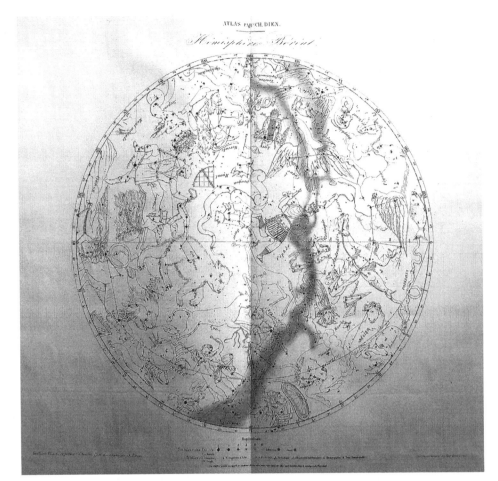

Figure 10.4. Chart A showing the northern celestial hemisphere, from the 11th edition of *Atlas Céleste* by Dien and Flammarion, published in 1904. 34.9-cm dia. hemisphere. Note that the images of the constellations of the northern celestial hemisphere are simply drawn in without boundaries or a coordinate system. Note also the presence of key asterisms (e.g., the Big Dipper, the Little Dipper, The Square of Pegasus) indicated by connecting lines that are superimposed on the constellations.

wrote with clarity, enthusiasm, and accuracy, but it was a flop with the general public. Nevertheless, he persisted with his scientific writing, despite having some financial difficulties, and gradually he began to make enough money to support his family as his reputation grew. He was very adept at projection and map drawing, and he participated in the charting of the stars for Argelander's great catalog (see Section 10.1.4.1). He also was keen on observing. In 1868 he produced a book entitled *Half-Hours with the Telescope*, which went through over 20 editions and was a primer for the observing amateur. In 1870 he published a geocentric atlas entitled *A Star Atlas*,

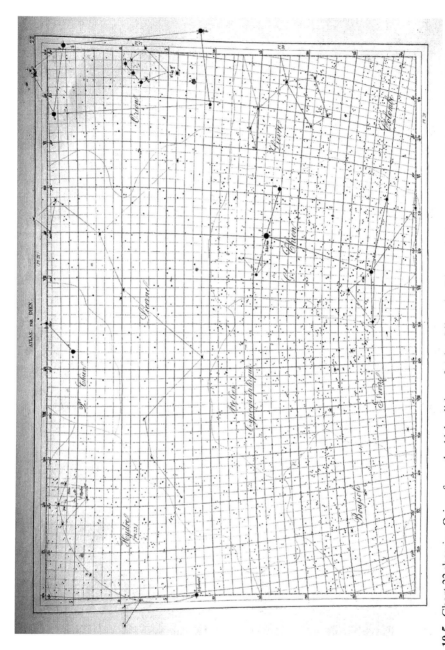

Figure 10.5. Chart 22 showing Orion, from the 11th edition of *Atlas Céleste* by Dien and Flammarion, published in 1904. 23.8 × 33.9 cm. Note the absence of constellation images, although there were connecting lines for the featured constellations. Note also the presence of loose and curving constellation boundaries and a detailed coordinate system.

in which the stars were labeled but unencumbered by constellation images or connecting lines. Stylistically, it resembled Norton's atlas (see Section 10.2.3.1). Proctor ultimately published some 57 books, drawing many of the figures and maps himself. He soon became well-known in England as one of the leading popularizers of astronomy in the Victorian period.

In 1866 he became a Fellow of the Royal Astronomical Society, and in 1872 he was elected as an honorary secretary, a position he left in 1873 when he departed for America to do a lecture tour. His subsequent original astronomical work included writing papers for the Society on the transits of Venus of 1874 and 1882; accurately determining the rotation of Mars by examining drawings of the planet going back to 1666; and describing a mechanism to explain the behavior of solar prominences. After his wife became ill in 1879 and subsequently died, he married again, this time to an American woman from St. Joseph, Missouri, in 1881. That same year, he founded the science magazine *Knowledge*, which he edited for several years. His family relocated to St. Joseph in 1884, although Proctor continued to travel and lecture abroad. After the death of two of his children in 1887, the family moved to Florida. While transiting through New York on a trip back to England, Proctor became ill, possibly of yellow fever, and he died on September 12, 1888. At the time of his death, he had been working on a textbook of astronomy, which he was never able to finish.

In 1869, Proctor published in London a book entitled *Half-Hours with the Stars*. This simple guide to observing was very popular, and it was still being printed after 1920. A special edition for American audiences (especially students) was also published in New York. The book consisted of 12 maps, each depicting the monthly night sky as seen from northern latitudes in a 19.2 cm diameter hemisphere centered on the zenith using a geocentric orientation. Accompanying each plate was an explanatory text describing the constellations. The major stars of the constellations were linked by connecting dashed lines and were labeled by Bayer letters. In some cases, prominent deep-sky objects were indicated, as well as the Milky Way. The stars were in white and were displayed against a black background, as they might appear in the night sky. In Figure 10.6, note Orion with its great nebula in the lower right, Leo in the middle left, and Ursa Major above Leo with its "pointer stars" indicating the way to the Pole Star. Note also that this image represents the sky at a number of different times and dates; i.e., 11 PM on February 6, 9:30 PM on March 1, and 9 PM on March 8. This convention is still used today in monthly popular astronomy magazines, as well as in portable celestial planispheres used by amateur astronomers and other people who are simply trying to orient themselves to the night sky (see Section 8.2.3). Since this was the major purpose for this popular atlas, there was no need for a coordinate system or for showing stars below naked eye visibility.

10.1.3.3 *Ball's Atlas of Astronomy*

Robert Stawell Ball was born on July 1, 1840 in Dublin, the second of seven children. His father, who was a naturalist and civil servant, died in 1857, leaving the family in some financial difficulties. Robert was given a scholarship to Trinity

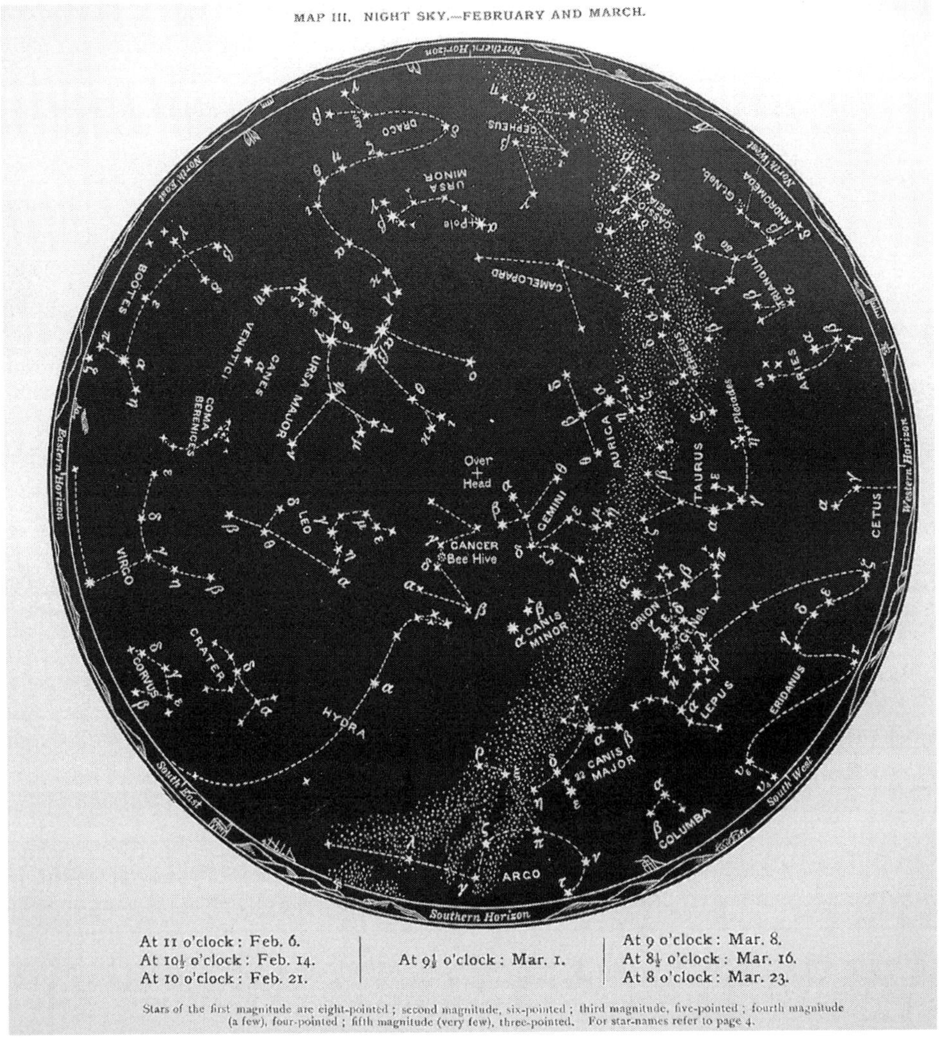

Figure 10.6. Map III showing Orion, from the 1887 edition of Proctor's *Half-Hours with the Stars*. 19.2 cm dia. Note that instead of images, the stars of the constellations are connected by simple dashed lines and that this map represents the appearance of the sky at a number of different times and dates, shown at the bottom.

College, Dublin, which he attended from 1857 to 1865. While there, he distinguished himself in mathematics and science, winning gold medals and prizes in these areas. Upon graduating, he spent two years tutoring the three sons of the 3rd Lord Rosse at Birr Castle. This happily gave him access to the 183 cm (72-inch) telescope there, which at the time was the largest reflector in the world. With this instrument, he discovered six nebulae and began writing papers in astronomy. In 1867, Ball was offered the position of Chair of Applied Mathematics and Mechanics at the Royal College of Science in Dublin. He married in 1868, ultimately having six children. He was elected a Fellow of the Royal Society in 1873 for his mechanical theories involving screws. In 1874 he was appointed Astronomer Royal of Ireland and Chair of the Astronomy Department at Trinity College, Dublin, positions he held for 18 years. Although he conducted research in astronomy, he was better known as a popular lecturer in this field, both in Great Britain and abroad in the United States and Canada. He wrote a number of books, including *The Story of the Heavens* in 1886, which went through several editions (see Figures 3.3 and 3.5). Also in 1886, he was knighted, and in subsequent years he served as president of many professional societies. In 1892 he was appointed Professor of Astronomy and Geometry at Cambridge, and in the following year he became the Director of the Cambridge Observatory. He died on November 25, 1913.

In 1892, Ball published a celestial atlas entitled *Atlas of Astronomy*. It contained 72 plates that used a detailed coordinate system and showed the stars in a geocentric orientation. There were no constellation figures, but the principal stars had Bayer letters and were connected by lines. The constellations were further demarcated by faint, curving boundary lines.

10.1.4   Star maps with no constellation images

10.1.4.1   Argelander's Bonner Durchmusterung

Friedrich Wilhelm August Argelander was born on March 22, 1799, in Memel, East Prussia. In 1817 he matriculated at the University of Königsberg, where he became interested in astronomy through his contact with Friedrich Wilhelm Bessel, who was a professor and directed the university's observatory. Argelander became Bessel's assistant in 1820 and received his Ph.D. in 1822. In 1823 he took a position at the Observatory of the University of Abo in Finland, where he subsequently became Director and Professor of Astronomy. Following a fire, the university was moved to Helsingfors (Helsinki) in 1832, and Argelander became Director of the Helsingfors Observatory. In 1837 he was appointed Professor of Astronomy at the University of Bonn and began to work on plans for an observatory there, which was inaugurated in 1845. In 1863 he received the Royal Astronomical Society Gold Medal. He died in Bonn on February 17, 1875. During his lifetime, he observed and wrote about a number of astronomical phenomena, including comets, the motion of the Sun and other stars in space, and aurorae.

Argelander also was interested in measuring the position of stars, which resulted in his publication in 1843 of *Uranometria Nova*. It was composed of 17 charts and an accompanying catalog of naked eye stars down to magnitude 6. The magnitudes were determined by a method Argelander developed for estimating the brightness of stars relative to each other. The charts still included constellation figures, which were patterned after those of Bayer, but they were subdued and de-emphasized in comparison with the stars. Although the charts were uncluttered with extraneous information, what was presented in terms of stellar positioning and magnitudes was precise and accurate. Although this atlas was a transitional one in terms of the presentation of the constellations, it was of high quality for the times in terms of positional astronomy.

Argelander also published a series of volumes entitled *Astronomische Beobachtungen auf der Sternwarte ... zu Bonn* (*Astronomical Observations from the Observatory at Bonn*) from 1846 to 1869. These contained tables of stellar positions in a number of northern and southern areas of the sky (Figure 10.7). As described by astronomer Joseph Ashbrook (1984b), Argelander's technique was to use a telescope whose eyepiece contained a vertical line that was divided into gradations that measured minutes of declination. Using this reticle, an observer would wait until the star being measured transited through the eyepiece and crossed the vertical line, and the exact time of this occurrence would be recorded. From this, a measure of the star's right ascension could be determined. In addition, the star's declination could be measured by the graduated vertical line. Argelander's measurements were quite precise, and this technique set the stage for his most important publication.

This was the *Bonner Durchmusterung*, which consisted of a 3-volume catalog of 324,198 stars down to 9th magnitude (and sometimes beyond) and a corresponding atlas of 40 plates (Figure 10.8). It was compiled from visual observations made through a 78 mm (3.1 inch) refracting telescope from 1852 to 1859 by Argelander and his assistants, E. Schoenfeld and A. Krueger. Publication of both the catalog and atlas was completed by 1863, and the total work covered the sky from the north equatorial pole to −2 degrees dec. (i.e., 2 degrees S dec.). By 1886, a version was published by Schoenfeld that extended the southern reach to −23 degrees dec. (i.e., 23 degrees S dec.). This was called the *Südliche Bonner Durchmusterung* and added 133,659 stars in 24 plates (Figure 10.9). By 1932, another version was produced from observations made in Cordoba, Argentina, that reached to the south equatorial pole and added 613,953 stars. This was the *Cordoba Durchmusterung*. Notably, the charts were devoid of constellation images. In addition, none of the stars were labeled or identified in any way except for their magnitudes (indicated by the size of the stellar image). For information on the plotted stars, the reader needed to consult the accompanying catalog, where they were listed by one-degree declination areas. In fact, the "BD" listings are still used today as a means of identifying stars down to 9th magnitude.

10.1.4.2 Dreyer's New General Catalogue

Although not accompanied by a dedicated star map, mention should be made of the *New General Catalogue*, since it is widely used today for cataloguing deep-sky objects and is familiar to many readers who are amateur astronomers. It was compiled by John Louis Emil Dreyer, who was of British descent but was born in Copenhagen on February 13, 1852. As a young student, he showed great academic talent, especially in physics, mathematics, and history. In 1874 he became the assistant to the 4th Lord Rosse at Birr Castle and its 183 cm (72-inch) telescope. Dreyer used this telescope to

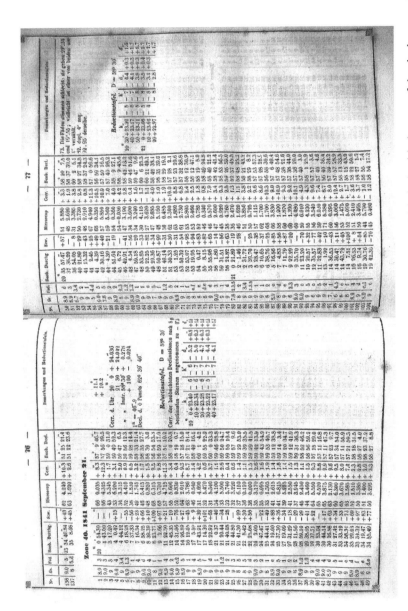

Figure 10.7. A table of star positions from Volume I of Argelander's *Beobachtungen*, which dealt with the area of the sky between 45 and 80 degrees N dec. (or +45 and +80 degrees dec.) and was published in 1846. 27 × 20.5 cm (page size). Note the entry for Zone 40, which was observed on September 21, 1841 and elsewhere is identified as covering the area of the sky from +57 to +60 degrees dec. and 20 hrs 1 min to 21 hrs 35 min right ascension. Information in the columns include the star number (col. 1), magnitude (col. 2), right ascension in hours/minutes/ seconds based on the time of transit as observed in the reticle of the telescope's eyepiece (col. 4), and the declination in degrees/minutes/seconds (col. 8).

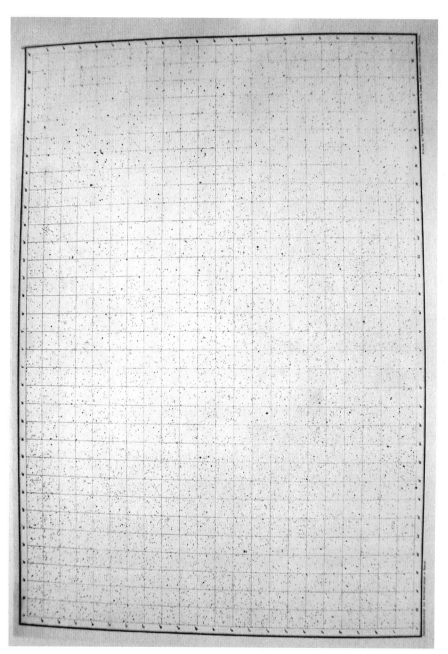

Figure 10.8. Photo of a lithograph published in 1898 from Argelander's *Bonner Durchmusterung* (BD), originally published in 1863. 15.1 × 21.8 cm. Note the lack of constellations and the stars comprising the belt of Orion at the bottom to left of center.

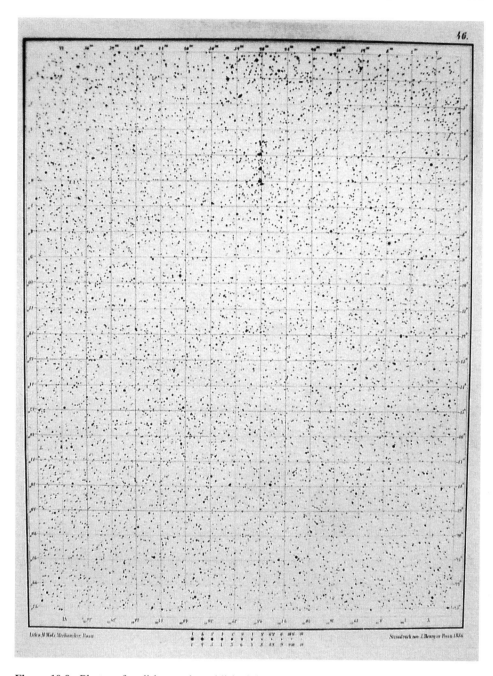

Figure 10.9. Photo of a lithograph published in 1898 from the *Suedliche Bonner Durch-musterung*, originally published in 1886 by Argelander's students as a southern supplement to the BD in Figure 10.8. 15.1 × 11.7 cm. Note the stars comprising part of the belt and the scabbard of Orion at the top center.

conduct a comprehensive survey of deep-sky objects (i.e., star clusters, nebulae, and galaxies). From 1878 to 1882, he was the assistant at Dunsink Observatory in Dublin, but in 1882 he was appointed Director of Armagh Observatory in Northern Ireland, where he remained until 1916. While there, he compiled a star catalog entitled *The Second Armagh Catalogue of Stars*. In 1890 he wrote an important biography in English of Tycho Brahe, followed by a 15-volume Latin compilation of Tycho's complete works. In 1906 he published his classic history of early astronomy entitled *The History of the Planetary System from Thales to Kepler* (which is still available under the title *A History of Astronomy from Thales to Kepler*—see Bibliography). He died at Oxford on September 14, 1926.

In 1888, Dreyer published in the *Memoirs of the Royal Astronomical Society* his important catalog of deep-sky objects with the complete and informative title: *A New General Catalogue of Nebulae and Clusters of Stars, being the Catalogue of the late Sir John F.W. Herschel, Bart., revised, corrected, and enlarged.* Now abbreviated "NGC", it contained 7,840 deep-sky objects that are seen from the Northern Hemisphere. It was an upgrade to the catalog published in 1864 by John Herschel (1792–1871) (see Section 4.3.2.6), which also contained a number of observations of deep-sky objects made by his father, William Herschel (1738–1822), the discoverer of Uranus (see Section 8.5.1). Additional objects were added by Dreyer in two supplementary *Index Catalogues*, or IC, published in 1895 and 1908, respectively. Together, they added 5,386 non-stellar objects to the NGC. As a result of these catalogs, the vast majority of deep-sky objects today have been given an NGC or IC number (even those that appear in Messier's catalog—see Section 8.5.4.3). The most recent version of the NGC/IC was generated in January 6, 2006 and lists the numbers and positions of 14,001 entries, along with their names (if any) and other data pertinent to these objects. It contains all of the objects in Dreyer's lists plus new objects discovered since then, including many in the Southern Hemisphere.

10.1.4.3   *Other star maps with no constellation images*

Winslow Upton was a Professor of Astronomy and the first Director of the Ladd Observatory at Brown University. In 1893, two years after its opening, he instituted a regular program of transit observations and time keeping at the observatory. In 1896, Upton published a celestial atlas, simply called *Star Atlas*. The plates had no constellation images or connecting lines, but the constellations were demarcated by curving, dashed lines. This atlas was popular and widely used in the United States, although astronomer George Lovi (1987) believes its plotting accuracy was flawed.

One of the most influential atlases to be published with no constellation images was *Norton's Star Atlas*, but since this was issued just after the turn of the century, it will be discussed in Section 10.2.3.1. Another group of star maps typically lacking constellation images were those taken from photographic plates, and we will now turn to these.

10.1.5   Astrophotography and its influence on star maps

10.1.5.1   *Early impact*

People began experimenting with photography early in the 19th Century, especially after a proper fixing agent was found in the late 1830s. In fact, some of the earliest pictures were crude daguerreotypes of the Moon taken by Louis Daguerre in 1838 and J.W. Draper in 1840. In 1842, Edmond Becquerel (1820–1891) used a daguerreotype plate to record the spectrum of the Sun. Astrophotography *per se* began in 1850, when W.C. Bond (1789–1859) at Harvard College obtained the daguerreotypes of the Moon and the star Vega through its observatory's 38 cm refracting telescope. Later in the 1850s, Warren de la Rue (1815–1889) in England used a collodion process and a mechanically driven telescope to take both lunar and solar pictures of higher quality, including stereoscopic images that allowed for a better characterization of sunspots. In the 1860s, spectroscopy was combined with astrophotography to obtain spectra of distant stars and nebulae. Photography was becoming a mainstay of astronomical activity, especially as improvements were being made in telescopic mechanical drives, lenses, and photographic techniques. As examples, from the 1870s on photographs were being made of the transit of Venus across the Sun's surface, features on Mars and other planets, comets and asteroids, deep-sky objects such as nebulae and star clusters, and the lunar surface (Figure 8.36). Astrophotography continues to be used today, both by professional and amateur astronomers.

Since more stars at greater distances could be seen in time-delayed photographs, and since permanent records could be made of their positions relative to each other, it was logical to think that astrophotography could be useful in celestial cartography as well, especially when it was realized that star coordinates could be measured directly from photographic plates. For example, about the same time that the *Südliche Bonner Durchmusterung* was being compiled in Argentina (see Section 10.1.4.1), there was a mapping of the southern stars called the *Cape Photographic Durchmusterung*. Covering the sky from −19 degrees dec. to the south equatorial pole, it contained the positions of nearly half a million stars measured from photographs taken from 1885 to 1889 at the Royal Cape Observatory, South Africa. Some atlases were actual blown-up photographic prints of the heavens. Others were reproductions of positive or negative photographic images. Enthusiasm for this method led to a prematurely ambitious undertaking.

10.1.5.2   *Carte du Ciel project*

This undertaking was planned at an international conference held in Paris in April, 1887, the purpose of which was to plan to catalog and map faint stars in the sky using photography, the *Carte du Ciel* project. According to Ashbrook (1984c), this project was stimulated by the development in 1885 of a 34.3 cm (13.5-inch) high-quality photographic lens made specially for astrophotography by two brothers, Paul and Prosper Henry, who worked at the Paris Observatory. As a result, 18 observatories throughout the world agreed to photograph a series of 2×2-degree assigned regions of the sky using telescopes with lenses of the Henry design. The plan demanded some 20,000 plates, with some 25 million stars being photographed down to the 14th magnitude and a catalog being developed giving the positions of some 1 million stars down to the 11th magnitude. However, the work soon bogged down, consuming decades of time and the valuable resources of the participants, especially the French. One of the problems was that the photographs were taken using rectangular coordinates, and converting these to the more conventional equatorial declination/right ascension system required the use of formulae and tables, a tedious task in the pre-computer era. Although a number of plates were produced, the project was never finished.

10.1.5.3   Other photographic star maps

More modest attempts at creating photograph-based celestial maps succeeded, however, and Lovi (1987) has reviewed many of these. Examples include: (1) the *Franklin–Adams* Charts of 1914, consisting of 206 prints made by a talented British amateur astronomer and recording stars down to the 15th magnitude; (2) the beautiful photographic atlases of the West German amateur astronomer and publisher Hans Vehrenberg, who produced his *Falkau Atlas* in the early 1960s (13th magnitude), followed by his *Atlas Stellarum 1950.0* (14th magnitude) and his *Atlas of Deep Sky Splendors*, which focuses on the Messier objects and is now in its 4th edition; and (3) the *True Visual Magnitude Photographic Atlas* of Chris Papadopoulos and Charles Scovil, which came out in 1979 and used a special lens and filter to correct for the tendency of red stars to appear fainter on photographs than they actually are visually.

But according to Lovi, the gold standard of photographic atlases is the *National Geographic Society–Palomar Sky Survey*, which was produced with photos made by the Mount Palomar Observatory 48-inch Schmidt telescope from 1949 to 1956. Each of the 935 pairs of 35.6×43.2 cm (14×17-inch) prints, one blue- and one red-sensitive, covers a 6×6-degree area of the sky. Stars down to magnitude 21 are shown, and the quality of the images is excellent. These prints are still used today by professional astronomers.

10.2   THE 20th AND 21st CENTURIES

Despite the success of photographic atlases, hand-plotted charts continued to be produced during the 20th Century and into the present Millennium. Improvements in the graphic arts and computer technology have led to advances in celestial cartography and the production of precise and sophisticated star atlases. Today, they rival photographic plates in accuracy and surpass them in utility and practical use. But before we move on to describe these great atlases, we need to discuss two new developments that occurred early in the 1900s.

10.2.1   Annie Jump Cannon and the *Henry Draper Catalog*

The *Henry Draper Catalog* was compiled at the Harvard College Observatory under the tutelage of its Director, Edward C. Pickering (1846–1919). Pickering had a distinguished career in his 42 years at the Observatory. For example, he established a station in Peru that took photographs of the southern skies, and he later published the first all-sky photographic map. He also was one of the discoverers of spectroscopic binary stars as well as the spectral lines of ionized helium.

In 1886 a memorial fund was set up by Anna Mary Draper, the wealthy widow of Henry Draper (1837–1882), to support photographic research in astronomy. Henry had been a physician, amateur scientist, and an early pioneer of astronomical photography, taking the first photographs of a stellar spectrum (the star Vega in 1872) and a nebula (the Orion Nebula in 1880). Pickering tapped into this fund to support a long-term project of obtaining stellar spectra and to ultimately classify the stars by their spectra. To do the routine data reduction and calculations involved, he hired a group of women later termed "Pickering's Women".

One of these individuals was Annie Jump Cannon, who was born on December 11, 1863 in Dover, Delaware. Her father was a prosperous ship builder and state senator, but as a little girl she learned about the constellations and astronomy from her mother, who encouraged her interest in this area. Cannon matriculated at Wellesley, where she studied physics and astronomy and learned to make spectroscopic measurements. She graduated in 1884 and returned to Delaware to help her mother run the large family household. After her mother died in 1894, she returned to Wellesley to work as a junior physics teacher, and then she went to Radcliffe to study astronomy until Pickering hired her in 1896.

Although one of her female colleagues had proposed a method of classifying stellar spectra, Cannon developed a simpler scheme based on spectral classes still used today (e.g., O, B, A, F, G, K, M). In 1911 she was appointed Curator of Astronomical Photographs at Harvard, and she began to systematically categorize the stars shown on photographic plates taken at the Harvard Observatory according to her spectral classification scheme. Between 1918 and 1924, she published this information in the *Henry Draper Catalog*, which included stars down to magnitude 9. From 1925 to 1949, she extended her work to reach fainter stars in selected areas of the sky. In the process, she also discovered and catalogued 300 variable stars. She received many honors for her work. She was the first woman to receive an honorary doctorate from Oxford (1925) and the first woman to win the Draper Gold Medal of the National Academy of Sciences (1931). She was also elected to the Royal Astronomical Society in 1931. She died in Cambridge, Massachusetts on April 13, 1941.

The *Henry Draper Catalog* continues to be useful today. Together with its extension, the positions and spectral types of 359,083 stars are recorded. Many of these stars do not have a Bayer or Flamsteed designation and so are known only by their "HD" number. In the original catalog, they are numbered 1 to 225,300, and in the extension, 225,301 to 359,083. The catalog does not include an associated star atlas.

10.2.2   Constellation boundaries and the IAU

Early in the 20th Century, two problems faced celestial cartographers. First, there was no agreed-upon set of constellations to include on star maps. Although most at the time showed 80 to 90 constellations, these would vary from atlas to atlas. The second problem was that as constellation images were being minimized or deleted, cartographers began to delineate the area of the stars belonging to a constellation with wavy-line boundaries. The limits of these boundaries varied from atlas to atlas, and even from edition to edition of the same atlas. There was no standardized method of showing which stars were to be included with which constellation. Something had to be done to rectify these two problems.

The matter was settled at the First General Assembly of the International Astronomical Union (IAU) in 1922. Founded in 1919, the I.A.U was set up to promote and safeguard standards in astronomy through international cooperation. Its members have included professional astronomers from all over the world who are active in research and education. As of August 2006, the IAU website lists a membership of 8,858 individuals from 85 different countries.

At its 1922 meeting, the IAU took up the issue of the number and boundaries of the constellations. A list of 88 constellations was drawn up that became the official recommendation of this astronomical governing body, and this grouping continues today as the standard number. Eighty-six of these constellations are indicated in bold type in the tables of Chapter 4; the remaining two members of the club are Crux and Coma Berenices. Furthermore, at the meeting the IAU directed Belgian astronomer Eugène Delporte (1882–1955) to draw up the official boundaries for these constellations, which he did. The results were published in his 1930 book entitled *Delimitation Scientifique des Constellations*.

Now, everyone could refer to the same standard number of constellations, and all of the stars in the heavens could be included within rectilinear constellation boundaries (including variable stars that were designated according to their constellation; e.g., RR Lyra and SS Cygnus). This greatly assisted astronomers in locating and naming new deep-sky objects that were being discovered by the huge telescopes that were being built. It also assisted celestial cartographers in planning and organizing their star maps. One of the first people to account for these actions was Arthur P. Norton, whose *Star Atlas* was the most widely used celestial atlas during the first half of the 20th Century and is still in print today. It is to this classic that we now turn.

10.2.3   Major atlases of the 20th Century

10.2.3.1   *Norton's Star Atlas*

Arthur Philip Norton was born in Cardiff, Wales (U.K.) on April 18, 1876. His father was a minister who served in several parishes, requiring the family to move frequently until they finally settled in Worcester, England in 1884. As a boy, Arthur became very interested in astronomy after he had been given an old telescope that had belonged to his great grandfather. After attending King's School and Barbourne College in Worcester, Norton left home to take a teaching post at a grammar school in 1896, but he returned after a year in order to study for an extension degree from Trinity College, Dublin, from which he obtained his B.A. in 1898. Although he continued with his teaching career, he also became more interested in astronomy, making several telescopes and grinding mirrors in his father's workshop. In 1910 he was elected to the British Astronomical Association. He also gave lectures to his students and to local groups on a variety of astronomical subjects, sometimes illustrated with lantern slides. He retired from teaching in 1936, but he continued to observe the heavens with his 11.5 and 26 cm refracting telescopes. He never married and died at home from cancer on October 13, 1955.

Norton's major claim to fame was his star atlas, which he completed in 1910. Known as *Norton's Star Atlas*, it was intended to be used as a companion to two popular books at the time: Webb's *Celestial Objects for Common Telescopes*, and Smyth's *Cycle of Celestial Objects*. Norton hand-positioned some 6,500 stars to magnitude 6 and 600 nebulae for the year 1920. According to writer Stephen James (1993), these objects were given letters or numbers according to 27 different listings (e.g., Bayer letters, Greenwich and Cape Observatory catalogs). There were two hemispheres centered on the north and south equatorial poles down to +50 and −50 degrees dec., respectively, using a geocentric orientation and polar projection. In addition, there were six gores, each covering 5 hours of right ascension and declinations from +60 degrees to −60 degrees. Norton devised his own projection (called a globular projection by Lovi, 1987) to minimize distortion over such a wide area, and like the hemispheres the orientation was geocentric. There were no constellation images or lines in any of the charts, but there were constellation boundaries that were marked by dashed lines. Initially, these boundary lines were curvy, but after the IAU boundaries were set, later editions of Norton's atlas showed rectilinear boundaries. For this and other differences, compare the 1919 edition image shown in Figure 10.10 with the later 1978 image in Figure 10.11. Besides the charts, there was much reference material in the atlas, covering such topics as time, the celestial sphere, observing tips, telescopes, a lunar map, the solar system, nebulae and star clusters, and useful tables; it was essentially a mini-textbook.

Norton's atlas was the most popular star atlas of the first half of the 1900s. Norton revised it to keep it current. For example, James (1993) points out that when the IAU adopted standardized constellation boundaries in 1930 (see Section 10.2.2), Norton included these in his 5th edition (1933), as well as redrawing the star disks to indicate half-magnitudes (showing stars down to magnitude 6.2), and adding 1,500 more objects. In this edition, stippling indicated the Milky Way, and the galactic poles and equator were added for the first time. Other major upgrades occurred in the 9th edition (1943) and the 18th edition (1989). The 19th edition (1998) was the first to use computerized typesetting. Currently, the atlas is in its 20th edition (2003).

10.2.3.2   *Becvar and his celestial atlases*

Antonin Becvar was born on June 10, 1901 in Stara Boleslav in the former Czechoslovakia. He received a degree of Doctor of Natural Sciences from Charles University in Prague. In 1937 he began working as a climatologist at the State Health Resort

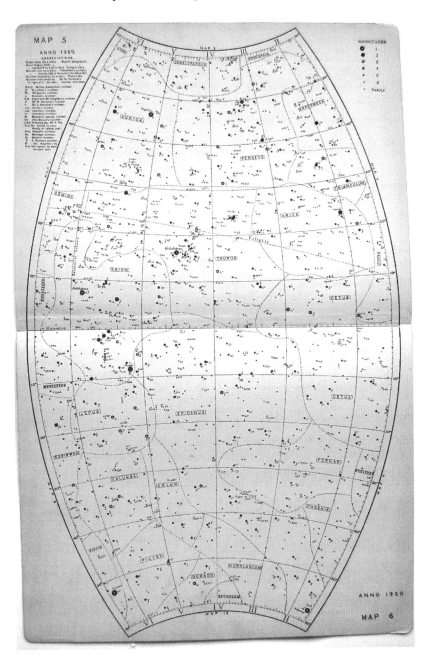

Figure 10.10. Maps 5/6 showing Orion, from the 2nd edition of *Norton's Star Atlas and Telescopic Handbook*, published in 1919. 44 × 28.2 cm. Note the absence of constellation images or lines, the presence of curvy constellation boundaries, stars and deep-sky objects shown for epoch 1920 (7,000 total), and no indication of the Milky Way. Compare with Figure 10.11 and note the slight change in the atlas title. Courtesy of Sky Publishing Corporation.

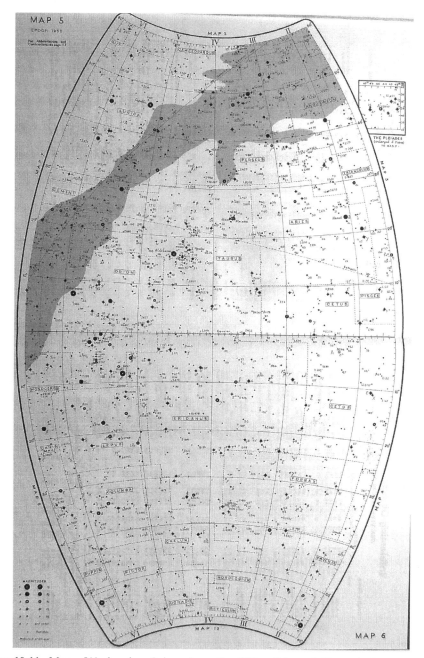

Figure 10.11. Maps 5/6 showing Orion, from the 17th edition of *Norton's Star Atlas and Reference Handbook*, published in 1978. 38.3 × 25.1 cm. Note the absence of constellation images or lines, the presence of rectilinear constellation boundaries, stars and deep-sky objects shown for epoch 1950 (9,000 total), and the Milky Way indicated in green (see Color Plate Gallery). Courtesy of Sky Publishing Corporation.

at Strbske Pleso in Slovakia. He subsequently founded and was the first director of the astronomical observatory at Skalnate Pleso in Czechoslovakia, a position he occupied from 1943 to 1950. His interests included comets, meteors, and atmospheric phenomena. He discovered two comets, in 1942 and 1947. He also made observations and recordings of the Ursid meteor showers. He wrote a number of papers and books on meteors, hydrology, and cloud formation. He died on January 10, 1965 in Stara Boleslav.

He is best known today for his four large astronomical atlases, which represented the state of the art for their time in terms of accuracy and sophistication. The most important was his *Atlas Coeli 1950.0*, which was published in 1948 but plotted the stars as they would appear in 1950. The atlas was created at the Skalnate Pleso Observatory by Becvar, his colleagues, and his students from a variety of astronomical catalogs and photographic atlases. It consisted of 16 celestial charts, showing some 32,500 stars in a geocentric orientation from pole to pole down to magnitude 7.75. It contained more information and labeling than many previous works, with specific markings for a variety of stellar and non-stellar objects, such as double and multiple stars, variable stars, star clusters, radio sources, diffuse and dark nebulae, galaxies, the Milky Way, etc. There were no constellation images or connecting lines, but there were dotted boundary lines following the IAU conventions. Soon after the atlas was published, an accompanying catalog was issued named *Atlas Coeli II*, which contained data on over 12,000 entries, with stars down to magnitude 6.25. In addition, an overseas copyright was purchased by the Sky Publishing Corporation, and an American edition of this atlas was issued in 1949 entitled *Skalnate Pleso Atlas of the Heavens 1950.0* and in 1958 entitled *Atlas of the Heavens: Atlas Coeli 1950.0* (see Figure 10.12). A number of editions were published in different formats, such as a multi-colored edition, a luxury edition, and an edition with white stars on a black background. The atlas was well received and was quite popular. However, it was not perfect. Lovi (1987) has pointed out that some of the conic projections between 20 and 65 degrees in both the north and south declinations were improperly constructed, squeezing the constellation patterns in an east–west direction. Nevertheless, for some 30 years this remained the premier star atlas for the general observer.

After he retired, Becvar produced three additional atlases: *Atlas Eclipticalis* (32 maps showing the sky between +30 and −30 degrees dec., issued in 1958); *Atlas Borealis* (24 maps showing the sky north of +30 degrees dec., issued in 1962), and *Atlas Australis* (24 maps showing the sky south of dec. −30 degrees, issued in 1964). These three atlases were produced for more specialized purposes related to photographic astrometry and photometry. Consequently, precise stellar position rather than magnitude was given preference in terms of which stars were included. Although the atlases were generally complete in showing stars down to the 9th magnitude, they also showed some stars down to the 13th magnitude, but not all. This discrepancy was due to the fact that the fainter stars came from specific catalogs that were not complete but were included because the stellar positions were known precisely. Also, there was a magnitude inconsistency, in that the *Atlas Borealis* adopted photographic magnitudes, whereas the other two atlases used visual magnitudes. These three atlases did not include any labeled deep-sky objects, but they did have one unique feature:

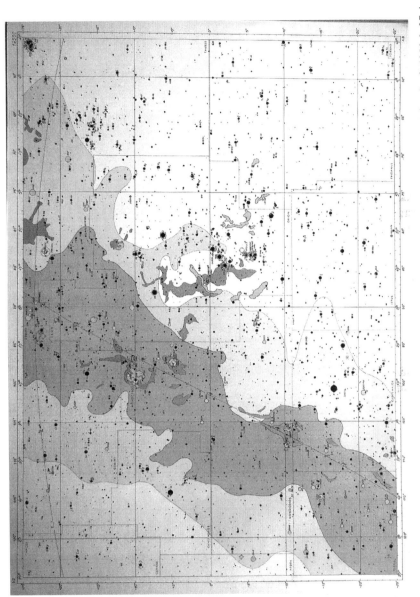

Figure 10.12. Map VII showing Orion, from the 1958 American edition of Becvar's *Atlas of the Heavens: Atlas Coeli 1950.0*. 37.4 × 52.4 cm. Note the absence of constellation images or lines, the presence of rectilinear constellation boundaries, stellar magnitudes indicated by clear variations in the size and design of the star images, and the Milky Way indicated in blue (see Color Plate Gallery). Courtesy of Sky Publishing Corporation.

each star was given one of six colors according to its spectral class. Again, this suited the special purpose for which these atlases were produced.

10.2.3.3   Tirion and his celestial atlases

The next advance in celestial cartography came from the Dutch uranographer (as he prefers to call himself) Wil Tirion, who was born on February 19, 1943. According to his website, he has been interested in the stars and in star maps for nearly all of his life. Trained as a graphic artist and designer, in 1977 he began working on his first star atlas, which consisted of five maps showing the stars down to magnitude 6.5. It was first published in the *Encyclopedia of Astronomy*, then as a separate set of charts by the British Astronomical Association entitled *B.A.A. Star Charts 1950.0*. He next began working on his classic *Sky Atlas 2000.0* (see below), which was published in 1981 and consisted of 26 star charts down to magnitude 8.0, with stars plotted as they would appear in the year 2000 in order to increase the atlas' time of usability. As a result, he was pursued by publishers to produce other star maps and atlases, and he has been engaged full-time in these activities since 1983. He has been honored for his work, including the naming of an asteroid for him by the IAU He is married and has two married children.

Over the years, Tirion has produced a number of star maps for atlases, books, and magazines, utilizing the latest computerized graphic programs. Most of these include deep-sky objects as well as stars. Some of his more popular titles include the *Collins Guide to Stars and Planets*, published with Ian Ridpath in 1984 and consisting of 88 constellation star maps down to magnitude 5.5; *The Monthly Sky Guide*, published with Ian Ridpath in 1987 and consisting of 12 monthly sky maps down to magnitude 5.0; *Uranometria 2000.0*, published with Barry Rappaport and George Lovi in two volumes in 1987 and 1988 and consisting of 473 star charts down to magnitude 9.5 (see below); the *Cambridge Star Atlas 2000.0*, originally published in 1991 and consisting of 20 star charts down to magnitude 6.5 (the current 3rd edition was published in 2001); *A Walk through the Heavens*, published with Milton Heifetz in 1996 and consisting of 40 star charts and four seasonal sky maps; and *How to Identify the Night Sky*, published with Storm Dunlop in 2001 and consisting of 45 constellation pages down to magnitude 6.0. Many of these titles have been produced in subsequent editions, including editions in other languages, deluxe editions, and field versions with white stars on black background. One of his more interesting contributions was to the book *Men, Monsters and the Modern Universe*, published together with George Lovi in 1989. In this book, 10 star charts down to magnitude 6.5 are reproduced from his *Bright Star Atlas 2000.0*, released in 1990 for the more casual observer, and they are contrasted with reproduced plates from Alexander Jamieson's 1822 *A Celestial Atlas*. It is interesting to compare and contrast the differences in cartographic styles that have developed over the course of nearly 170 years.

Two celestial atlases with Tirion's geocentric star maps stand out for their quality and impact in the field. The first of these is *Sky Atlas 2000.0*, which became an instant classic when it appeared in 1981 (see Figure 10.13). Lovi (1987) mentions several

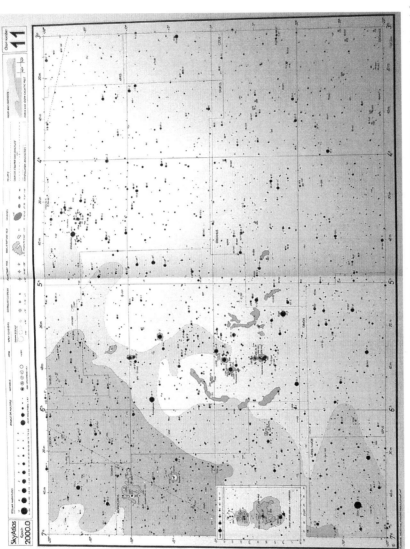

Figure 10.13. Chart 11 showing Orion, from the 1981 1st deluxe edition of Tirion's *Sky Atlas 2000.0*, the first atlas plotted for the epoch of 2000 AD. 31.1 × 45.4 cm. Note the absence of constellation images or lines, the presence of rectilinear constellation boundaries, stellar magnitudes indicated by clear variations in the size and design of the star images, the Milky Way indicated in blue (see Color Plate Gallery), and an insert on the left showing the Orion Nebula. Courtesy of Sky Publishing Corporation.

reasons for this. First, unlike previous atlases that were plotted for the year 1950 (which had already passed), Tirion's stars were plotted ahead for coordinate epoch 2000, which made his stellar positions even more accurate as time progressed to this date and extended the usability of the atlas. Second, the *Sky Atlas 2000.0* showed some 43,000 stars plotted down to magnitude 8.0, along with 2,500 deep-sky objects, all of which were labeled. Third, the map projections were better done in Tirion's atlas, without the distortion error found in Becvar's atlas. Finally, the 26 charts were clearly and beautifully presented. As in the Norton and Becvar atlases, there were no constellation images or connecting lines, but there were dotted boundary lines following the IAU conventions. In the same year that the black-on-white desk edition of *Sky Atlas 2000.0* was published, both a white-on-black field edition and a multi-colored deluxe edition were issued. A 2nd edition of all three came out in 1998 that included over 80,000 stars to magnitude 8.5 and around 2,700 deep-sky objects. A companion to the atlas entitled *Sky Catalogue 2000.0* was published by Alan Hirshfeld and Roger W. Sinnott in 1982 (Volume I on stars) and 1985 (Volume II on double and variable stars and non-stellar objects). A second edition of Volume I was published in 1991.

The other significant Tirion atlas was the *Uranometria 2000.0*, which was produced with Barry Rappaport and George Lovi in two volumes in 1987 and 1988. Unlike the atlases mentioned above that were hand-plotted, this one was computer-generated. In this regard, it was preceded by the computer-plotted *Smithsonian Astrophysical Observatory Star Atlas*, which came out in the 1960s and had an accompanying catalog that contained nearly 258,997 stars down to the 9th magnitude. Like the specialized atlases of Becvar, the SAO atlas and catalog were incomplete in that they only included stars that had accurate positional data. This was because they were not designed for the general observer but for the specialized purpose of assisting space vehicle trackers in producing accurate orbits using precise stellar reference points. In contrast, the *Uranometria 2000.0* was designed for the sophisticated observer and had the precision that one finds in a computer-plotted star atlas, as well as showing 332,556 stars down to magnitude 9.5 and 10,300 deep-sky objects, all spread over 473 charts. Like the atlases mentioned above, there were no constellation images or connecting lines. Tirion's input was used to create a beautiful and user-friendly atlas with labels and IAU-approved constellation boundaries. In terms of numbers of labeled objects and precision of plotting, this atlas is an important advance in the field of celestial cartography. The current 2nd edition of *Uranometria 2000.0* was produced in 2001 and consists of three volumes: I (northern hemisphere to −6 degrees), II (southern hemisphere to +6 degrees), and III (deep-sky field guide) (Figure 10.14).

10.2.4   Other contemporary star maps and atlases

There are a number of other star maps and atlases produced in addition to those described above. Some are found in monthly astronomy-oriented magazines, such as *Sky & Telescope*, *Astronomy*, *The Griffith Observer*, and *Mercury* (bimonthly). Others are located in general texts, such as encyclopedias, world atlases, or astron-

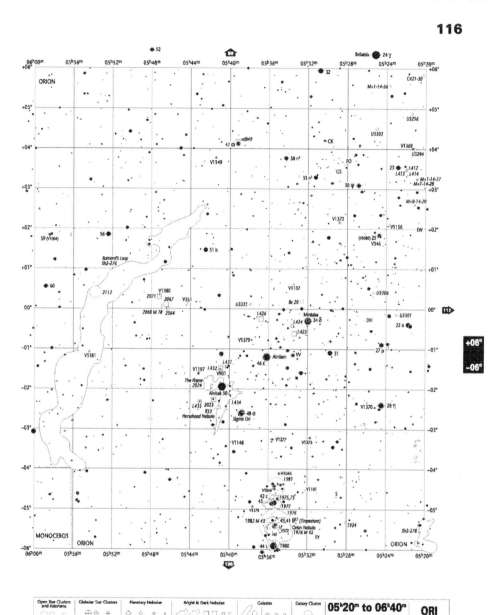

Figure 10.14. Chart 116, showing Orion, from the 2001 edition of Volume 1 of *Uranometria 2000.0* (Tirion, Rappaport, and Remaklus). 22.1 × 18.5 cm. Note the absence of constellation images or lines, the presence of rectilinear constellation boundaries, stellar magnitudes indicated by clear variations in the size and design of the star images, and the nebulae of Orion and Barnard's loop. Courtesy of Willmann-Bell Inc., Copyright 1987-2001.

omy books. One example of the latter is the classic *Burnham's Celestial Handbook: An Observer's Guide to the Universe beyond the Solar System*, by Robert Burnham, Jr. Published in a complete 3-volume set by Dover Publications in 1978, this book is still available in hardbound and paperback versions and contains a wealth of astronomical information organized alphabetically by constellations. Although there are no star map plates in the book, there is a catalog of stars and non-stellar objects for each constellation, as well as the occasional map for selected stars. Another useful book was *A Field Guide to the Stars and Planets, 2nd Edition, Revised*, by Donald H. Menzel and Jay M. Pasachoff, published in 1983. This was a small compact guide to astronomy for use in the field that introduced 52 star charts drawn by Wil Tirion, which were useful for the casual observer, especially with binoculars in hand. The latest 1999 edition continues this tradition. Also, *The Astronomical Companion* (last published in 2000) and *The Astronomical Calendar* (published yearly) by Guy Ottewell are useful books that present the universe in a lucid text and a series of sketchy but understandable three-dimensional maps. The *Calendar* also includes star maps and events for the current year. *The Observer's Handbook* is another annual book published by the Royal Astronomical Society of Canada that includes a number of tables as well as information and maps of events for the year. There are a number of other general astronomical textbooks that include star maps, but listing them all would go beyond the reach of this book.

Other contemporary star maps are found in atlases designed for special purposes or for the more casual observer. One such atlas for the beginner interested in seeing non-stellar objects was entitled *1000 + : The Amateur Astronomer's Field Guide to Deep Sky Observing*, by Tom Lorenzin and Tim Sechler, published in 1987 (with a paperback version in 1992), which is currently out of print. This book focused on over 1,000 deep-sky objects that could be seen by an observer in the Northern Hemisphere using a 20 cm (8-inch) telescope. Following written material on telescopes and observing tips, the authors provided a catalog of these objects, giving their location, NGC or other number, and information on their classification and other helpful comments. This was followed by a set of 8 star maps, 2 north circumpolar charts down to +50 degrees dec., and 6 gores centered at 0, 4, 8, 12, 16, and 20 hours of right ascension and covering the sky from +60 to −50 degrees dec. The charts were computer-generated and showed the deep-sky objects against a background of stars down to magnitude 6.0, plotted for epoch 2000.0 (see Figure 10.15). Although the stars and constellation areas were labeled, there were no constellation figures, lines, or boundaries, and the coordinate grid was fairly simple. However, the atlas was not designed for the detailed plotting of faint stars but for finding the deep-sky objects listed in the catalog by "star-hopping", using the stars in the atlas as a guide.

Another book for beginners interested in non-stellar objects is *The Observer's Sky Atlas*, by Erich Karkoschka, which came out in 1988 but currently is available in a year 2007 paperback version. It has 50 sky charts showing constellation regions, with stars down to magnitude 6, and smaller finder charts for the deep-sky objects, with stars down to magnitude 9. Each chart is accompanied by a listing of interesting objects in the area. The atlas highlights 250 deep-sky objects (including all of the

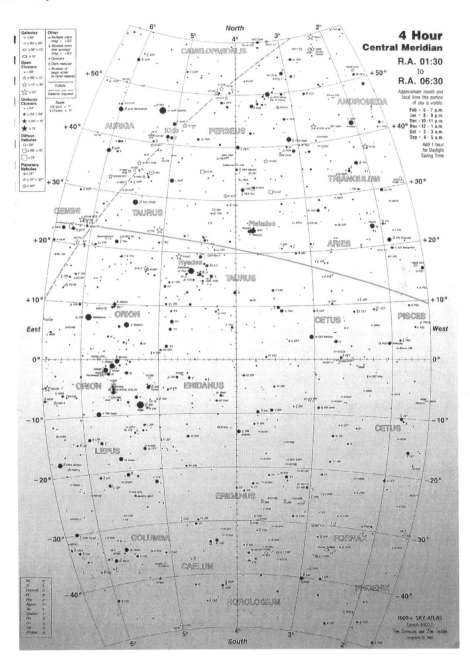

Figure 10.15. "4 Hour" chart showing Orion, from the 1987 1st edition of *1000+: The Amateur Astronomer's Field Guide to Deep Sky Observing*, by Tom Lorenzin and Tim Sechler. 35 × 23.9 cm. Note the absence of constellation images, lines, and boundaries in this atlas, designed primarily for finding the deep-sky objects listed in the catalog by "star-hopping", using the stars in the atlas as a guide. Courtesy of Tom Lorenzin.

Messier objects), along with a number of binary and variable stars. There is also introductory material that is helpful for the observer, such as how to use celestial coordinates, sidereal time, stellar magnitudes, nebulae, meteors, and a calendar of astronomical events through 2018.

One notable specialized atlas is the *AAVSO Variable Star Atlas*, compiled in 1980 by Charles E. Scovil and now in its 1990 2nd edition. This atlas contains the chart boundaries for the 178 maps of the American Association of Variable Star Observers. The maps show the locations of a number of variable stars along with nearby comparison stars whose magnitudes are marked. Stars down to magnitude 9.5 are included. Also included are all Messier objects as well as other deep-sky objects, making this atlas of value to general observers as well as to those mainly interested in following the brightness course of variable stars.

The most comprehensive star atlas published today is *The Millennium Star Atlas*, by Roger W. Sinnott and Michael A.C. Perryman. This 3-volume book, covering right ascension regions of 0–8 hours, 8–16 hours, and 16–24 hours, respectively, was published in 1997, with a paperback version produced in 2006 (Figure 10.16). It has 1,548 charts showing over 1 million stars down to 11th magnitude and some 10,000 non-stellar objects. It is so precise that even the galaxies are pictured with their correct orientation. It is an excellent atlas for the advanced amateur and professional observer. Notably, many of its stars were compiled from data gathered by satellite, such as the European Space Agency's *Hipparcos* satellite. Celestial cartography is now a part of the space age!

10.2.5   Computerized star maps

As we have seen above, computers have been used to generate accurate star placements on paper since the specialized SAO atlas of the 1960s, and they have helped to construct more general star atlases since *Uranometria 2000.0* in 1987/88. Although computers have been used for decades to guide professional telescopes to preferred locations, amateur astronomers formerly had been limited to star-hopping, whereby they located faint objects with the use of a paper atlas by pointing their telescope to a bright star, then swept to fainter and fainter landmark stars located progressively nearer to their target using their telescope's finderscope, and finally arrived at the desired object via their main telescope eyepiece. This activity encouraged the use of star atlases and produced a group of observers who had an intimate knowledge of the sky and its constellations. However, in the 1980s computer-controlled telescopes entered the amateur market. These telescopes contained a database of celestial objects that the observer could locate by simply dialing up the coordinates. Such "go-to" systems now have become common. The result is that star-hopping has become less used, decreasing the need for a paper star atlas and sadly causing many of today's amateur astronomers to become less facile in navigating their way through the night sky than in previous generations.

However, all is not lost. As personal devices (PDs) like iPhones, iPads, and Android systems have become more common, a number of applications have developed that display the sky in dynamic detail and in some cases allow the observer to

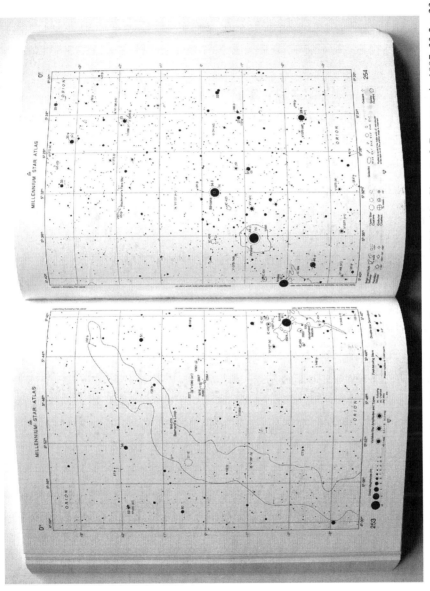

Figure 10.16. Maps 253/254 showing Orion from *The Millennium Star Atlas*, published by Sinnott & Perryman in 1997. 33.2 × 23.4 cm (page size). Note more stars than in any other printed atlas (over one million), the absence of constellation images or lines, the presence of rectilinear constellation boundaries, galaxies shown with their correct orientation, and computer-generated magnitude symbols.

drive a go-to telescope using the apps. Much like the star map division introduced in Chapter 1, these apps can be divided into two categories: general star maps and solar system maps. An example of the former is Star Charts by Wil Tirion, which contains 18 charts that are based on the maps of this famous uranographer. Another is Star Walk, which displays constellations, stars, and planets in the area of the sky to which you point your PD. Another is DSS Browser, which contains a database of over 45,000 celestial objects. Examples of solar system maps include Solar Walk, which offers flybys of our planetary system, and Exoplanet, which provides updates on planetary discoveries around distant stars. There are more novel astronomy-related apps as well, including one that simulates collisions between galaxies and another that displays the size of a crater an asteroid might make if it hit the Earth. Thus, star maps have gone hi-tech. This limits their value as collectibles, but for someone with a PD and a few dollars for the app, the sky is no longer the limit.

10.3 BIBLIOGRAPHY

Ashbrook, J. (1984a) Chapter 31. An episode in early astrophotography. *The Astronomical Scrapbook*. Cambridge, U.K.: Cambridge University Press, pp. 160–165.

Ashbrook, J. (1984b) Chapter 80. How the BD was made. *The Astronomical Scrapbook*. Cambridge, U.K.: Cambridge University Press, pp. 427–436.

Ashbrook, J. (1984c) Chapter 81. The brothers Henry and the *Carte du Ciel*. *The Astronomical Scrapbook*. Cambridge, U.K.: Cambridge University Press, pp. 436–440.

Ashworth, W.B. Jr. (1997) *Out of this World—The Golden Age of the Celestial Atlas*. Kansas City, MO: Linda Hall Library.

Batten, A.H. (1991) Argelander and the Bonner Durchmusterung. *Journal of the Royal Astronomical Society of Canada*, **85**, 43–50.

Brown, Basil (1932) *Astronomical Atlases, Maps and Charts*. London: Search Publishing.

Dreyer, J.L.E. (1953) *A History of Astronomy from Thales to Kepler*, 2nd edn. New York: Dover Publications.

James, S. (1993) Arthur Philip Norton (1876–1955): The man and his star atlas. *Journal of the British Astronomical Association*, **103**, 289–293.

Johnston, P.A. (1985) *Celestial Images—Astronomical Charts from 1500 to 1900*. Boston: Boston University Art Gallery.

Kanas, N. (2002) Mapping the solar system: Depictions from antiquarian star atlases. *Mercator's World*, **7**, 40–46.

Kanas, N. (2005) Are celestial maps really maps? *Journal of the International Map Collectors' Society*, **101**, 19–29.

Lovi, G. (1987) Uranography yesterday and today. In: W. Tirion, B. Rappaport, and G. Lovi (eds.), *Uranometria 2000.0*. Richmond, VA: Willmann-Bell.

North, J. (1995) *The Norton History of Astronomy and Cosmology*. New York: W.W. Norton & Co.

Ridpath, I. (1997) Bonner Durchmusterung/Cordoba Durchmusterung. *A Dictionary of Astronomy*, www.encyclopedia.com

Shubinski, R. (2011) There's an astro app for that. *Astronomy*, August, 58–59.

Warner, D.J. (1979) *The Sky Explored: Celestial Cartography 1500–1800*. Amsterdam: Theatrum Orbis Terrarum.

11

Color plate gallery

This chapter contains 91 full-page color reproductions of important figures that have appeared earlier in this book. These plates have been computer-enhanced to improve their clarity and to make it easier for the reader to observe the fine details that are mentioned in the text and in the legend to each figure. Care has been taken to match the colors shown in these plates with those in the original prints from which they were photographed. Through these efforts, it is hoped that the reader will not only gain a better understanding of what the images are trying to convey but also of their beauty and artistry.

Figure 1.1. View of several northern constellations, taken from the 1795 edition of Fortin's *Atlas Céleste de Flamsteed*. 15.7 × 20.6 cm. Note the constellations of Lacerta ("le Lezard", which looks more like a dog than a lizard); Cygnus (flying in the Milky Way); Lyra (with the bright star "Wega"); and Vulpecula ("le Renard") clutching in its jaws Anser ("l'Oye", which no longer exists).

Figure 1.2. A plate produced by Doppelmayr for Homann Publications, ca. 1720, which also appeared in Homann's 1742 *Atlas Coelestis*. 48.2 × 56.8 cm, 43.6 cm dia. hemisphere. It depicts the state of astronomical knowledge in the early 1700s. Note the Copernican cosmological system in the center, complete with the planets and their moons and textual and numerical information on the proportionate diameters of the planets. See text for a description of the other images.

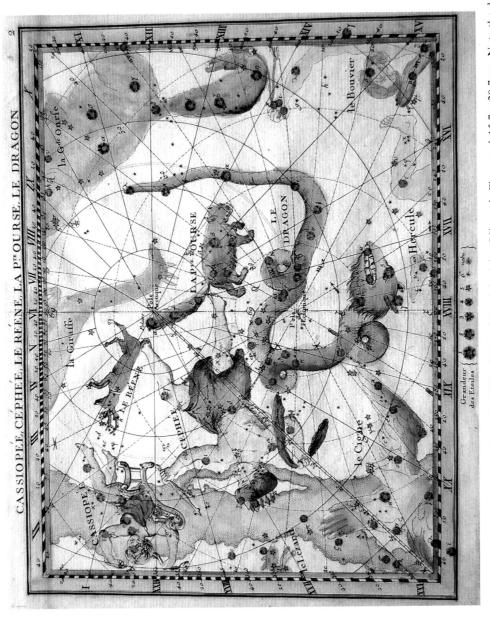

Figure 1.5. View of the north pole, taken from the 1776 edition of Fortin's *Atlas Céleste de Flamsteed*. 15.7 × 20.7 cm. Note the double-grid system centered on the poles of the ecliptic and the celestial equator.

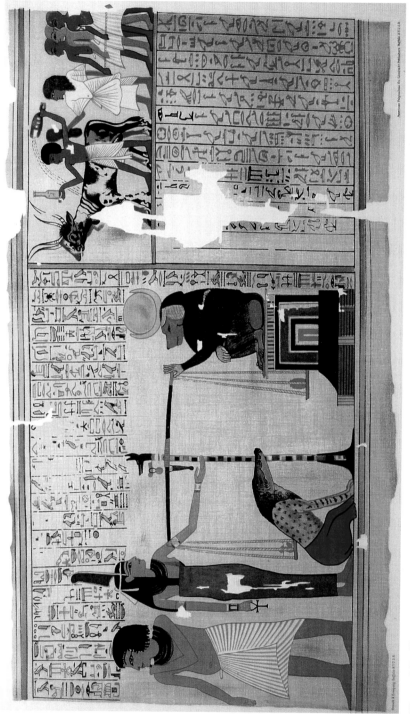

Figure 2.5. Chromolithograph of an Egyptian papyrus "Judgment of the Dead", from Binion's (1887) *Ancient Egypt or Mizraim.* 21 × 43.6 cm. Note the goddess of truth, Maat, presiding over the weighing of the heart of the deceased man on the left. Thoth, the god of wisdom and writing, is ready to record the outcome, and the "devourer" is waiting to destroy the heart if there is an unfavorable outcome. Being a lunar deity, Thoth has a Moon over his head.

377

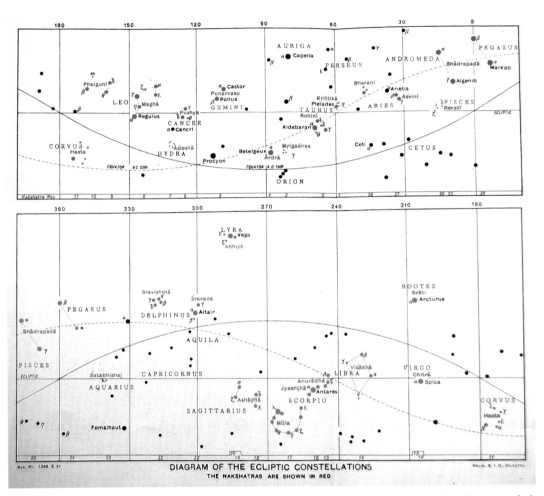

Figure 2.8. The 27 *naksatra* constellations from Vedic mythology, from G.R. Kaye's *Memoirs of the Archaeological Survey of India, No. 18: Hindu Astronomy*, published in Calcutta in 1924. 22.7 × 27.4 cm. Note that they include both constellations and individual stars, some of which are familiar (e.g., *Krittika*, the Pleiades; *Svati*, the star Arcturus).

Figure 2.10. "Planisphere Egyptien" representing the Egyptian sky according to Athanasius Kircher, from Charles-François Dupuis' *L'Origine de tous les Cultes ou Religion Universelle*, ca. 1795. 18.1 cm dia. Note the syncretic mixture of western and eastern zodiacal images around the periphery and the more traditional western constellations in the center.

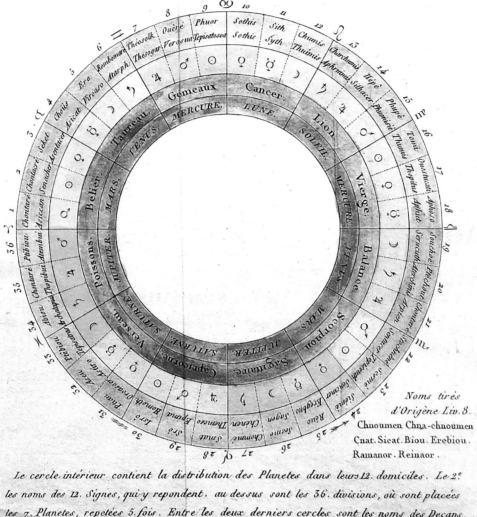

Figure 2.11. A figure from Charles-François Dupuis' *L'Origine de tous les Cultes ou Religion Universelle*, ca. 1795, which summarizes important astrological information and is reminiscent of a Greco-Roman astrological board from Egypt (see text). 16.2 cm dia. Note the concentric rings that represent (from the outside in) the names of the 36 decans and the symbols for their planetary associations, and the names of the 12 zodiac constellations and the names of their planetary associations.

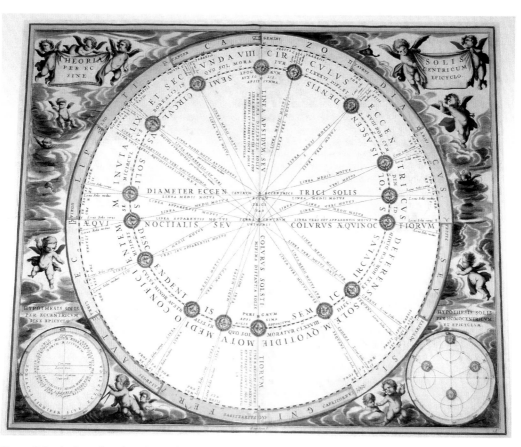

Figure 3.2. A plate showing the orbit of the Sun around the central Earth according to Hipparchus and adapted by Ptolemy, from Cellarius' *Harmonia Macrocosmica*, ca. 1661. 42.1 × 50.4 cm, 38.5 cm dia. hemisphere. Note that the eccentric orbit accounts for the unequal period of time between the equinoxes, with the lower part (autumn to spring) being shorter than the upper part (spring to autumn).

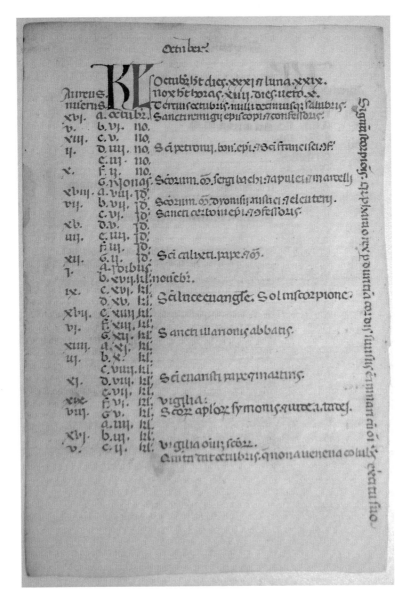

Figure 3.8. Anonymous calendar leaf for October from a Book of Hours, ca. 1350. This manuscript was written on vellum and is probably a page from a Psalter from Italy. The margins have been cut and are uneven but measure about 12.7 × 8.6 cm (page size). Note from left to right that the columns depict: Roman numerals indicating the year in the 19-year solar/ lunar calendar where the full Moon falls on the given date; letters representing the days of the week, with the red "a's" probably being the dominical letter for Sunday; the Roman calendrical system, including "idib" for the "ides" of the month in the middle; and descriptions of various saints' and feast days on the right. Note also at the top the large "KL" for "Kalendarium", and the indication that October had XXXI days and XXIX lunar days that year.

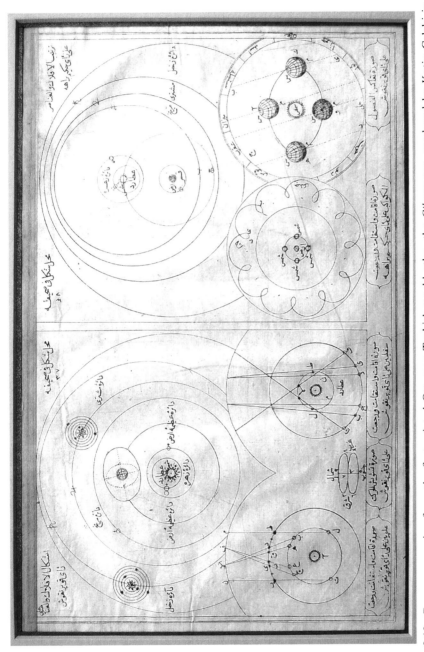

Figure 3.10. Copper engraving from the first printed Ottoman Turkish world atlas, the *Cihannuma*, produced by Katip Celebi in 1732. 16.5 × 26.1 cm. Note the Western influences: in the upper part are illustrations of the cosmologies of Copernicus (*left*) and Tycho Brahe (*right*), and in the lower part are diagrams showing the mechanism for the retrograde motion of a superior and inferior planet, the looped appearance of a superior planet's orbit in the sky, and the 3rd motion of the Earth according to Copernicus.

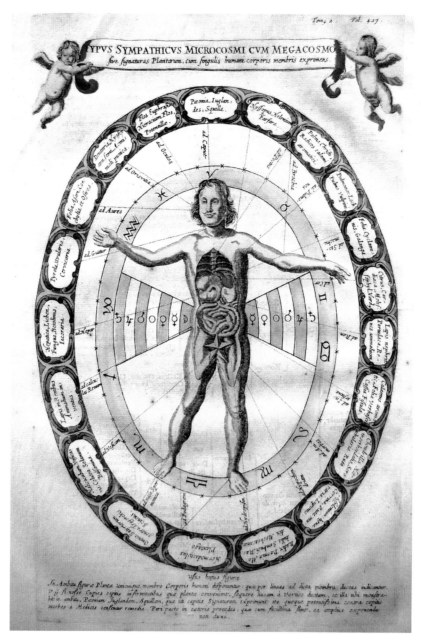

Figure 3.13. An engraved plate from the first Dutch edition (1682) of *Mundus Subterraneus*, first published in 1665 by Athanasius Kircher. 35.6 × 20.8 cm (plate size). This shows the correspondence between the microcosm (the "microcosmic" or "zodiac" man in the center) and the macrocosm (symbols of planets and zodiac in the periphery). Note the dotted lines linking the sympathies of the parts of the human body and various medicinal plants (identified in the Baroque-styled labels in the outer circle).

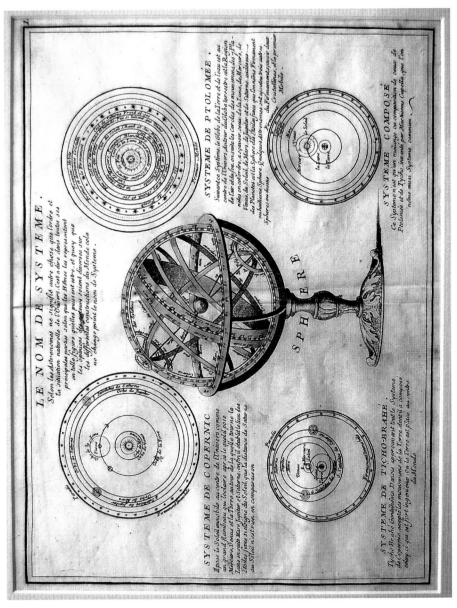

Figure 3.17. Print entitled "Le Nom de Systeme", from Le Rouge's *Atlas Nouveau Portatif à L'Usage des Militaires et du Voyageur*, ca. 1761. 19.4 × 26.1 cm. Note the central armillary sphere surrounded by cosmographic models taken from Copernicus, Ptolemy, Tycho Brahe, and Martianus Capella, which is labeled "Compose". Note that the models of Tycho Brahe and Capella are both hybrids of Copernicus and Ptolemy, allowing the Earth to remain in the center of the universe but having all of the other planets (Tycho Brahe) or just Mercury and Venus (Capella) revolve around the Sun.

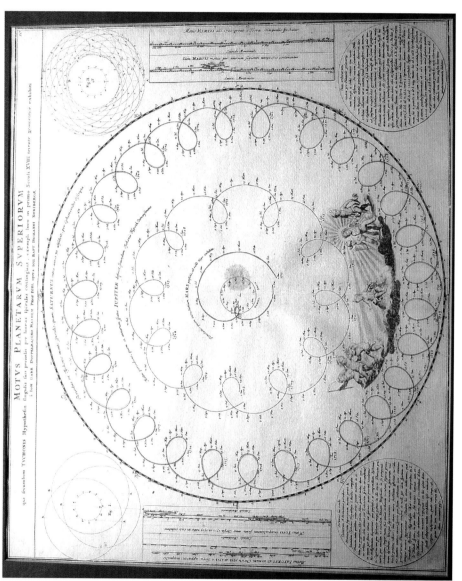

Figure 3.18. Print entitled "Motus Planetarum Superiorum Qui Secundum Tychonis Hypothesin . . .", from the 1742 edition of Doppelmayr's *Atlas Coelestis.* 48 × 57.2 cm, 43.8 cm dia. hemisphere. Note the movement of the outer planets in the early 1700s according to the view of Tycho Brahe. The loops represent the retrograde actions of the planets as seen from the Earth. Note also that Mars sometimes passes through the sphere of the Sun's orbit (here labeled *Orbita Solis*), demonstrating that it is not made of solid crystalline material, which was proposed by Aristotle.

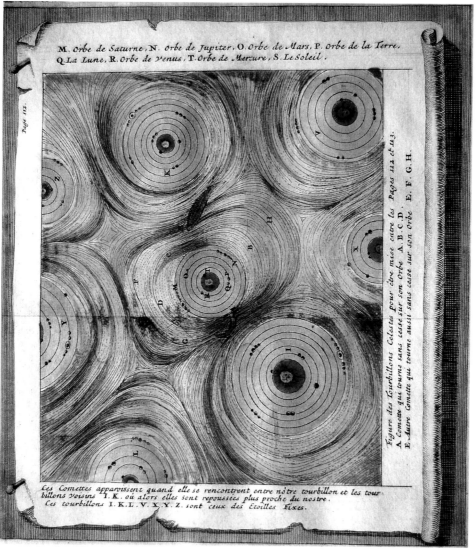

M. *Orbe de Saturne*, N. *Orbe de Jupiter*, O. *Orbe de Mars*, P. *Orbe de la Terre*,
Q. *La Lune*, R. *Orbe de Venus*, T. *Orbe de Mercure*, S. *Le Soleil*.

Figure des Tourbillons Celestes pour estre mise entre les Pages 112 et 113.
A *Comette qui tourne sans cesse sur son Orbe* A. B. C. D.
E *Autre Comette qui tourne aussi sans cesse sur son Orbe* E. F. G. H.

*Ces Comettes apparoissent quand elle se rencontrent entre nôtre tourbillon et les tour-
billons voisins* I. K. *où alors elles sont repoussées plus proche du nostre.*
Ces tourbillons I. K. L. V. X. Y. Z. *sont ceux des Etoilles Fixes.*

Figure 3.19. Copper engraving from Bion's *L'Usage des Globes Celestes et Terrestres . . . ,* which was published in Amsterdam and bound into the 1700 edition of Nicolas Sanson's *Description de tout l'Univers en Plusieurs Cartes.* 17.7 × 19.8 cm. Note the depiction of how comets enter into orbits around a star according to the cosmological system of Descartes, which postulates swirling vortices surround the stars.

Figure 4.4. Two constellations from a manuscript copy of al-Sufi's *Book of Fixed Stars*, ca. 1009AD. 29.7 × 17.8 cm (page size). Note the vivid colors and the clothing contemporary with the times rather than classical Greece. On the left is Perseus holding the head of a male demon instead of Medusa. On the right is Andromeda with two fish in front of her, reflecting a Bedouin tradition. Courtesy of the Bodleian Library. University of Oxford (MS. Marsh 144 p. 111 & 167).

Figure 4.5. The southern celestial hemisphere ("Hemisphere Austral") centered on the south equatorial pole, from the 1795 edition of Fortin's *Atlas Céleste de Flamsteed*. 17.5 × 22.4 cm image, 16.9 cm dia. hemisphere. Note the additional constellations from those in Figure 4.2, but still the presence of a blank area below and to the left of the pole. See text for details.

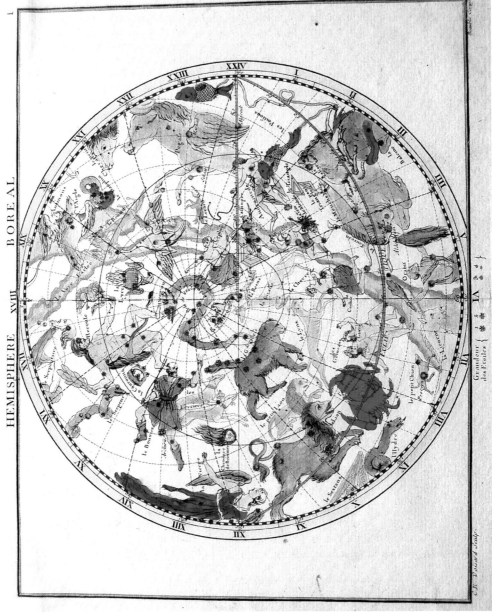

Figure 4.6. The northern celestial hemisphere ("Hemisphere Boreal") centered on the north equatorial pole, from the 1795 edition of Fortin's *Atlas Céleste de Flamsteed.* 17.5 × 22.4 cm image, 16.9 cm dia. hemisphere. Note the additional constellations to those in Figure 4.1. See text for details.

Figure 4.7. Lacaille's famous map of the south celestial polar region, from the 1776 edition of Fortin's *Atlas Céleste de Flamsteed.* 19.4 × 23.9 cm image, 18.4 cm dia. hemisphere. Note the new constellations invented by Lacaille that picture scientific instruments and tools of the period. In comparison with Figure 4.5, the whole area has now been filled in with constellations. See text for details.

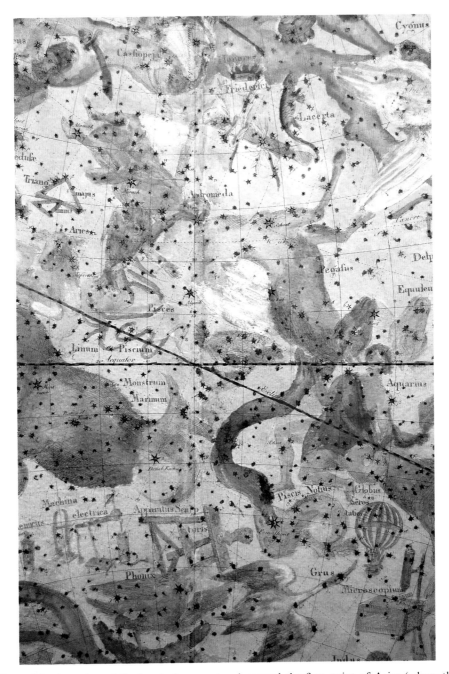

Figure 4.9. A section of the hemisphere centered around the first point of Aries (where the ecliptic and celestial equator lines cross), from Bode's *Uranographia*, published in 1801. 56.7 cm dia. hemisphere. Note the new constellations that appear here for the first time: Honores Friederici (*top right*), Machina Electrica (*bottom left*), and Globus Aerostaticus (*bottom right*).

Figure 5.1. The constellation of Aries, from a 10th-Century manuscript written on vellum used to illustrate the text of Aratus' *Phaenomena*. 38 × 29.9 cm (page size). Note the text in the constellation, which is an excerpt from the writing of Hyginus. Photograph taken from *Celestial Charts* (C. Stott, Studio Editions, 1995). Courtesy of Ms. Stott and the British Library (Harley MS 647).

Figure 5.4. The northern celestial hemisphere produced by Albrecht Dürer in 1515. Approximately 42.9 × 42.9 cm. Note the radial lines resulting in a primitive coordinate system, allowing the stars to be located in the heavens. This woodcut is the first printed star map. Courtesy of the Collection of Robert Gordon.

Figure 5.5. The northern celestial hemisphere produced by Johannes Honter in 1541. Approximately 27.9 × 27.9 cm. Note the geocentric orientation, the addition of lines for the Tropics and the celestial equator, and the medieval clothing on some of the figures. Courtesy of the Collection of Robert Gordon.

Figure 5.8. The northern celestial hemisphere of Hood's 1590 *The Use of the Celestial Globe in Plano* (facsimile). 53.7 × 54.4 cm. Note the accurate latitude and longitude coordinate system centered on the ecliptic pole, the beautiful detail, and the text information written beside each constellation. Photograph taken from a facsimile copy from the British Library and published as *The English Experience* #533 (Theatrum Orbis Terrarum, 1973). Courtesy of the British Library (Maps 184.h.1).

Figure 6.1. A later-colorized image of the constellation Bootes, from a mid-1600 edition of Bayer's *Uranometria*. 27.5 × 37.8 cm. Note the geocentric orientation, the use of Greek letters to indicate stellar magnitude, the use of an accurate grid system in the margins, a bundle of wheat instead of Coma Berenices to the lower right of the image, and the five bright stars of the handle of the Big Dipper to the upper right.

Figure 6.2. The Christianized constellation St. Sylvester (a.k.a. Boötes), from the 1627 edition of Schiller's *Coelum Stellatum Christianum*. 23.8 × 30.6. Note the external orientation, the five bright stars of the Big Dipper to the upper left of the image of the saint, and the wand with ribbons instead of Coma Berenices to the lower left.

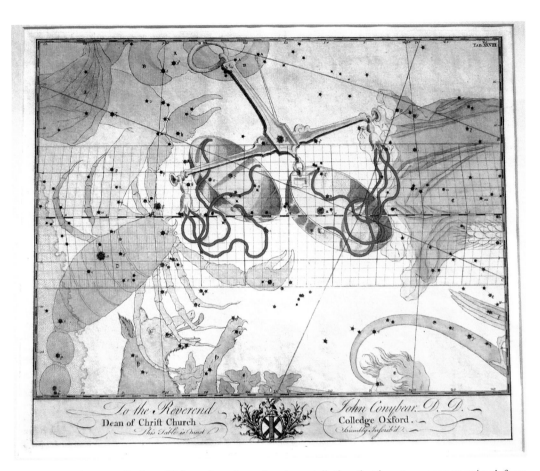

Figure 6.4. The constellation of Libra, the only zodiacal constellation that is not a person or animal, from Bevis' *Atlas Celeste*, ca. 1786. 26.8 × 36.3 cm. Note the central grid area that represents the zodiac and the dedication at the bottom to the Dean of Christ Church, Oxford.

Figure 6.5. The northern celestial hemisphere from Hevelius' *Firmamentum Sobiescianum*, first published in 1687. Approximately 50.8 × 49.5 cm, 46.5-cm dia. hemisphere. Note the external orientation and the nicely drawn constellation images. Courtesy of the Collection of Robert Gordon.

Figure 6.7. The northern celestial hemisphere from Zahn's *Specula Physico-Mathematico-Historica . . .*, published in 1696. 35.2 × 40.7 cm, 31.9 cm dia. hemisphere. Note that this is a near-perfect copy of Hevelius' northern hemisphere map (compare with Figure 6.5).

Figure 6.8. The southern celestial hemisphere from Zahn's *Specula Physico-Mathematico-Historica ....*, published in 1696. 35.2 × 40.9 cm, 31.9 cm dia. hemisphere. Note that this is a near-perfect copy of Hevelius' southern hemisphere map and likewise incorporates the work of Halley.

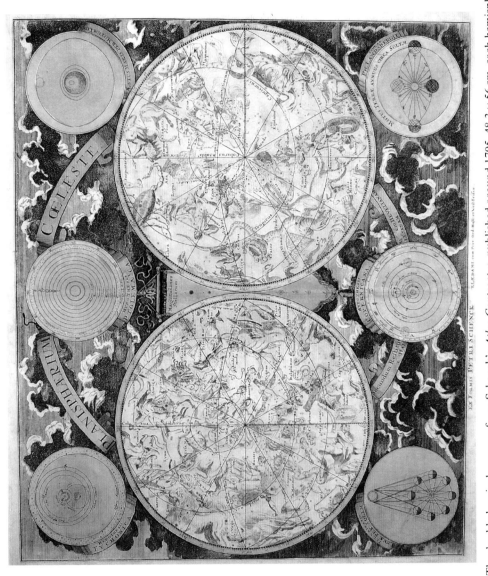

Figure 6.9. The double-hemisphere map from Schenck's *Atlas Contractus*, published around 1705. 48.3 × 56 cm, each hemisphere 26.3 cm dia. Note that the constellations are taken from Hevelius. The format of this map is nearly identical to maps by Eimmart and Lotter and quite similar to the double-hemisphere map by Seutter.

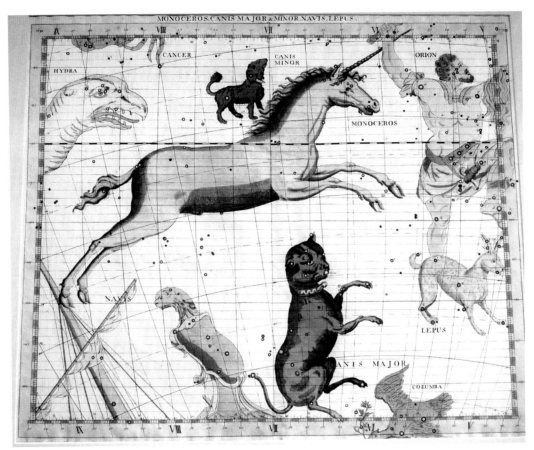

Figure 6.10. The map labeled "Monoceros, Canis Major & Minor, Navis, Lepus", from Flamsteed's *Atlas Coelestis*, published in 1729. 47.3 × 58.2 cm. Note the absence of constellations (e.g., Lacaille's "Pyxis") between Monoceros and Navis and the double-grid system, the major one centered on the celestial equator and the other centered on the ecliptic.

Figure 6.11. The map showing the region around Monoceros (the unicorn), from Reissig's *Constellations Represented on XXX Tables*, published in 1829. 19.1 × 19.2 cm. Note the similarity to the map shown in Figure 6.10 from Flamsteed's atlas, although the constellations are labeled in Cyrillic.

405

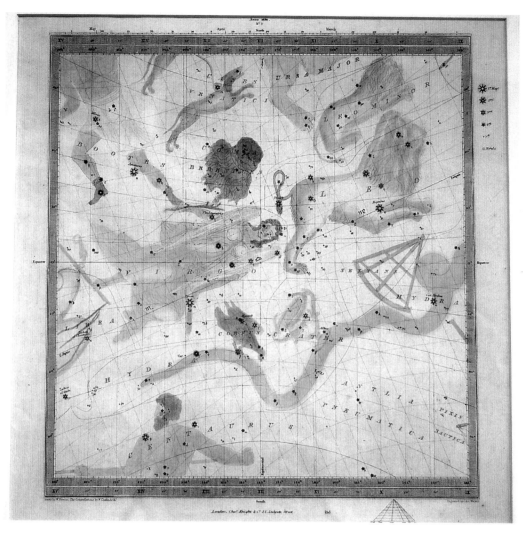

Figure 6.12. The constellations around the location of the Sun in the zodiac at the time of the autumnal equinox, from the 1833 edition of *The Constellations*, published by the Society for the Diffusion of Useful Knowledge. 25.2 × 24.9 cm. Note the influence of Flamsteed and the original pastel colors on this inexpensive, practical English celestial map.

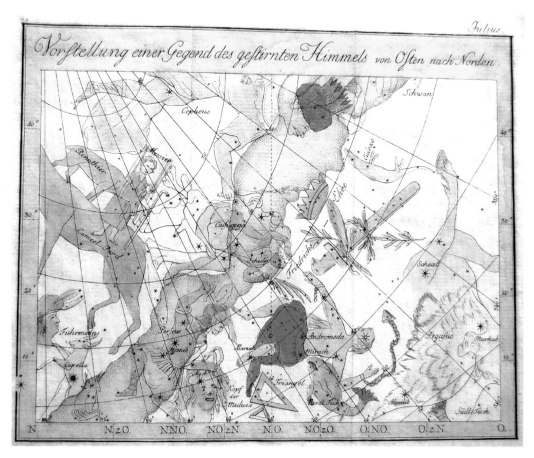

Figure 6.13. The constellations for "Julius" (July), from a ca. 1787 edition of Bode's *Anleitung zur Kenntniss des Gestirnten Himmels*. 15.5 × 19.3 cm. Note the crudely drawn constellation figures; the now obsolete constellations of the Reindeer and Messier; and the new constellation of "Friedrichs Ehre" (e.g., Frederick's glories), introduced by Bode in 1787 to honor the death of his patron, Frederick the Great.

Figure 6.14. Plate III showing Perseus and Andromeda, from the 1782 edition of Bode's *Vorstellung der Gestirne* ... 15.8 × 20.6 cm. Note the similarity of this map to the maps in the atlases of Flamsteed and Fortin.

Figure 6.15. The hemisphere "Coelum Stellatum Hemisphaerium Arietis", centered around the location of the Sun in the zodiac at the time of the vernal equinox, from Bode's 1801 *Uranographia*. 56.7 cm dia. hemisphere. Note the large number of stars and constellations, more than in any previous atlas. Note also the added colorization, whose poor quality suggests that it was done by non-professionals, perhaps children.

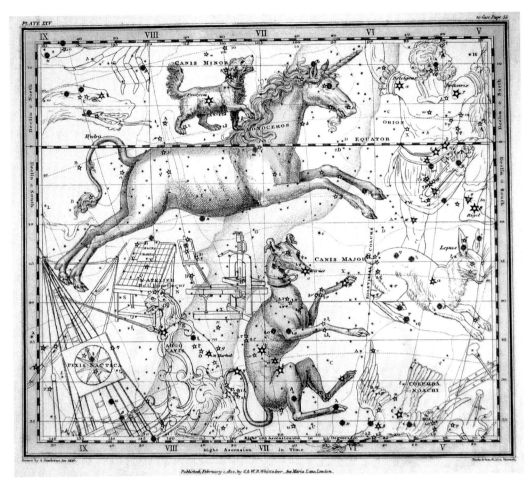

Figure 6.16. Plate XXV featuring the region of Monoceros, Canis Major and Minor, from Jamieson's 1822 *A Celestial Atlas*. 17 × 20.1 cm. Note the stylistic similarity to Bode's *Vorstellung*. Note also the constellation below Monoceros' rear legs labeled here (in French) as *l'Atelier de l'Imprimeur*, which Bode introduced in his 1801 *Uranographia* as "Officina Typographia", the printing press, to honor Gutenberg.

Figure 6.17. An original hand-painted card of the constellation "Orion", from a later edition (ca. 1840) of Leigh's *Urania's Mirror*. 18.3 × 12.4 cm (image size). Note the holes punched in the brightest stars that give a representation of the constellation pattern when the card is held up to the light.

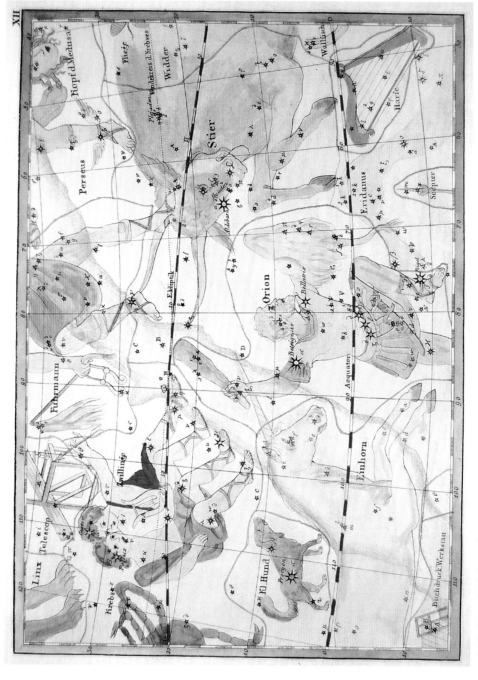

Figure 6.18. Plate XII featuring the region around Orion, from Riedig's *Stern-Karten in 20 Blättern*. 14.4 × 20.5 cm. Note the attractive constellation images, the ecliptic and celestial equator lines, and the constellation of the harp, depicted in Bode's *Uranographia* as Harpa Georgii (honoring King George III of England).

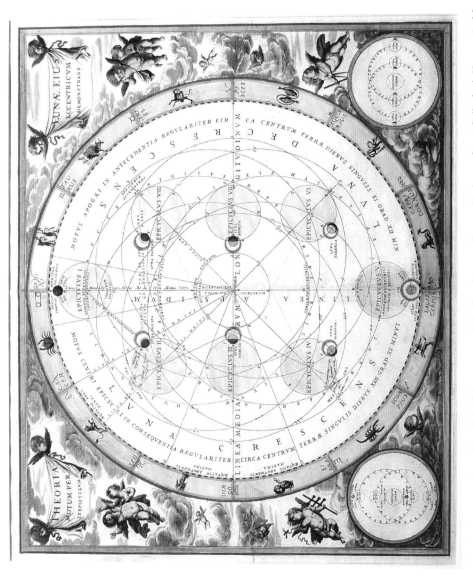

Figure 7.1. A diagram showing the orbit of the Moon around the Earth according to Ptolemy, from Cellarius' *Harmonia Macrocosmica*, ca. 1661. 42 × 50.4 cm, 36.8-cm dia. hemisphere. Note the locations of the epicycle of the Moon as it revolves along its deferent orbit. Note also that the Moon's epicycle is pulled in toward the Earth by a "crank" mechanism hinged on a central epicycle (dotted line), which would correctly account for its location in the sky ("save the phenomena") but would result in an erroneous increase in its apparent diameter, especially at the quarter phase.

413

Figure 7.2. A diagram showing a planisphere centered on the north ecliptic pole down to about 20 degrees in the southern hemisphere, from Cellarius' 1660 edition of *Harmonia Macrocosmica*. Approximately 43.2 × 50.8 cm, 40.7 cm dia. planisphere. Note the beautiful constellation images and the Baroque elements (clouds, putti) in the periphery. Courtesy of the Collection of Robert Gordon.

Figure 7.4. A print from Mallet's 1683 *Description de l'Univers*. 14.5 × 10.1 cm. Note the five constellation images, in which are shown the locations of supposed novae (see text). Also, note the accompanying rural scene at the bottom, meant to aesthetically enhance the astronomical part of the print.

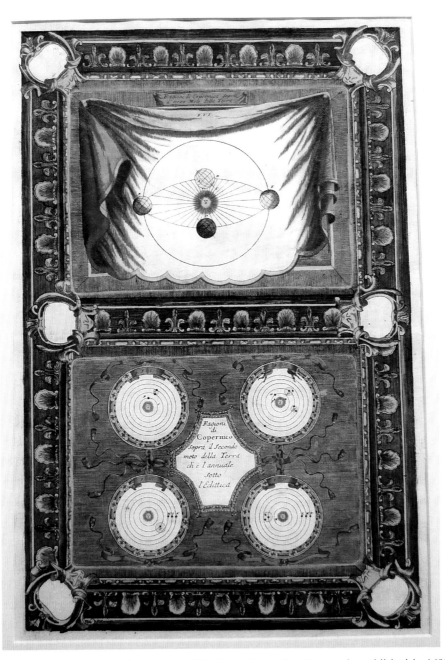

Figure 7.5. A double print from Coronelli's *Corso Geografico Universale*, published in 1692. 38.4 × 25.5 cm. Note the diagram of the third motion of the Earth at the top, the second motion of the Earth at the bottom, and the beautiful Baroque margin uniting the diagrams into one print.

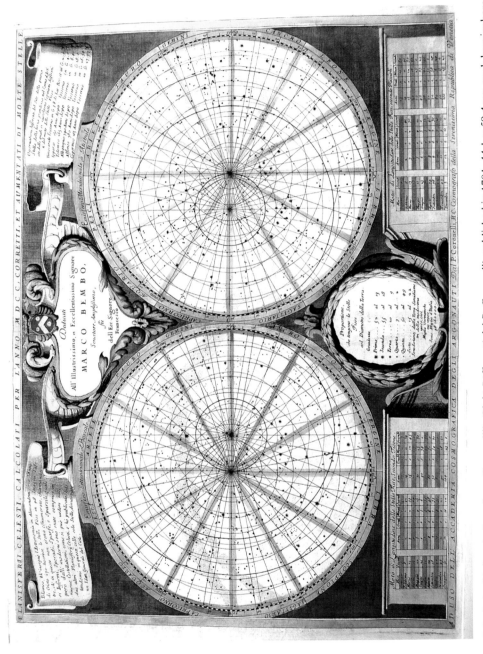

Figure 7.6. A double hemisphere from Coronelli's *Globi Differenti del P. Coronelli*, published in 1701. 44.1 × 59.4 cm, each hemisphere 28.3 cm dia. Note the lack of constellation figures and the double-grid system, showing that Flamsteed was not the first person to use this in a single chart.

Figure 7.7. A double hemisphere from Seller's *Atlas Coelestis*, published ca. 1680. Each hemisphere is approximately 7 cm in diameter. Note the traditional constellation images, the signs of the zodiac above and below, and outlines of the planets with their symbols at the bottom. Courtesy of Jonathan Potter Ltd.

Figure 7.8. One of two pages depicting the zodiac in three long horizontal strips, produced by Senex in 1718. Approximately 53 × 64.5 cm (each page). Note the prominent grid system. Courtesy of Jonathan Potter Ltd.RED

419

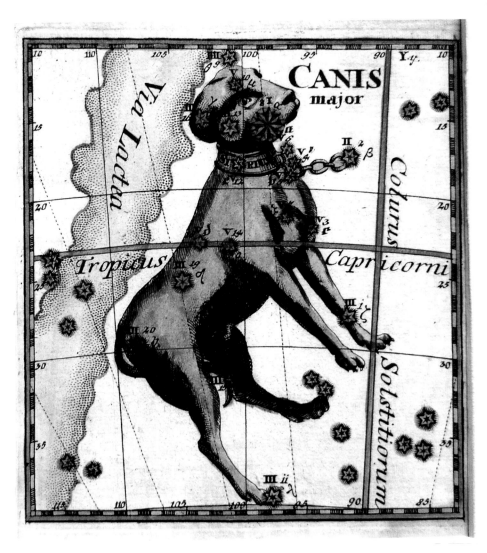

Figure 7.9. The constellation of Canis Major, from Thomas' 1730 *Mercurii Philosophici Firmamentum*. 11.9 × 11.2 cm. Note the use of Bayer Greek letters for stars, Roman numerals for stellar magnitudes, and Arabic numerals for references to the star catalog. Note the Via Lactea (Milky Way) to the left of the image, and the crossing lines representing the Tropic of Capricorn and the Colurus Solstitiorum.

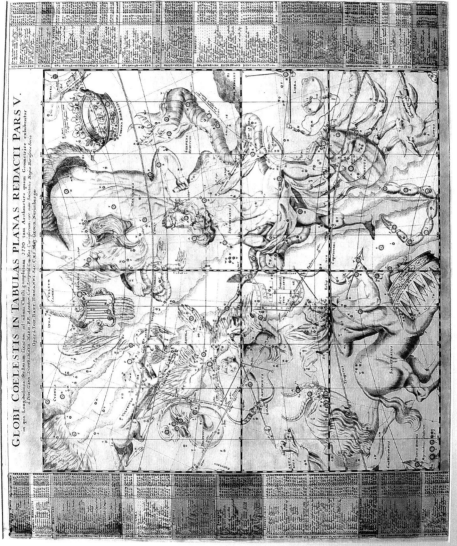

Figure 7.10. The stars and constellations that are centered around the location of the Sun in the zodiac during the winter solstice, according to Doppelmayr (from his *Atlas Coelestis*, 1742). 48.5 × 57.4 cm. Note the influences of Hevelius (including the presence of his constellation Scutum Sobiescianum) and the paths of several comets, including Kepler's 1607 viewing of what was later to be called Halley's comet (to the right of Serpens Ophiuchi).

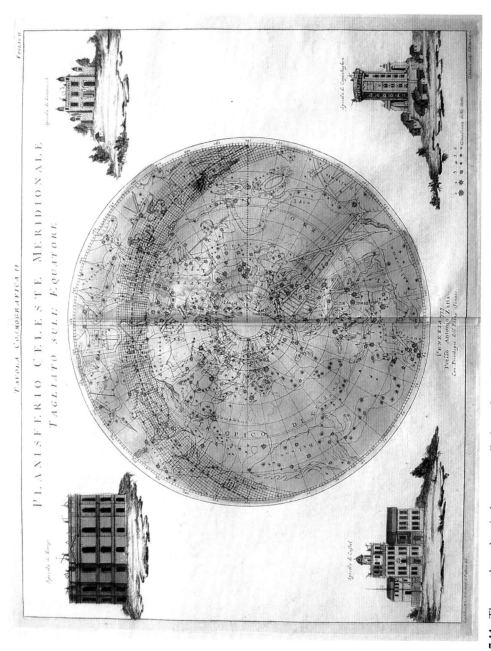

Figure 7.11. The southern hemisphere constellations, from Zatta's 1779 *Atlante Novissimo*. 31 × 40.3 cm, 24-cm dia. hemisphere. Note the Magellanic Clouds and the beautiful drawings of famous European observatories in the corners (at Paris, Greenwich, Cassel, and Copenhagen).

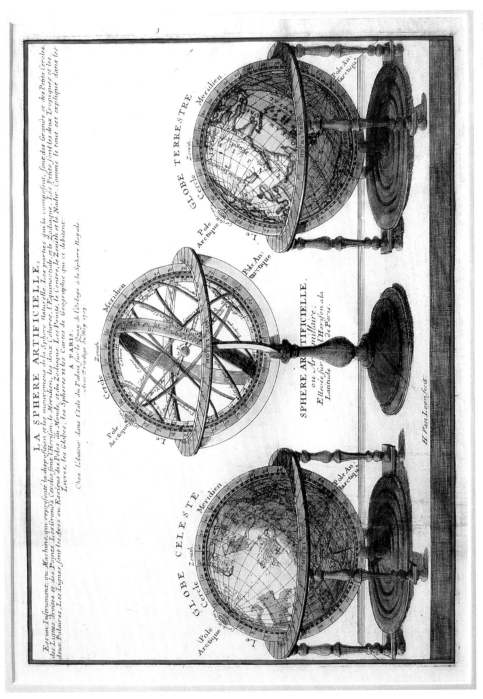

Figure 7.13. The depiction of three important globes ("La Sphère Artificielle"), from Nicolas De Fer's 1703 *L'Atlas Curieux*. 22.5 × 32.1 cm. Note the central armillary sphere and the celestial and terrestrial globes to either side.

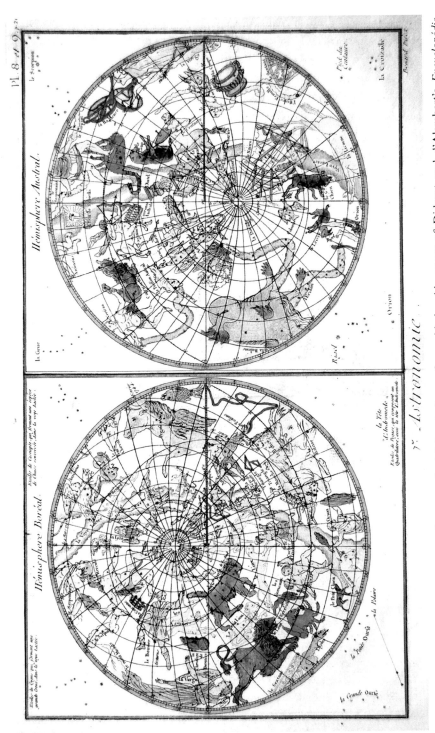

Figure 7.15. Copies of Le Monnier's famous celestial hemispheres, from side-by-side pages of Diderot and d'Alembert's *Encyclopédie*, published in Lausanne and Berne ca. 1780. 19.9 × 35.1 cm, each hemisphere 16.7 cm dia. Note the beautiful detail in the hemispheres and the constellation star patterns in the corners.

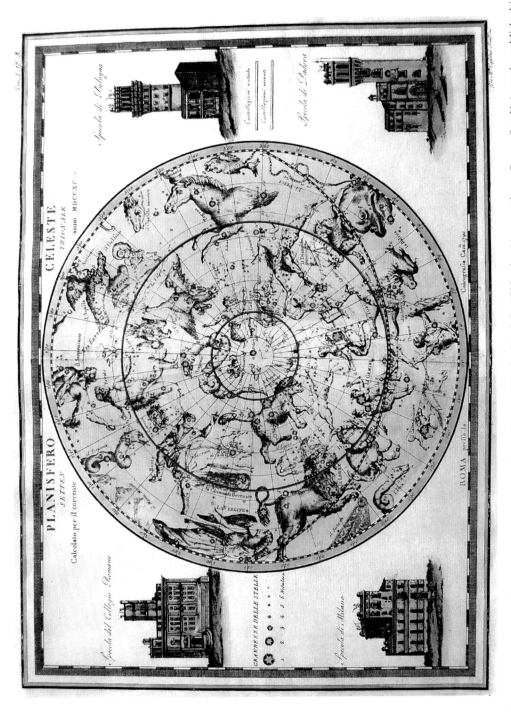

Figure 7.16. The northern celestial hemisphere, from Giovanni Maria Cassini's 1792 atlas *Nuovo Atlante Geografico Universale*, published in Rome. 29.1 × 41.3 cm, 27.3-cm dia. hemisphere. Note the fine images of four Italian observatories in the corners.

Figure 8.1. Nine inch celestial globe on a Bakelite stand, produced by Rand McNally in 1936. Note the stars in yellow, the mirror-reversed constellation images in light blue, and the dark blue background, which simulates the night sky. Note also the movable, vertical calibrated meridian ring on which the globe is attached; the hour ring near the north equatorial pole of the globe; and the fixed horizon circle that is connected to the base. Copyright Rand McNally. Reproduced with permission, R.L.07-S-22.

Figure 8.2. Close-up view of a 9 inch celestial globe on a Bakelite stand, produced by Rand McNally in 1936. Note the vertical gore lines. Note also the place where the lines representing the ecliptic and the celestial equator cross (i.e., the place in the zodiac where the Sun is located at an equinox, in this case the vernal or spring equinox). The vertical calibrated declination line that passes through this point connects the north and south equatorial poles. Copyright Rand McNally. Reproduced with permission, R.L.07-S-22.

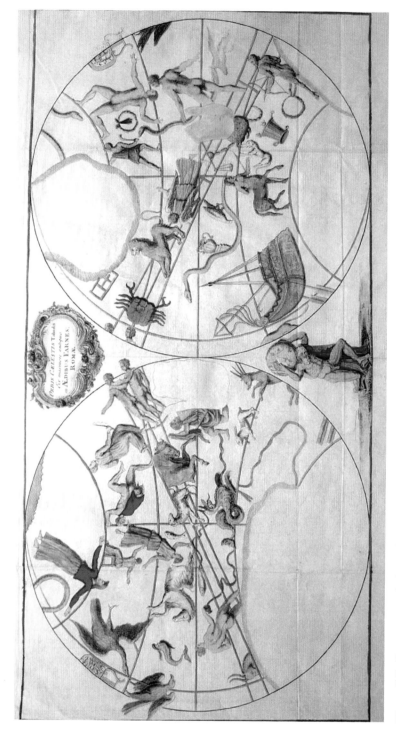

Figure 8.3. Paper transcription of two hemispheres drawn from the celestial globe that is part of the 2nd Century AD Farnese Atlas. 25.9 × 52.2 cm. This print appeared in Richard Bentley's 1739 edition of Manilius' *Astronomicon ex Recensione*. Note the classical Greek constellations, the gaps representing areas of damage or places where Atlas holds the globe, and a picture of the statue in the lower center.

428

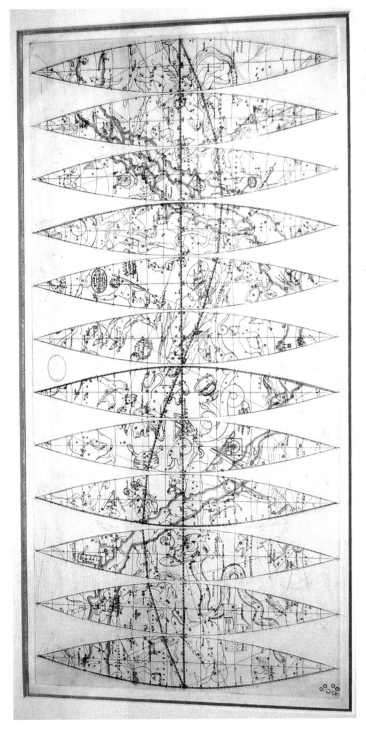

Figure 8.4. A finely engraved set of 12 gores for a 5 inch diameter celestial globe by Bales, 1845. 20.3 × 41.2 cm. Note the ecliptic and equator lines and the typical constellation figures reversed left to right.

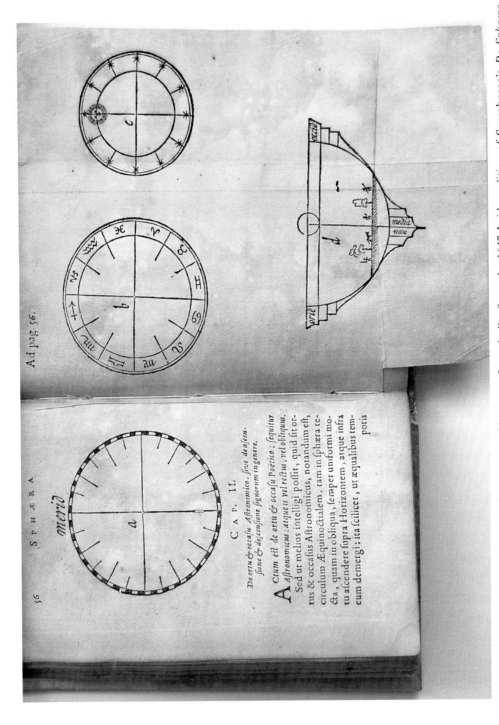

Figure 8.6. An insert (*right side*) containing the movable parts of a volvelle, from the 1647 Leiden edition of Sacrobosco's *De Sphaera*. 17.8 × 13.1 cm insert. Note that these pieces can be cut out and attached to the printed page on the left on to the 5.5 cm dia. printed circular scale.

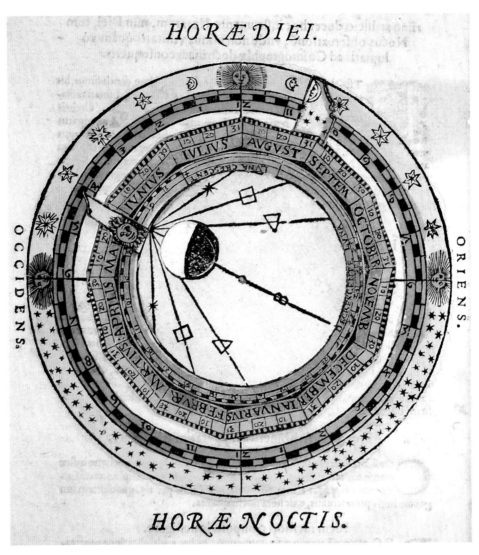

Figure 8.7. A beautifully colored volvelle from the 1584 edition of Peter Apianus' *Cosmographia*. 13.4 cm dia. circular scale. Note that by moving the inner and outer disks in the manner described in the text, the relative locations of the Moon and Sun in the sky can be demonstrated for different lunar phases.

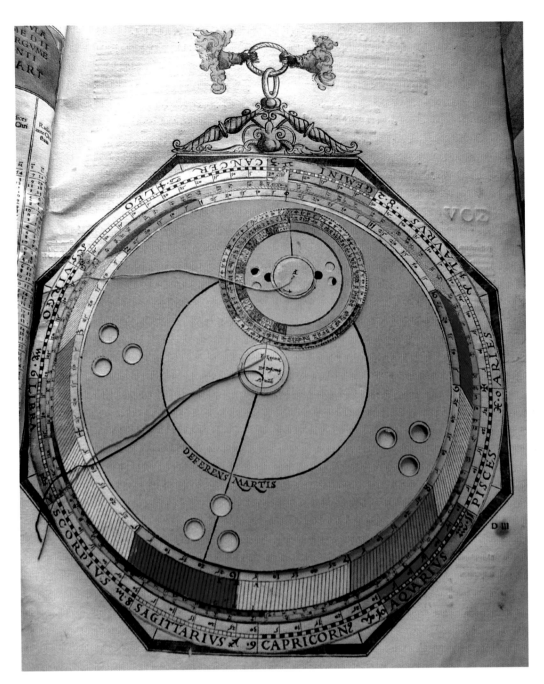

Figure 8.8. An equatorium type of volvelle from Peter Apian's *Astronomicum Caesarium*, published in 1540. Approximately 47 × 31.8 cm, 29.8 cm dia. hemisphere. Note that by manipulating the various disks and strings, the location and movement of the planet Mars in the sky can be determined as it moves through the ecliptic. Courtesy of the Collection of Robert Gordon.

Figure 8.10. A Victorian planisphere, ca. 1890, constructed under the direction of A. Klippei from Dortmund. Note the 22.2 cm square black heavy paper envelope with four Victorian-decorated pointers. This holds the 25.7 cm dia. revolving wheel, on which are printed the date and the night sky constellations (seen through the window of the envelope).

Figure 8.11. A French planisphere, ca. 1900, constructed under the direction of Camille Flammarion. Note the 18 cm square heavy paper base, on which are printed the date and night sky constellations. This is overlayed by a 14.5 cm dia. revolving wheel with clear plastic over the window.

Figure 8.12. The front of a brass astrolabe made by Badr ibn ʿAbdallah ca. 1130 probably in Baghdad. 18.8 × 13.4 cm. Note the rete star map and the tympan engraved with circles of altitude for the latitude of Persia. Courtesy of the Adler Planetarium & Astronomy Museum, Chicago, Illinois, A-84.

Figure 8.13. The back of a brass astrolabe made by Badr ibn 'Abdallah ca. 1130 probably in Baghdad. 18.8 × 13.4 cm. Note the Persian writing and the absence of an alidade sighting bar. Courtesy of the Adler Planetarium & Astronomy Museum, Chicago, Illinois, A-84.

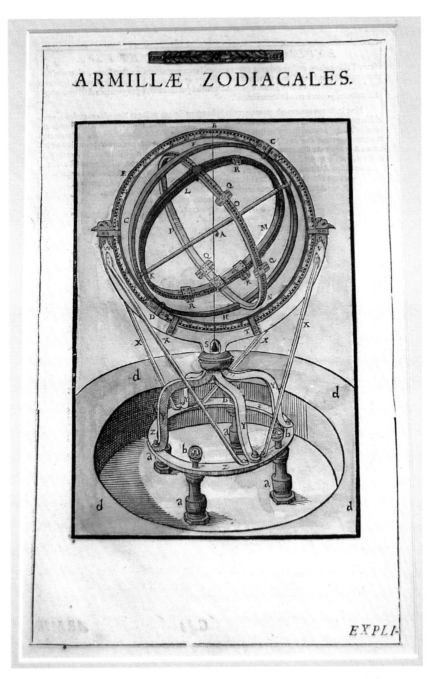

Figure 8.14. Page depicting an armillary sphere from an edition of Tycho Brahe's *Astronomiae Instauratae Mechanica*, first published in 1598. 27.7 × 16.3 cm. The instrument pictured was used by this famous Danish astronomer at his observatory, Uraniborg. Note the letters indicating the parts of the instrument that were keyed to a description in the accompanying text.

Figure 8.18. A beautiful frontispiece from volume 4 of a mid-1700s French book. 16.5 × 9.5 cm (page size). Note Galileo describing the telescope (termed here "The Dutch Spying-glass") and his discoveries of 1609 to Lord Sagredo and the Venetian nobility on the tower of St. Mark's in Venice.

Figure 8.20. Small spyglass-type refracting telescope made by Semitecolo, a well-known Venetian telescope maker, ca. 1800. It is made out of pasteboard, with bony rings separating each section. The main tube is 25.5 cm long. Note the beautiful floral and Greek key design and the screw-on eyepiece cap.

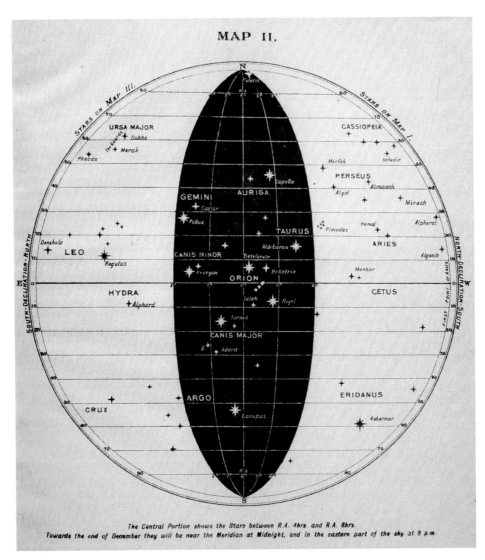

Figure 8.22. Map II from the 1921 edition of *Brown's Star Atlas*. 28.3 × 24.6 cm (page size), 20.7 cm dia hemisphere. Note the focus on bright stars and the highlighting of the area near the meridian (here emphasizing the star Sirius and the constellation Orion), which reflects the use of this atlas for navigation at sea.

440

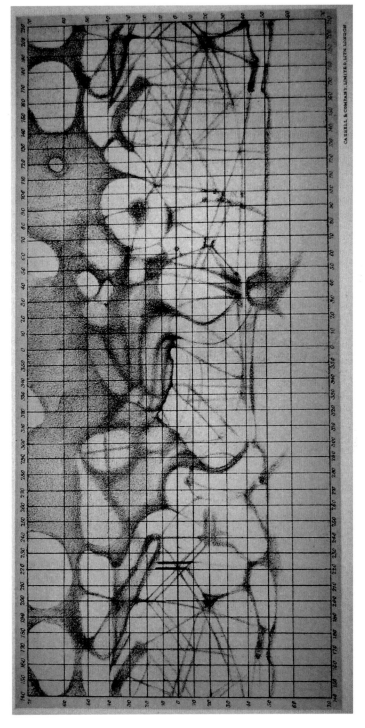

Figure 8.28. A reproduction of Schiaparelli's map drawing of Mars, 1881–1882, taken from Sir Robert Ball's *The Story of the Heavens*, published in 1897. 23 × 15.3 cm (page size). Note the prominent system of canals, some of which are in doubles.

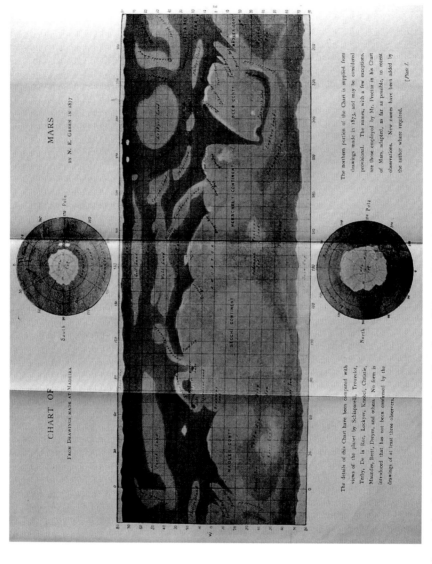

Figure 8.29. A reproduction of N.E. Green's map drawing of Mars, 1877, taken from the 1894 American edition of Flammarion's *Popular Astronomy.* 23.2 × 31.1 cm (pull-out page size). Note the absence of a prominent canal system and the statement that this is a compilation of the drawings of more than ten observers (including Schiaparelli) and that no item is introduced that has not been confirmed by three observers.

Figure 8.30. A contemporary view of "Mars and Syrtis Major", *http://grin.hq.nasa.gov/*, image #PR99-27D. Compare with older images shown in Figures 8.28 and 8.29. Courtesy of NASA.

Figure 8.31. A contemporary "Solar System Montage" showing the planets from Mercury (top) to Neptune (bottom), *http://grin.hq.nasa.gov/*, image #PIA01341. Compare with older drawings shown in Figures 8.24 and 8.25. Courtesy of NASA.

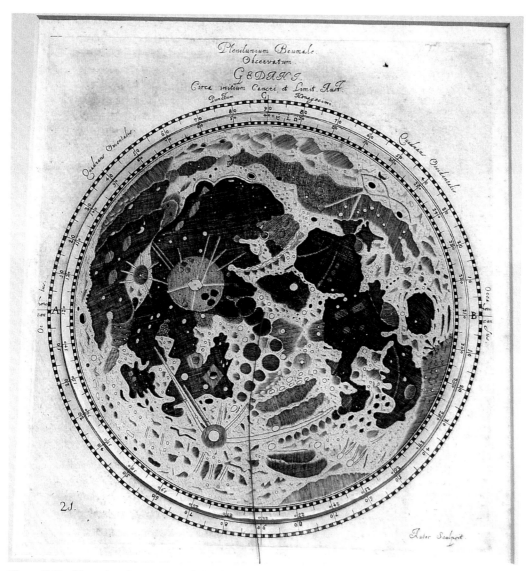

Figure 8.32. Plate showing a map of the full Moon from the first true lunar atlas, *Selenographia*, by Hevelius, which was published in 1647. 16.4 cm dia. inner rotating disk, 17.6 cm dia. outer circular ring. Note that this plate is actually a volvelle (the only one in the atlas), with a revolving Moon and the original measuring string that allowed for the angular measurement of the lunar axis with respect to the background stars.

Figure 8.33. This image is from a copper engraving by Johann Doppelmayr and published by Homann Publications, ca. 1730. 47.9 × 57.2 cm, each hemisphere 27.7 cm dia. This print depicts the lunar nomenclature system developed by Hevelius on the left and by Riccioli on the right. Hevelius named the features of the Moon after geographical features on Earth (which can be seen by turning the image 90 degrees counterclockwise), whereas Riccioli named them for famous people and scientists. Note in the Hevelius map *Sicilia* with *Mt. Aetna* in the center, which Riccioli named *Copernicus*, a name that persists to the present day.

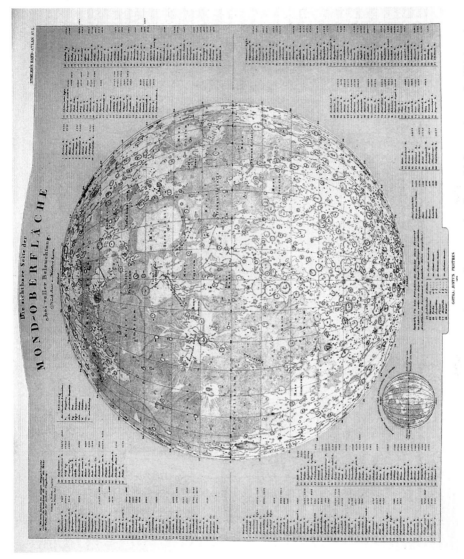

Figure 8.35. Map of the Moon printed in 1876, from the popular *Stieler's Hand-Atlas*, which was begun in 1816 and went through many editions until well into the mid-20th Century. 33.6 × 41.1 cm, 28.3 cm dia. lunar disk. Note the careful attention to detail and the almost photographic appearance of the lunar surface. Note also that the heading states that this image was taken from the famous lunar map produced by Beer and Mädler (in 1837).

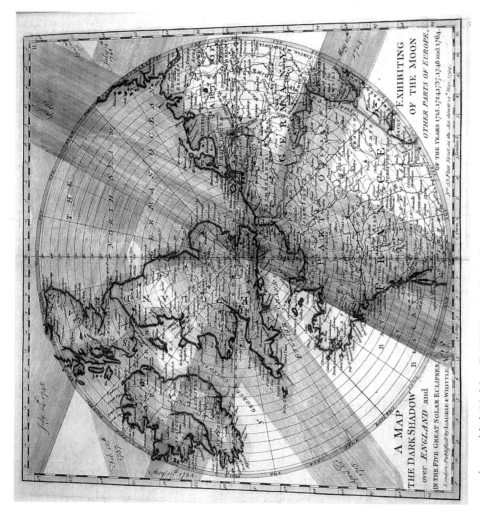

Figure 8.38. Copper engraving entitled "A Map Exhibiting the Dark Shadow of the Moon . . .", produced by Laurie and Whittle in 1794. 29 × 28.7 cm. Note the shadow paths from the five total solar eclipses seen from Britain in the 18th Century (in 1715, 1724, 1737, 1748, and 1764), which are superimposed on a high-quality terrestrial map of the U.K. and northern Europe.

Figure 8.43. A contemporary image labeled "Magnificent Details in a Dusty Spiral Galaxy", which is NGC4414 as shown by the Hubble Space Telescope, *http://grin.hq.nasa.gov/*, image #PR99-25. Compare with the older engraving of a spiral galaxy shown in Figure 8.42. Courtesy of NASA.

Figure 8.44. A celestial-themed card from a set of rare playing card maps by Hegrad entitled *Jeu des Cartes Geographiques*, published in 1785. 11.3 × 6.4 cm (card size). Note the image of an armillary sphere on an ace (the non-aces had geographical maps), and instead of a suit mark there is a red colored square.

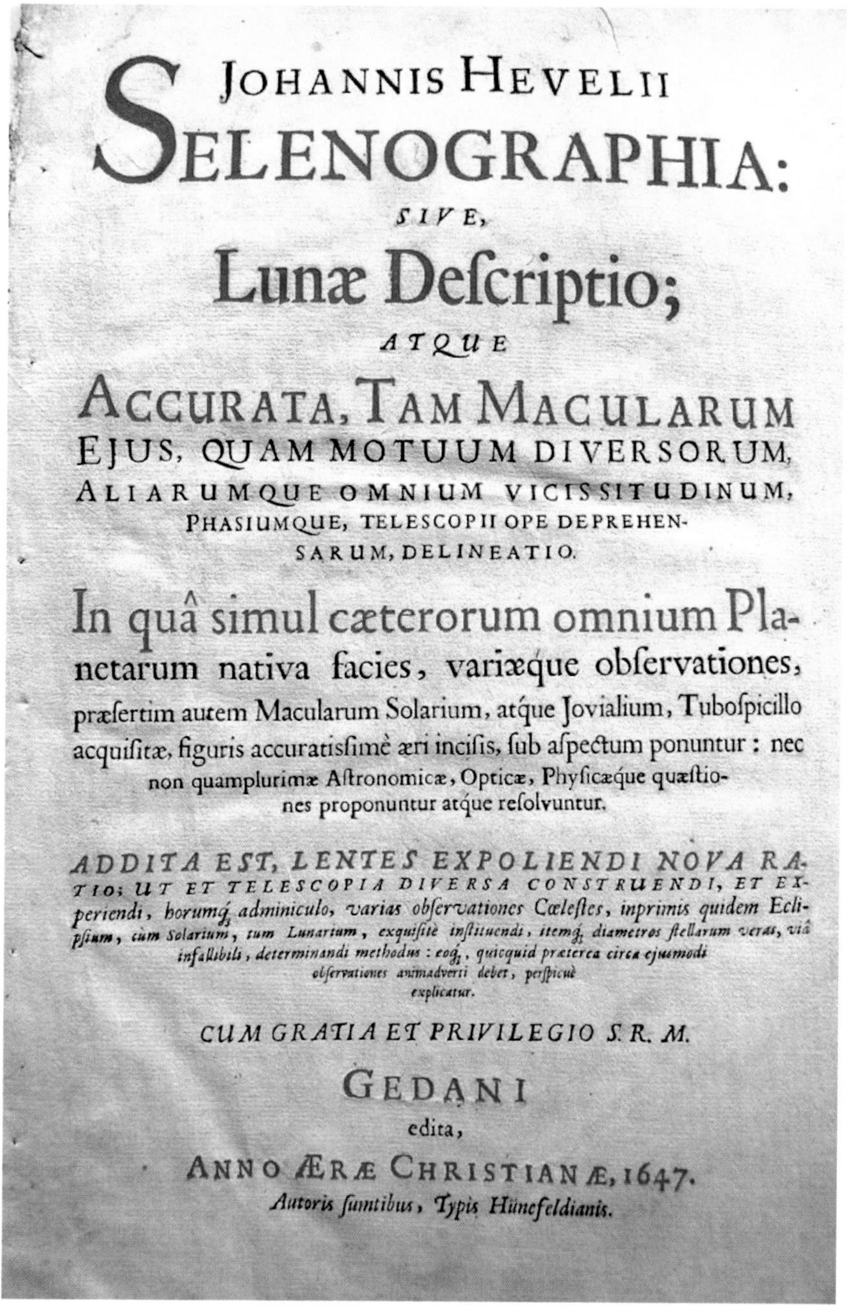

JOHANNIS HEVELII

SELENOGRAPHIA:

SIVE,

Lunæ Defcriptio;

ATQUE

ACCURATA, TAM MACULARUM
EJUS, QUAM MOTUUM DIVERSORUM,
ALIARUMQUE OMNIUM VICISSITUDINUM,
PHASIUMQUE, TELESCOPII OPE DEPREHEN-
SARUM, DELINEATIO.

In quâ simul cæterorum omnium Planetarum nativa facies, variæque obfervationes,
præfertim autem Macularum Solarium, atque Jovialium, Tubofpicillo
acquifitæ, figuris accuratisfimè æri incifis, fub afpectum ponuntur : nec
non quamplurimæ Aftronomicæ, Opticæ, Phyficæque quæftio-
nes proponuntur atque refolvuntur.

*ADDITA EST, LENTES EXPOLIENDI NOVA RA-
TIO; UT ET TELESCOPIA DIVERSA CONSTRUENDI, ET EX-
periendi, horumq̃ adminiculo, varias obfervationes Cæleftes, inprimis quidem Ecli-
pfium, cùm Solarium, tum Lunarium, exquifitè inftituendi, itemq̃ diametros ftellarum veras, viâ
infallibili, determinandi methodus : eoq̃, quicquid præterea circa ejufmodi
obfervationes animadverti debet, perfpicuè
explicatur.*

CUM GRATIA ET PRIVILEGIO S.R.M.

GEDANI

edita,

ANNO ÆRÆ CHRISTIANÆ, 1647.
Autoris fumtibus, Typis Hünefeldianis.

Figure 8.47. Title page from the first true lunar atlas, *Selenographia*, by Hevelius, which was published in 1647. 34.9 × 22.5 cm (page size). Note the amount of information given, including comments that not only the Moon but also the planets will be shown as they appear naturally through the telescope.

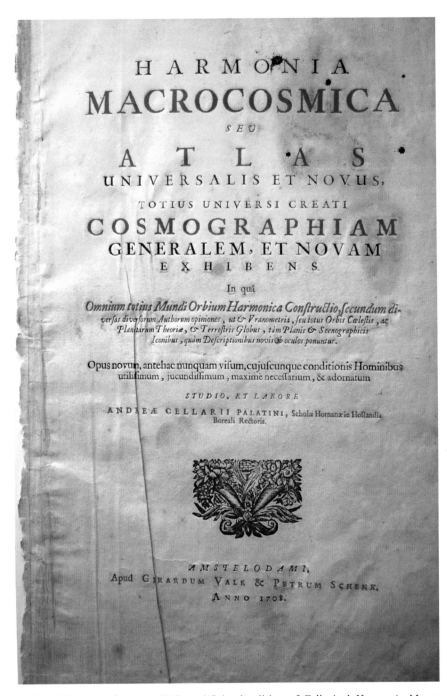

HARMONIA
MACROCOSMICA
SEU
ATLAS
UNIVERSALIS ET NOVUS,
TOTIUS UNIVERSI CREATI
COSMOGRAPHIAM
GENERALEM, ET NOVAM
EXHIBENS.

In quâ

Omnium totius Mundi Orbium Harmonica Conſtructio, ſecundum diverſas diverſorum Authorum opiniones , ut & Vranometria , ſeu totus Orbis Cœleſtis , ac Planitarum Theoriæ, & Terreſtris Globus , tàm Planis & Scenographicis Iconibus , quàm Deſcriptionibus novis ab oculos ponuntur.

Opus novum, antehac nunquam viſum, cujuſcunque conditionis Hominibus utiliſimum , jucundiſſimum , maxime neceſſarium, & adornatum

STUDIO, ET LABORE

ANDREÆ CELLARII PALATINI, Scholæ Hornanæ in Hollandia Boreali Rectoris.

AMSTELODAMI,
Apud GIRARDUM VALK & PETRUM SCHENK,
ANNO 1708.

Figure 8.49. Title page from the Valk and Schenk edition of Cellarius' *Harmonia Macrocosmica*, 1708. 51.2 × 30.7 cm (page size). Note that like the example shown in Figure 8.47, there is a lot of information given in the title page about the contents of the atlas.

Figure 8.50. Frontispiece from the Valk and Schenk edition of Cellarius' *Harmonia Macrocosmica*, 1708. 43 × 26.2 cm. Note the six famous astronomers at the bottom surrounding Urania, the muse of astronomy. They are pointing at the heavens or at important astronomical instruments, models, and books of their times.

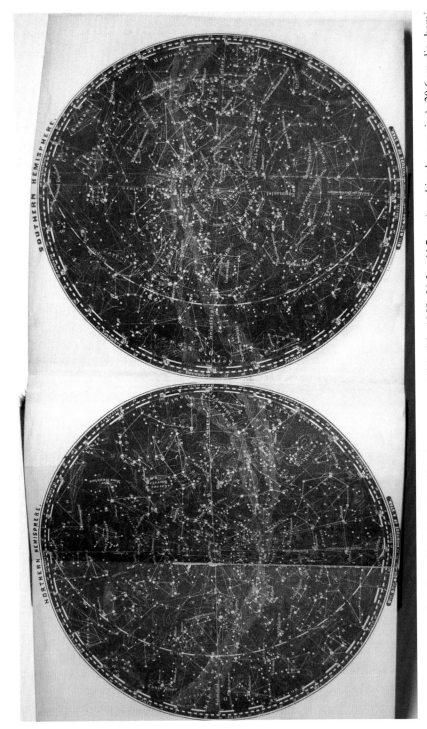

Figure 8.56. Frontispiece from Bouvier's *Familiar Astronomy*, published in 1857. 21.3 × 44.7 cm (combined page size), 20.6 cm dia. hemispheres. Note the very faint images but the prominent stick figures of the constellations in these northern and southern celestial hemispheres, which were meant to be consulted by the reader as he or she perused the text.

454

Figure 8.57. Combined frontispiece and title page from the 1661 edition of Bayer's *Uranometria*. 28.7 × 19 cm. Note the addition of a small printer's mark at the bottom and the names of various owners of the atlas handwritten in the margins, including a monastery entry at the bottom.

DECEMBER hath 31 days, beginning on Wednesday. [1847.

The year is died! I turn back thine eye
Along the path of life, and mark the way
O'er which thy soul, with many a tear and sigh,
Hath reached the dying year's departing day.

New Moon 7d 13h 42m after. Full Moon 21d 5h 20m after.
First qr. 14d 10h 3m after. Last qr. 29d 9h 0m morn.

M	W	CALENDAR, &c.	⊙ rise	⊙ sets	⊙dec south.	Full sea.	☽'s pl'c	☽ rise
1	4	7's sou.11h Cold and	7 23	4 37	21 48	4 16	♐	2
2	5	☽ Apog. windy	7 24	4 36	21 57	4 58	R.	1 50
3	6	♃ rises 7h.11m. ♂ ☽ D.	7 25	4 35	22 6	5 41	♑	2 54
4	7	☉ so.9h Pleasant, then	7 25	4 35	22 15	6 25	S.	3 51
5	C	♂ in Adv. comes	7 26	4 34	22 22	7 9	♒	4 48
6	2	♭ set 11h. snow	7 26	4 34	22 29	7 57	♒	5 44
7	3	● Fri. 3h. 15m M. or	7 27	4 33	22 36	8 46	T. sets.	
8	4	Day 6.5h.43m. ⓵low.	7 28	4 32	22 43	9 37	♈	5 57
9	5	♀ 6h.35m. rain or	7 28	4 32	22 49	10 29	K.	6 45
10	6	♭ 1h 57m M. both	7 29	4 31	22 55	11 21	♈	7 38
11	7	☉ fast cl 6m. Some	7 29	4 31	23 0	aft 14	L.	8 37
12	C	☉ in Ad. pleasant days,	7 29	4 31	23 5	1 6	♉	9 42
13	2	then comes snow or	7 29	4 31	23 9	1 57	♊	10 50
14	3	♑ se.10h.30m. ♂ ☽ D.	7 29	4 31	23 14	2 48	L.	11 55
15	4	♀ ri.3h.23m M. rain	7 30	4 30	23 17	3 39	♋	morn
16	5	♀ Perihe. again	7 30	4 30	23 20	4 31	H.	1 2
17	6	♀ ♂ ☽ Pleasant	7 30	4 30	23 22	5 25	♌	2 10
18	7	♀ ☽ Perigee. for the	7 30	4 30	23 24	6 20	N	3 18
19	C	♃ in Ad. season	7 30	4 30	23 26	7 17	♍	4 25
20	2	☽ ri. 5h 56m Cold	7 30	4 30	23 27	8 15	A.	5 31
21	3	full tides. ♂ so. 5h.	7 30	4 30	23 27	9 13	♎	rise.
22	4	◐ high. frosty weather.	7 30	4 30	23 27	10 9	B.	5 52
23	5	♀ se.10h ♃ ☽ D. Pleas.	7 30	4 30	23 27	11 0	♏	6 63
24	6	♀ ri.3h.34m M.ant ente-	7 30	4 30	23 26	11 54	H.	7 53
25	7	Christmas. rings for	7 29	4 31	23 25	morn	♐	8 53
26	C	Christmas visits, and	7 29	4 31	23 23	0 42	B.	9 63
27	2	now I will end the farce	7 29	4 31	23 21	1 29	♑	10 49
28	3	♀ se.1h 35m M. of the	7 29	4 31	23 18	2 11	♑	11 40
29	4	☽ so.0h 38m M weather	7 29	4 31	23 15	2 54	R	morn
30	5	for it is no better than	7 28	4 32	23 11	3 36	♒	0 42
31	6	guess work altogether.	7 28	4 32	23 7	4 19	S.	1 37

A LIST OF SHIPS BELONGING TO NEW-LONDON, STONINGTON, MYSTIC, SAG-HARBOR, AND GREENPORT EMPLOYED IN THE WHALE FISHERY.

VESSELS NAMES.	TONS	MASTERS.	AGENTS.	SAILED.
NEW-LONDON.				
Alert	398	Middleton	Williams & Haven	June 16,1845
Armata	413	Fitch	Abner Bassett	July 24, 46
Atlantic	700	Beck	Miner, Lawrence, Co	Aug 5, 45
Black Warrior	231	Chappell	Williams & Haven	May 3, 45
Benjamin Morgan	407	Bellows	Perkins & Smith	June 25, 46
Bengal,	304	Prink,	Thomas Fitch, 2d	May 21, 44
Brookline	360	Jeffrey	Perkins & Smith	July 7, 45
Chas. Carroll	404	Long	Perkins & Smith	Aug 26, 45
Candace	319	Bolles	Williams & Haven	June 2, 45
Clematis	311	Bailey	Williams & Barns	June 17, 45
Charleston	373	Bailey	N. & W.W. Billings	June 26, 44
Columbia	492	Kelley	Perkins & Smith	June 18, 44
Columbia	279	Lano	Miner, Lawrence, Co	July 15, 46
Clem, Perry, bark	270	Bailey	Frink, Chew & Co	Aug 20, 44
Connecticut, bark	398	Town	"	Aug 23, 45
Columbus, brig	159	Forsyth		June 3, 45
Charles Henry,	265	Allot	Williams & Barns	July 15, 45
Ceras, bark	176	Harris	Perkins & Rogers	Oct 12, 44
Catharine	384	Smith	Weaver & Rogers	Sept 3, 45
Carolina	395	Prentis	Thomas Fitch, 2d	July 1, 45
Columbus	344	Buchman	Stoddard & Learned	July 2, 46
Corea	365	Hempsted	Lyman Allyn	Oct 20 45
Dove, bark	145	Douglass	Frink, Chew & Co	
Dromo	430	Jeffrey	Williams & Haven	Aug 12, 45
Electra	306	Steel	Benjamin Brown	June 20, 44
	348	Ward	Thomas Fitch, 2d	April 22, 45
Exile, schr.	70	Church	Williams & Barns	June 18, 44
Fanno, bk.	258	Mitchell	Learned,& Stoddard	April 9, 46
Flora	338	Potter	William Tate	June 4, 46
Franklin schr.	119	Butler	N & W. Billings	July 28, 44
Gratland, schr.	60	Nichols	Benjamin Brown	June 17, 44
Gen. Williams	446	Ward	Perkins & Smith	June 21, 45
Gen. Scott	620	Holt		July 29, 45
Georgia,	350	Sistare	Williams & Barns	June 23, 46
George & Mary	344	Full	Weaver Rogers	June 2, 45
Hibernia	356	Bailey	Thomas Fitch, 2d	Aug 13, 44
Hand, schr.	550	Smith	Lyman Allyn	June 5, 44
Henry Thompson	915	Holmes	Frink, Chew & Co	Sept 11, 44

Figure 9.1. This figure shows two pages from Nathan Daboll's 1847 edition of *The New-England Almanac, and Farmer's Friend*. 15.8 × 10.7 cm (page size). Note on the left an ephemeris for December showing the daily rising and setting of the planets, the Sun, and the Moon, and the predicted weather for the month (but with the ending caveat: "now I will end the farce of the weather for it is no better than guess work altogether"). On the right is a table containing information about whaling ships belonging to several northeastern seaports.

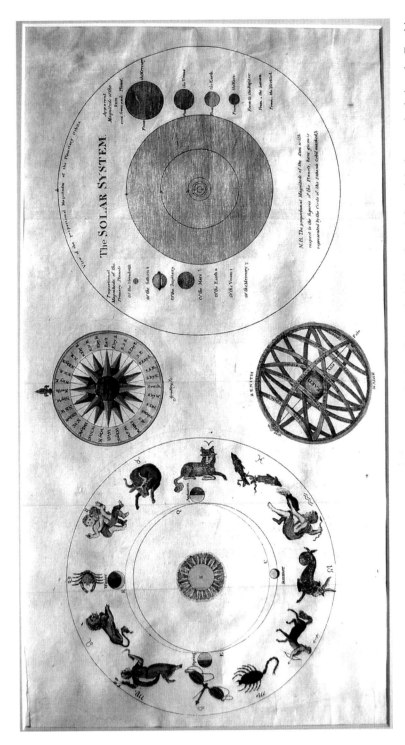

Figure 9.3. Print engraved by Enoch G. Gridley, ca. 1800. 18.4 × 35.4 cm (image size). Note from left to right images depicting the Earth's orbit surrounded by the figures and symbols of the constellations of the zodiac (14.5 cm dia.); a compass rose (*top*) and an armillary sphere (*bottom*); and information concerning the diameters and orbits of the known planets of the solar system, giving the name Herschel for what later was called the planet Uranus (15.5 cm dia.).

457

Figure 9.5. This figure is from a colored version of the 1835 edition of Elijah Burritt's *Atlas Designed to Illustrate the Geography of the Heavens*, showing the constellations in the Virgo/Leo region of the sky. 33.3 × 31.7 cm. Note the now-extinct constellation of Noctua the Owl, which is perched at the tip of Hydra's tail at the bottom left.

Figure 9.7. A double-page star chart showing the region involving Orion, Taurus, Musca, Eridanus, and George's Harp, from Mitchel's *Atlas Designed to Illustrate Mitchel's Edition of the Geography of the Heavens* . . . 23.3 × 32.7 cm. Note the differences from Burritt's atlas (Figure 9.5), including the white stars on black background and the faint constellation outlines.

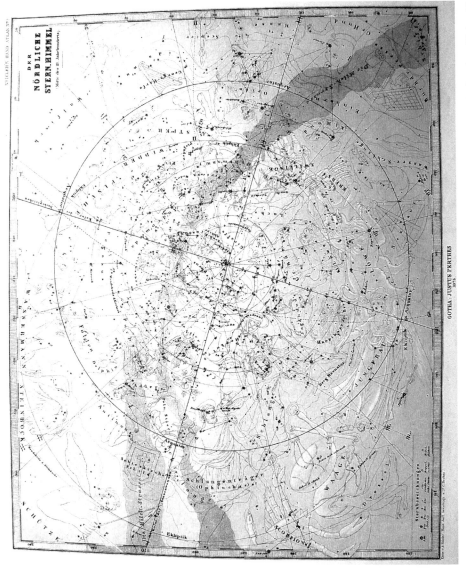

Figure 10.1. A map of the northern celestial region printed in 1875 from the 6th edition of *Stieler's Hand-Atlas*. 27.1 × 33.6 cm. Note the presence of a number of faint constellation images, constellation boundaries that are not very pronounced, stellar magnitudes indicated by slight variations in the size and shape of the star images, and an overall cluttered look.

460

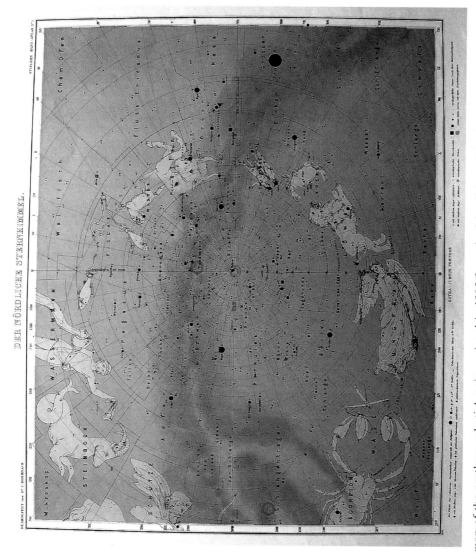

Figure 10.2. A map of the northern celestial region printed in 1905 from the 9th edition of *Stieler's Hand-Atlas*. 32.3 × 39.4 cm. Note, in contrast to Figure 10.1, only the zodiac constellations are imaged, the constellation boundaries are more pronounced, stellar magnitudes are more differentiated, and nebulae and novae are indicated by hexagonal symbols.

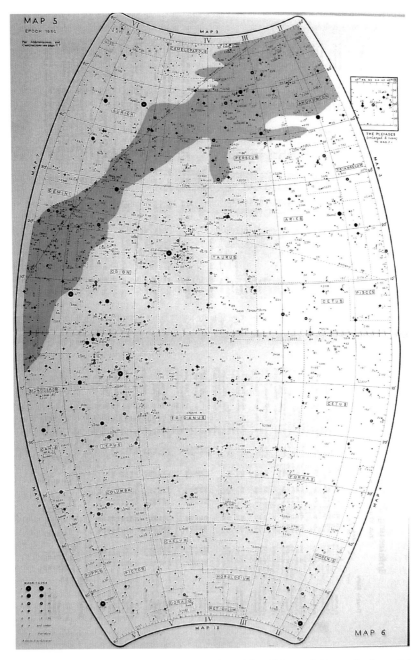

Figure 10.11. Map 5/6 showing Orion, from the 17th edition of *Norton's Star Atlas and Reference Handbook*, published in 1978. 38.3 × 25.1 cm. Note the absence of constellation images or lines, the presence of rectilinear constellation boundaries, stars and deep-sky objects shown for epoch 1950 (9,000 total), and the Milky Way indicated in green. Courtesy of Sky Publishing Corporation.

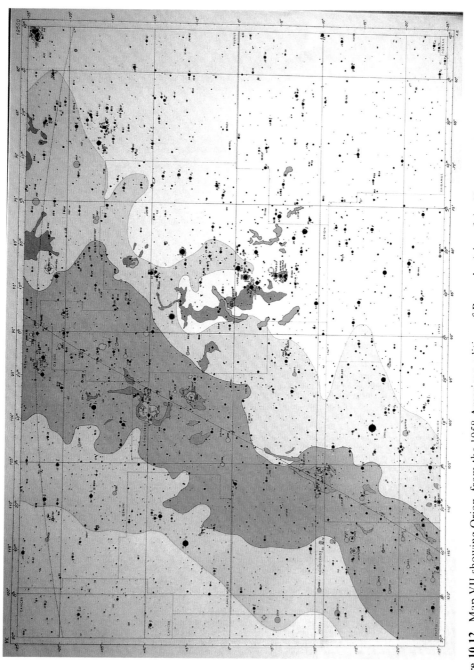

Figure 10.12. Map VII showing Orion, from the 1958 American edition of Becvar's *Atlas of the Heavens: Atlas Coeli 1950.0.* 37.4 × 52.4 cm. Note the absence of constellation images or lines, the presence of rectilinear constellation boundaries, stellar magnitudes indicated by clear variations in the size and design of the star images, and the Milky Way indicated in blue. Courtesy of Sky Publishing Corporation.

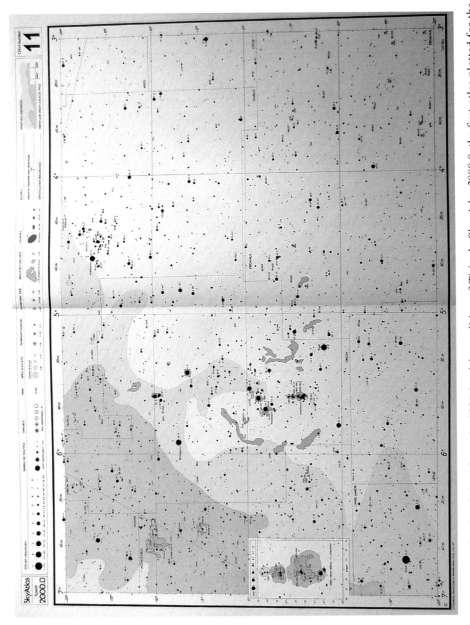

Figure 10.13. Chart 11 showing Orion, from the 1981 1st deluxe edition of Tirion's *Sky Atlas 2000.0*, the first atlas plotted for the epoch of 2000 AD. 31.1 × 45.4 cm. Note the absence of constellation images or lines, the presence of rectilinear constellation boundaries, stellar magnitudes indicated by clear variations in the size and design of the star images, the Milky Way indicated in blue, and an insert on the left showing the Orion Nebula. Courtesy of Sky Publishing Corporation.

Appendix A

Collecting celestial maps and prints

Collecting antiquarian celestial maps is actually a specialized area of map collecting in general, similar to other specialized areas (e.g., county or state maps of a country, maps of a country or continent, road maps, sea charts, maps of military campaigns, town plans, maps of an individual mapmaker, maps of a particular period of time). In all of these cases, one is concerned with such issues as authenticity, condition of the paper, type of colorization (if any), displaying the map, preservation, and price. Much of what will be said below could be said of any antiquarian map, be it celestial or terrestrial. I will only touch on some of the highlights specifically related to collecting celestial maps; more information is given in the references at the end.

One issue that sometimes comes up has to do with the ethics of buying any kind of antiquarian map that was once part of a bound atlas. Since it is sometimes more profitable to sell a series of individual pieces than to sell a bound collection, some dealers have been known to buy an atlas and then cut it up into individual pieces. Many book dealers consider this to be unethical. However, many antiquarian maps were originally produced in a loose state and were never bound up in an atlas. In addition, many atlases that are purchased by dealers are already damaged or incomplete to begin with and would have no value as a bound source. Finally, most collectors cannot afford to buy a complete atlas but can afford to buy an individual piece, and in this way antiquarian maps remain in circulation for the enjoyment of people rather than being stored away somewhere unseen and unappreciated.

A.1 THE MATERIALS IN PRINTED CELESTIAL MAPS

Hand-made celestial manuscripts on vellum or papyrus are exceedingly rare and expensive. Consequently, they are beyond the reach of most collectors, who are more likely to acquire printed woodblock maps or engravings made after the 1450s. All maps printed from an original woodblock or copper plate are considered to be original maps. Over the years, however, the plate may have been damaged and repaired, or perhaps new or updated information has been carved or engraved on it. In this case, the resulting map is called a second, third, or higher "state", depending on the number of times changes were made between printings. Some maps are produced for a specific atlas or book; others are produced loose to be used separately or to be bound later. Because paper size may vary from printing to printing, the most reliable indicator of a map's dimension is the image itself. Some people measure the height and width of the block mark (for woodblocks) or plate mark (for engravings). Others (as in this book) favor the dimensions of the innermost margins shown on the image.

Some maps are produced and left uncolored, whereas others are colored at the time of production or shortly thereafter to enhance their beauty and value. Such coloring is called "original", whereas if it is done some time after production it is called "later" coloring. Original coloring generally is uniformly and artistically done, may show some fading with age, uses more traditional colors rather than garish or modern "in vogue" colors, and generally matches the coloring of other maps made at the time. The green pigment verdegris used in the past had a copper and acid base that oxidized to brown over time and often leeched through the paper to show markings on the other side; both of these characteristics can be used to distinguish old from new coloring. However, there are some exceptions to these color rules. For example, to save money some maps were originally hand-colored by amateurs or even children in assembly-like fashion, resulting in variations in quality and colors that crossed into or over boundaries. On the other hand, some modern colorists are true artists who are able to add color that is difficult to distinguish from that used in past centuries.

A.2 CELESTIAL MAP PRESERVATION AND CONSERVATION

The ideal map should have a clear and strong image printed on clean and flawless paper with sufficient margins all around. However, antiquarian maps rarely come pristine, and condition often is a relative term. In some maps, the paper may be brittle or discolored due to overexposure to light, the acid found in the paper itself (rare in old maps prior to 1800 that were made of rag-pulp), or from contact with framing material using modern acidic wood-pulp paper (resulting in a clearly demarcated brownish "mat burn"). Other maps have mildew damage or "foxing", which are brownish spots due to mould caused by too much moisture in the air (see Figure 9.1 for example). Trauma and misuse can cause tearing, which is more serious if it impinges upon the image itself rather than being confined to the margins. Some maps that were originally issued across two pages in books may show excessive wear or splitting on the centerfold through repeated opening and closing of the facing pages, or discoloration from the glue used to attach a binding stub to the back (verso) of the map. Insects, bookworms, and other vermin can produce wormholes or bite marks on maps. Dirt, smoke, and water can discolor or stain any product made out of paper.

Maps are occasionally washed and bleached to remove dirt and stains, and this may cause them to feel stiff and acquire an unnatural whiteness.

Many of these problems can be prevented. Stored maps should be placed in areas that are free from direct light, moisture, dirt, and insects and vermin. They can be inserted in Mylar pockets that are acid-free and allow the maps to be seen but still protected from the elements. A room with a moderate temperature and dehumidifier or drying agent can help. Maps that are displayed should not be placed in direct light. They should be mounted on acid-free paper using acid-free tape. Ultra-violet ray blocking glass or plastic on the front of the frame can help to preserve the color and paper. As with stored maps, care should be taken to provide moderate temperatures and low humidity in areas where the maps are hung.

Where damage has already been done to a map, it can sometimes be restored using the services of a professional conservator. Of course, the decision to pay for this service depends on the value of the map and the extent of the damage. It is important to keep in mind that antiquarian celestial maps look old because they are old, and minor wear and tear is to be expected in most cases. Occasionally a map may be "backed" with a thin sheet of tissue or linen or even "laid down" on a piece of heavy cardboard. This is done as a conservation measure to give strength to an otherwise fragile sheet of paper. To see if the map has been so treated, as well as to fully assess the quality of the paper, it is important to be able to examine the verso when buying a celestial print, especially if it is presented to you already framed. Other conservation measures are beyond the scope of this book and are best left to be discussed with a professional.

A.3   BUYING CELESTIAL MAPS

Celestial maps tend to be cheaper than their terrestrial counterparts. However, one can still spend over $10,000 for a print and over $100,000 for a complete atlas. But if one is careful, the vast majority of antiquarian celestial prints can still be purchased for less than $100 up to $2,000. Although most map collectors focus on terrestrial maps, celestial maps are growing in popularity, and their prices will probably increase. Constellation maps are especially popular, although if one understands their meaning, cosmological maps also can be very satisfying.

Some people buy celestial maps as investments or to decorate an office or boardroom, but most buy them for their beauty and historical value. Particularly undervalued are maps from the 19th and 20th Centuries. An inexpensive alternative to an authentic map is a nicely printed reproduction, and to an atlas is a facsimile edition. One should be careful to buy from a reputable source that will provide a statement of authenticity. This helps guard against fakes, which may be hard to identify. One should be suspicious, however, if a supposed antiquarian map is too inexpensive, looks too clean and bright, lacks block or plate marks, and shows a solid appearance in the image where one would expect to find fine engraved lines. In addition, the paper used in antiquarian maps tends to be rougher than modern paper and often has a watermark or chain line impressions that are left by the grid on which the pulp was left to dry.

Most map dealers carry antiquarian celestial maps, although they also may be found in a second-hand bookshop. There is nothing like visiting a map dealer's show room, talking with the staff, and seeing the actual maps in person before buying. If you are a serious collector and have dealt with the dealer before, many will give you a small discount on their published price. Always ask the dealer what his or her "best price" is. Keep in mind that the ability of dealers to give you a discount will depend on a number of factors, such as the rarity and condition of the piece, the price they paid to purchase it, and their cost for doing business. For example, map prices tend to be higher in some high-cost urban areas, such as New York and London, which is unfortunate since these areas often have a number of dealers and a high volume of material. If you have a good framer, it is usually cheaper to buy a map unframed from a dealer and have it framed later yourself.

Once you have established a relationship with a dealer, you can check to see if he or she has a website, especially if you live far away. As new items are acquired by the dealer and are posted, you can do your shopping online. Many websites include photos of the items, and most dealers have a policy where you can return a map within a period of time for a refund (paying just the return shipping and insurance costs). Other sources for celestial maps include book and map fairs, where dealers come together to display their wares. These are wonderful opportunities to compare items and prices and to speak with a variety of people who know the field. Although not all of a dealer's items can be displayed, one can still make contacts and check out a dealer's stock online or through a mail-order catalog.

Many fine celestial maps can be purchased through auction, either in person or through the Internet. Some auction houses specialize in the Internet trade and show photos online of their offerings. There are some risks to buying at auction. For example, it is more difficult to get a refund of your money from an auction house than a dealer, even if there is a misrepresentation or a legitimate problem with the purchase. Also, there is usually an auction house buyer's premium added to the hammer price that needs to be factored in when you make a bid. Nevertheless, some rare pieces can be found at good prices in auctions, and they represent a viable source for more sophisticated and disciplined collectors.

A.4 BIBLIOGRAPHY

Clancy, R. (1995) *The Mapping of Terra Australis*. Macquarie Park, NSW, Australia: Universal Press.

Manasek, F.J. (1999) *Collecting Old Maps*. Norwich, VT: Terra Nova Press.

Moreland, C. and Bannister, D. (2000 [1993]) *Antique Maps*, 3rd edn. London: Phaidon Press.

Potter, J. (1999) *Collecting Antique Maps: A New Edition*. London: Jonathan Potter.

A.5 SOURCES OF ANTIQUARIAN CELESTIAL MAPS

Below is a list of map dealers and Internet auction houses that I have found to be reputable and reasonable in their pricing and service. In most cases there is a website or Internet address as well as the mailing address and phone number. Also included below is a list of map societies and journals. The dealer list is meant to be suggestive, not totally inclusive, and the absence of a dealer does not necessarily mean that he or she does not engage in good business practices. Also, this list reflects my personal experiences alone, although I have obviously spoken with other collectors over the years about dealers with whom they have had good (or bad) experiences and have factored their opinions into my own activities. I am receiving no remuneration from any of the sources listed below (all of whom have given me their written permission to print their names). I am providing these lists with the hope that it will give potential collectors who read this book some sources to begin with as they begin their journey into a new and exciting hobby.

A.5.1 Dealers who carry celestial maps

1. Altea Antique Maps & Rare Charts; 35 Saint George Street; London W1S 2FN; U.K.; 44-(0)20-7491- 0010; *www.alteagallery.com*
2. Antiquariat Nikolaus Struck; 10178 Berlin; Spandauer Strasse 29; Germany; 49-30-24-27-261; *www.antiquariat-struck.de*
3. Antiquariat Reinhold Berg; Wahlenstrasse 6; 93047 Regensburg; Germany; 49-941-52229; *www.bergbook.com*
4. The Antiquarium; 3021 Kirby Drive; Houston, Texas 77098; U.S.A.; 713-622-7531; *www.theantiquarium.com*
5. Barry Lawrence Ruderman Antique Maps; 7463 Girard Avenue; La Jolla, California 92037; U.S.A.; 858-551-8500; *www.raremaps.com*
6. George Glazer Gallery; 28 East 72nd Street, Suite 3A; New York, New York 10021; U.S.A.; 212-535-5706; *www.georgeglazer.com*
7. Gowrie Galleries; 4/380 Matcham Road; Matcham NSW 2250; Australia; 612-4365-6399; *maps@sydney.net*
8. Hanno Schreyer Buch- und Kunstantiquariat; Bonn-Endenich; Euskirchener Strasse 57-59; 53121 Germany (by appointment only); 49-(0)-22-86-21-059; *SchreyerBonn@t-online.de*
9. Iris Antique Globes and Maps; Weverweg 11; 6961 KM Eerbeek; The Netherlands; 31-(0)575-439440; *www.irisglobes.nl*
10. Jonathan Potter Limited; 52A George Street; London W1U 7EA; U.K.; 44-(0)-20-7491-3520; *www.jpmaps.co.uk*
11. Librairie Le Bail; 13 rue Frédéric Sauton; 75005 Paris; France; 33-1-43-29-72-59; *www.librairie-lebail.fr*
12. Libreria Antiquaria Perini; via A. Sciesa 11; 37122 Verona; Italy; 0039-045-8030073; *www.libreriaperini.com*
13. Librairie Loeb-Larocque; 31 rue de Tolbiac; 75013 Paris; France; 33-1-44-24-85-80; *www.loeb-larocque.com*

14. The Map House of London; 54 Beauchamp Place; Knightsbridge, London SW3 1NY; U.K.; 44-(0)-20-7589-4325; *www.themaphouse.com*
15. Martayan Lan; The Heron Tower, 6th Floor; 70 East 55th Street; New York, New York 10022; U.S.A.; 800-423-3741; 212-308-0018; *www.martayanlan.com*
16. Murray Hudson Antiquarian Books, Maps, Prints & Globes; P.O. Box 163; 109 South Church Street; Halls, Tennesee 38040; U.S.A.; 731-836-9057; *www.murrayhudson.com*
17. Old Print Shop; 150 Lexington Avenue; New York, New York 10016; U.S.A.; 212-683-3950; *www.oldprintshop.com*
18. Peter Harrington; 100 Fulham Road; Chelsea, London SW3 6HS; U.K.; 44-(0)-20-7591-0220; *www.peterharrington.co.uk*
19. Philadelphia Print Shop; 8441 Germantown Avenue; Philadelphia, Pennsylvania 19118; U.S.A.; 215-242-4750; *www.philaprintshop.com* (also gallery in Denver)
20. Robert Putnam Antique Maps and Books; P.O. Box 70084; 1007 KB Amsterdam; The Netherlands; *www.putmap.com*
21. Tim Bryars Ltd.; 8 Cecil Court; London WC2N 4HE; U.K.; 44-(0)20-7836-1901; *www.timbryars.co.uk*
22. Arader Galleries. San Francisco: 435 Jackson Street, San Francisco, California 94111, U.S.A., 415-788-5115, *www.aradersf.com*; New York: 29 East 72nd Street and 1016 Madison Avenue, New York, New York 10021, U.S.A., 212-628-3668, *www.aradernyc.com*. Also galleries in Philadelphia, Houston, and elsewhere.

A.5.2   Internet auction houses and fairs with a focus on celestial maps

1. Old World Auctions; *www.oldworldauctions.com*
2. Paulus Swaen Internet Map Auction and Gallery; *www.swaen.com*
3. Virtual Antique & Rare Art—Book & Map Fair; *www.antiquebookprintmap-fair.com*

A.5.3   Map societies

1. British Cartographic Society; *admin@cartography.org.uk*; *www.cartography.org.uk*
2. California Map Society; *www.californiamapsociety.org*
3. International Map Collectors' Society; *www.imcos.org*
4. Philip Lee Phillips Society; *www.loc.gov/rr/geogmap/phillips.html*
5. Washington Map Society; *www.washmap.org*
6. Society for the History of Discoveries; *www.sochistdisc.org*

A.5.4   Journals

1. *The Cartographic Journal* (the journal of the British Cartographic Society); Maney Publishing; Suite 1c; Joseph's Well; Hanover Walk; Leeds LS3 1AB; U.K.; 44-(0)-113 243 2800; *www.maney.co.uk*
2. *Imago Mundi; the International Journal for the History of Cartography* (c/o The

British Library Map Library, *http://www.maphistory.info/imago.html*); Taylor & Francis Publications; 4 Park Square; Milton Park; Abingdon, Oxfordshire OX14 4RN; U.K.; 44-(0)-207-017-6000; *http://www.tandf.co.uk/journals/imagomundi*

3. *IMCoS Journal* (the journal of the International Map Collectors' Society); London; U.K.; *www.imcos.org*

4. *The Portolan* (the journal of the Washington Map Society); U.S.A.; *www.portolan.washmap.org*

5. *Terrae Incognitae* (the journal of the Society for the History of Discoveries); U.S.A.; *www.sochistdisc.org*

Appendix B

Supplementary reference catalog
(see also Index for names and sources discussed in the chapters)

In addition to the people and maps described in the main text, the collector may run across additional antiquarian sources of relevance for collection or background research. For reference purposes, it is useful to briefly mention some of the involved individuals and their contributions, which is done below. Although I have directly seen examples of some of these works, I am dependent upon other sources of information for many others. Five such catalogs have been referenced. One of these, the catalog by Ashworth, is still in print and available online or in hard copy from the Linda Hall Library in Kansas City, MO. The Johnston catalog, which accompanied a traveling exhibition sponsored by the Boston University Art Gallery in 1885, is out of print, but it has morphed into a new version by McCarroll that is available and accompanied its own traveling exhibition. Sadly, the classic surveys by Basil Brown and Deborah Warner are out of print, and one only hopes that they will become available again (especially the Warner book, arguably the best detailed catalog in the field).

The complete references and the codes used in the subsequent summaries are:

A = Ashworth, W.B. Jr. (1997) *Out of this World—The Golden Age of the Celestial Atlas*. Kansas City, MO: Linda Hall Library.

B = Brown, B. (1932) *Astronomical Atlases, Maps and Charts*. London: Search Publishing.

J = Johnston, P.A. (1985) *Celestial Images—Astronomical Charts from 1500 to 1900*. Boston: Boston University Art Gallery.

M = McCarroll, S. (2005) *Celestial Images: Antiquarian Astronomical Charts and Maps from the Mendillo Collection*. Boston: Boston University Art Gallery.

W = Warner, D.J. (1979) *The Sky Explored: Celestial Cartography 1500–1800*. Amsterdam: Theatrum Orbis Terrarum.

The following entries are in alphabetical order by a person's last name, along with a brief description of their most important celestial maps and books. This material supplements (but does not repeat) the entries from earlier chapters. For books or atlases, the first edition is generally described. In some cases, subsequent editions will be mentioned, particularly if there are important differences. Readers wanting to research later editions can find these by accessing the online search catalog of books: *worldcat.org*.

Too late for inclusion in this edition is an excellent illustrated catalog of 130 celestial atlases (116 after 1800) by Robert W. McNaught entitled: *A Catalogue of Celestial Atlases*, 2011 (order through *www.lulu.com*).

Francisco de Aerden (also **Frans Afferden**) (1653–1709). Cartographer from Antwerp who in 1696 compiled an atlas entitled *El Atlas Abreviado* that contained a print labeled *Plani-spherium Coeleste*. It measured 19×30.3 cm (total page size) and included two 13 cm diameter celestial hemispheres. They were centered on the ecliptic poles using a polar stereographic projection with external orientation. The maps were influenced by Joan Blaeu and were bound in later editions of Afferden's atlas (until 1725) and an atlas by Verdussen. (J, M, W)

Carel Allard (1648–1709). Cartographer and publisher from Amsterdam who ca. 1706 produced two 50×57.6 cm prints labeled *Planisphaerii Coelestis Hemisphaerium Septentrionale* or *Meridionale* that each featured a 40.6 cm diameter celestial hemisphere. They centered on the ecliptic poles using a polar equidistant projection with geocentric orientation. Insets and tables also were included. The maps were influenced by Coronelli and were bound in atlases by Visscher, Allard, and Covens & Mortier. (W)

Noel André (see **R.P. Chrysologue de Gy, Capuchin**).

Johann Ludwig Andreae (active in early 1700s). Celestial map and globe maker who in 1724 produced two celestial hemispheres with no labels but attributed to Andreae. They were centered on the equatorial poles using a polar equidistant projection with geocentric orientation. The maps were influenced by Coronelli and could be bent into cones and viewed from the inside to give an illusion of the celestial vault. (W)

Remmet Teunisse Backer (active 1600s). Dutch cartographer who ca. 1684 produced a 48×58 cm print with a 28.5 cm $\times$ 56 cm rectangular map of the heavens whose long label began *Sterre Kaert of Hemels Pleyn* … It covered the area from the north to the south equatorial pole with external orientation. This and subsequent variations appeared in atlases by Backer; Danckerts or de la Feville (under the name of Ioannes de Ram); and Ottens or Elwe (under these names). (W)

Francis Baily (active mid-1800s). Author of the *Revised Charts of the Stars*, which were used in the production of the *Catalogue of Stars of the British Association for the Advancement of Science* in 1845. In the charts, lines were drawn around the figures to better define the constellation boundaries. (B)

John Bainbridge (1582–1643). The first Savilian Professor of Astronomy at Oxford who in 1619 produced an 18.5 × 16.5 cm celestial map entitled *Cometae qui effulsit Anno 1618 descriptio a 18 Nov.; ad 16 Decemb*. It showed the path of the Nov.–Dec. 1618 comet through constellations using a geocentric orientation. The map was printed by Edward Griffin for John Parker. (W)

James Barlow (active late 1700s). British cartographer who ca. 1790 produced two prints that each featured a celestial hemisphere labeled *N:* (or *S:*) *Coelestial Hemisphere*. They were centered on the ecliptic poles using a polar stereographic projection with geocentric orientation. Insets were included. The constellation figures resembled those from Seller. (W)

Jacob Bartsch (1600–1633). Influential mathematician, astronomer, and physician who was born in Silesia but traveled widely for his schooling. He spent much of his time in Strasbourg, where he was appointed Professor of Mathematics and published several books on astronomy and astrology. From 1626 to 1627, he worked in Augsburg with Julius Schiller and supervised the publication of his atlas after Schiller died. He also helped Kepler calculate his ephemerides based on the Rudolphine Tables, and in 1630 he married Kepler's daughter, Susanna. In 1624 he published a book entitled *Usus Astronomicus Planisphaerii Stellati seu Vice-Globi in plano ...* The book was popular, went through 1661 and 1674 editions, and contained astronomical tables, constellation lore, and biblical references. It also included three celestial maps labeled *Planisphaerii Stellati sue Vice-Globi Coelestis ...* One map was 32 × 26.5 cm and featured a planisphere centered on the north equatorial pole down to below the Tropic of Cancer using a polar equidistant projection with geocentric orientation. The other two maps were 16 × 37 cm, and each showed half of the zodiac from 35 degrees N to 55 degrees S dec. using a rectangular projection and geocentric orientation. (B, W)

James Bassantin (ca. 1504–1568). Scottish astronomer and mathematician who taught at the University of Paris and in 1557 published a book entitled *Astronomique Discours*, which contained a celestial planisphere labeled *Instrument pour trouver les longitudes & latitudes des Etoiles fixes ...* It was folio size and centered on the north ecliptic pole down to 60 degrees S lat. in polar equidistant projection with external orientation. The map was reprinted in the 1613 edition of his book. It was copied from Apian's *Astronomicum Caesareum*. (W)

George Baxter (see **Isaac Frost**).

Nicolaus Bazelius (active ca. 1578). Flemish physician who in 1578 published a treatise entitled *Prognostication nouvelle de cest An calamiteux 1578. Avec description de la comète vue le 14 de Novembre en l'an passé* (and a Latin version entitled *Prognosticon novum ...*), which contained a celestial map showing the path of the comet of 1577. (W)

Nicolas Bion (ca. 1652–1733). Official engineer for mathematical instruments to the King of France who in 1699 wrote a book entitled: *L'Usage des Globes Célestes et Terrestres* that included a number of celestial diagrams and maps. Later editions included a plate with two celestial hemispheres labeled *Planisphère Céleste*. Each hemisphere was 12.2 cm in diameter and was centered on the equatorial poles using a polar equidistant projection with geocentric orientation. These hemispheres were derived from those of de la Hire. Around 1708, he constructed a movable planisphere overlaid with a metal meridian and horizon circle centered on the north equatorial pole to 26 degrees S dec. using a polar projection with geocentric orientation, but no copies of this are known to exist. (W)

John Blagrave (ca. 1558–1611). British mathematician, surveyor, and instrument maker who in 1596 published a book and celestial planisphere that both began with the title *Astrolabium Uranicum Generale* ... The planisphere was about 25 cm in diameter and was centered on the north equatorial pole down to 30 degrees S. dec. using a polar stereographic projection with external orientation. It was patterned after Mercator's 1551 celestial globe. (W)

Joan Blaeu (1596–1673). Cartographer in Amsterdam (and son of Willem Janszoon Blaeu) who in 1648 produced a 205 × 299 cm map of the world in two hemispheres labeled *Nova Totius Terrarum Orbis Tabula*. In the upper corners were two ca. 35 cm diameter celestial hemispheres centered on the ecliptic poles using a polar stereographic projection with external orientation. In his 1662 book entitled *Atlas Maior*, Blaeu also showed a series of observing instruments that were enlarged versions (42.2 × 27.6 cm total page size) of those shown in Tycho Brahe's 1598 book *Astronomiae Instauratae Mechanica*. (J, W)

Richard Bloom (1641–1705). English engraver who produced a number of prints and maps that sometimes were plagiarized from other sources, such as Speed, Sampson, and de Sauvigny. They ended up in periodicals such as *The Gentlemen's Recreation*. (M)

Zacharias Bornmann (active 1596). German cartographer who in 1596 published an atlas of the Ptolemaic constellations labeled *Astra. Alle Bilder des Himmels* ... Each map was 13 × 16.5 cm and showed several constellations with a star showing the direction to the equatorial pole. The constellations were geocentric and resembled those of Dürer. (W)

Georg Brentel (active ca. 1573). Bavarian artist and sundial and instrument maker who in 1573 produced a celestial hemisphere labeled *Georgius Brentel has imagines formabat* ... that was centered on the north ecliptic pole using a polar projection with exterior orientation. Although zodiacal figures are shown, only four circumpolar constellations with stars are mapped, along with the positions of the Sun, Moon, and planets on Dec. 1, 1572. (W)

R. Brook (active 1799). British engraver who in 1799 produced two 34 cm diameter hemispheres labeled *Synopsis of the Universe. Northern* (or *Southern*) *Hemisphere*. They were centered on the ecliptic poles using a polar projection with external orientation (but no constellation figures). The stars are white against a black background, and there are figures depicting the Sun, Moon, and planets. (W)

Henry Brooks (active early 1800s). Author of an atlas on navigation in 1820 entitled *Guide to the Stars in either Hemisphere*, which contained 12 planispheres. (B)

Francesco Brunacci (1640–1703). Italian astronomer who in 1687 produced a print with two celestial hemispheres labeled *Planisfero del Globe Celeste*. Each was 26.7 cm in diameter and was centered on the ecliptic poles using a polar stereographic projection with external orientation. Insets of the Sun, Moon, and planets were included, along with text. (W)

Margaret Bryan (active 1797–1815). English schoolmistress and natural philosopher who wrote several books on astronomy for the lay public. In her book *A Compendious System of Astronomy*, first published in 1797 with follow-up editions, there appear two celestial hemispheres labeled *The Northern* (or *Southern*) *Celestial Hemisphere*. Each is 15.7 cm in diameter in a pull-out measuring 22.9 × 18.7 cm total page size, and each was centered on an equatorial pole using a polar stereographic projection with external orientation. The maps were influenced by Senex but reversed left-to-right. (J, W)

Jacques Buot (died ca. 1675). French astronomer, geometrician, and original member of the Académie Royale des Sciences who in 1657 produced a 44.5 cm diameter celestial planisphere with no label (but which included a statement *Par son très humble Servitor Buot*). It was centered on the north equatorial pole to the Tropic of Capricorn using a polar projection with exterior orientation (but no constellation figures). (W)

Georg Busch (died ca. 1590). Artist and astronomer who lived in Nuremberg and Erfurt who in 1577 published a document entitled *Beschreibung von zugehörigen Eigenschafften* ... that contained a picture on the title page and two maps of the path of the comet of 1577. One map was a planisphere 12 cm in diameter and centered on the north ecliptic pole down to 20 degrees S lat. using a polar projection and external orientation (but with no constellation figures). The other map was a 11.5 cm-square quadrant of the northern sky using an ecliptic polar projection with external orientation (and including constellation figures seen from the rear). (W)

Friedrich Büthner (1622–1701). Professor of mathematics at the Danzig Gymnasium who in 1681 published a document entitled *Cometa Scepticus. Oder Kurtzer Discursz Über Jungst-Erschienenen Cometen MDCLXXXI*, which contained two celestial maps. One measured 10 × 17.5 cm and showed the path of a comet in Nov. 1680 through constellations using a geocentric orientation. The other measured 9.2 × 27 cm and showed the path of the comet from Dec. 1680–Feb. 1681 through constellations using a geocentric orientation. (W)

Claude Buy de Mornas (active mid-1700s). Prominent French mapmaker who in 1761 produced *Atlas Méthodique et Elementaire de Géographie et d'Histoire*, which contained several celestial maps and diagrams. Typically, his plates were double-paged with a central, highly decorated image and text written on either side. (M).

Christoph Cellarius (1638–1707). (No relation to Andreas Cellarius, described in Section 7.1.) German historian and geographer who in 1705 published a document whose long label began *Elementa Astronomia* ... It contained two 13.5 cm diameter celestial planispheres centered on the equatorial poles to 35 degrees N dec. using a conical projection with geocentric orientation. The constellation figures were copied from the conical maps of Wilhelm Schickard (see below) but were more primitive in style. (W)

G.F. Chambers (active mid-1800s). Author of the *Cycle of Celestial Objects* in 1860, an observer's guide that consisted of many celestial plates and a catalog and went through at least one additional edition in 1881. (B)

R.P. Chrysologue de Gy, Cupuchin (1728–1808). Astronomer and geographer, also known as Noel André, who lived in a Capuchin convent in Paris adjacent to Le Monnier's observatory. In 1778 he produced two prints that each featured a 58 cm diameter celestial hemisphere labeled *Planisphère Céleste Boréal* (or *Austral*) *Projeté sur le Plan de L'Equateur*. Each was centered on an equatorial pole using a polar stereographic projection with external orientation. The maps contained over 4,400 stars from observations of Flamsteed, Hevelius, Halley, Le Monnier, Lacaille, and others, and they used Bayer letters and boundary lines around the constellations. The maps were influenced by Senex and were bound in atlases from chez Merigot l'aine and chez Perrier &Verrier. The hemispheres were approved by the Académie Royale des Sciences, and according to Warner they represented the epitome of large celestial hemispheric maps. Chrysologue de Gy also produced smaller celestial planispheres covering the area from the north equatorial pole to 33 degree S dec. (W)

A. Cottam (active late 1800s). Author of *Charts of the Constellations*, published in 1889 and containing 36 star charts down to 35-40 degrees S. dec. Additional editions and reissues were also published. (B)

Lucas Cranach the Elder (1472–1553). Painter and engraver from Wittenberg who designed a paper astrolabe with constellation figures derived from Dürer. The rete was a cut-out map of the constellations rather than the customary star pointers. This astrolabe was published by de Nova Domu in Wittenberg in 1529. (W)

Augustin Darquier de Pellepoix (1718–1802). French astronomer who in 1771 published in Paris an introductory atlas of the brightest stars entitled: *Uranographie, ou contemplation du ciel, à la portée de tout le monde*. A second edition followed in 1780. (W)

Cunradus Dasypodius (ca. 1530–1600). Humanist and professor at Strasbourg who edited several ancient mathematical and astronomical texts and helped facilitate the construction of the astronomical clock in the Strasbourg cathedral. In 1573 he produced a broadsheet labeled *Brevis et succincta descriptio corporis luminosi ...* showing the location of the 1572 nova in Cassiopeia in a celestial hemisphere centered on the north ecliptic pole using a polar projection with external orientation. A German version was also published. (W)

Heinrich Decimator (ca. 1544–1615). German theologian and author who in 1587 published a poetical discussion of astronomy with woodcut maps of the Ptolemaic constellations labeled *Libellus de Stellis Fixis et Erraticis ...* in octavo size with external orientation. The figures were shown from the rear and resembled those of Dürer.

Louis Charles Desnos (active ca. 1768). French cartographer who served both the French and Danish kings who produced a papier-mâché nautical device labeled the *Cosmo-Plane and Planisphere* in Paris in 1768. It was similar to an astrolabe and consisted of a 16.5 cm diameter revolving star finder over a 25.4 cm diameter planisphere. (J)

Thomas Digges (ca. 1546–1595). British mathematician and early supporter of the Copernican system who produced a celestial map of Cassiopeia that appeared on the verso of the title page of Marsh's 1573 book *Alae Seu Scalae Mathematicae ...* It was 17 × 13 cm and contained the 13 Ptolemaic stars in a geocentric orientation. On the next page was a catalog of the stars giving their magnitude, latitude, and Copernican longitude. (W) He also proposed his own heliocentric model where our solar system is surrounded by stars that are not bounded in a shell but extend out to infinity. A famous image of this world view was published in his 1576 *A Perfit Description of the Caelestiall Orbes ....*

Edwin Dunkin (active mid-1800s). British Superintendent of Altazimuth at the Greenwich Observatory and a Fellow of the Royal Astronomical Society who in the 1860s published 38 star maps in *Leisure Hours* and again in a book for students entitled *The Midnight Sky*. The maps showed white stars on a black background and were accompanied by an index and other astronomical information. The book underwent several later editions. (B)

Charles-François Dupuis (1742–1809). French author who in 1795 published *L'Origine de tous les Cultes ou Religion Universelle*, a book on the beliefs of religion and astrology. It contained 22 engravings, including depictions of Egyptian constellations and their division of the sky into 36 segments called decans, which could be used for astrological purposes. Subsequent editions were published in 1822 and 1835. (M)

Paul Fabricius (ca. 1525–1589). Viennese mathematician and Imperial counselor who according to Warner produced two celestial maps showing the paths of comets. One was labeled *Der Comet in Mertzen ...* and showed the comet of March 1556. The other was labeled *Cometae qui Anno 1577. a die 10. Nouemb: ad 22. diem Decemb: conspectus* ... and showed this comet in a planisphere centered on the north ecliptic pole down to 10 degrees S lat. using a polar projection with external orientation. (W)

James Ferguson (1710–1776). British teacher and popularizer of science who in 1757 published a volvelle he called *A New Astronomical Instrument* ... Warner describes this as being a paper instrument that "consists of a calendrical circle overlaid with a rotating star map that, in turn, is overlaid with a cutout plate of alt-azimuth coordinates and charts of Dominical Letters" (p. 79). The star map is about 18 cm in diameter and is centered on the north equatorial pole down to 40 degrees S dec. using a polar equidistant projection and geocentric orientation. Although he did not publish a star atlas, Ferguson wrote several books on astronomy that included plates of astronomical instruments and star charts, including his *Astronomy Explained upon Sir Isaac Newton's Principles* ..., which according to Brown went through some 13 editions between 1756 and 1811 and introduced Sir William Herschel to astronomy. (B, W)

M. Flecheux (1738–1793). French astronomer who in 1778 produced a 42 cm diameter celestial hemisphere labeled *Planétaire ou Planisphère Nouveau, Inventé et Dessiné par M. Flecheux*. It was centered on the north equatorial pole using a polar equidistant projection with external orientation. It included a number of coordinates and was intended to show the paths of the Sun and Moon, the daily equation of time and the time of lunar meridian passage, the locations of stars and their transits for a given location, and the longitude at sea. Its use was explained by a booklet entitled *Planétaire, ou Planisphère nouveau, inventé par M. Flecheux*, published in 1778 and 1780. (W)

Frisius (see **Isiberandius Middoch**).

Isaac Frost (active mid-1800s). Scientist, artist, and member of the Muggletonians, a religious sect formed after the English Civil War in the 1650s by two cousins, John Reeve and Lodowick Muggleton. They believed that Christ was in fact the true God, that Heaven and Hell existed on Earth rather than in the afterlife, and that formal religious ceremonies and evangelism were useless. Frost was instrumental in helping to refine the sect's astronomical theory of a geocentric universe during the Victorian period and was involved with the publication of *Two Systems of Astronomy* in 1846. This book contained a series of 22.9 × 30.5 cm planetary engravings drawn by Frost and printed in London by George Baxter, who had developed a method of printing using oil pigments that produced a glowing effect in the image. (M and other sources)

Christlieb Benedict Funke (1736–1786). Professor of Physics at the University of Leipzig who in 1770 published a book entitled *Anweisung zur Kenntniss der Gestirne auf zwei Planiglobien und zween Sterntegeln, nach Doppelmayers Himmelscharten*, which contained four celestial maps derived from the work of Doppelmayr. In 1777 he published a similar book entitled *Anweisung zur Kenntniss der Gestirne auf zwei Planiglobien und zween Sterntegeln, nach Bayern und Vaugondy*, which contained four celestial maps derived from the work of Bayer and de Vaugondy. Two of these, labeled *Planiglobium boreale* or *australe*, consisted of 49 cm diameter celestial hemispheres centered on the equatorial poles using a polar stereographic projection with geocentric orientation. The other two, labeled *Coniglobium boreale* or *australe*, consisted of 49 cm diameter celestial hemispheres centered on the equatorial poles using a conical projection with geocentric orientation. Both sets of maps included constellations from Hevelius and southern stars with Greek and Roman letters as assigned by Lacaille. (W)

Rev. James Gall (active mid- to late 1800s). Author of *An Easy Guide to the Constellations* in 1857, which contained 6 star maps with white stars shown against a sky-blue background. This went through several editions into the 1900s. (B)

Adam Gefugius de Vellendorph (active ca. 1565). Cartographer who in 1565 produced a northern celestial hemisphere (and possibly a southern one as well) centered on the ecliptic pole and labeled *Speculum firmamenti sive coelum stellatum* ... on a 26.1 × 26.6 cm print that used a polar projection with geocentric orientation. It was modeled after Honter. (W)

Cornelius Gemma (1535–1577). Physician and astronomer at Louvain who in 1578 published a book entitled *De Prodigiosa Specie, Naturaq. Cometae* ... that contained a celestial map of the path of the comet of 1577 in a 13.5 cm radius quarter circle centered on the north ecliptic pole using polar stereographic projection with external orientation. It was modeled after Dürer. (B,W)

Jacob de Gheyn (1565–1629). Dutch artist who in 1586 engraved a portrait of Tycho Brahe and ca. 1600 engraved 39 maps of the Hyginus constellations for Germanicus Caesar's *Arati Phaenomena in Latinis versibus conversa* ... The maps were 15.5 × 13.5 cm with external orientation and first appeared in his *Hug Grotii Syntagma Arateorum* ... in 1600 and in *Aratea sive signa coelestia in quibus Astronomicae* ... in 1621 and 1652. Warner states that the woodcuts were derived from Ratdolt's 1482 prints and later influenced Bayer. (W)

Simon Girault (active ca. 1592). French author of children's books who in 1592 published a book entitled *Globe du Monde* ... that contained a print of two celestial hemispheres labeled *Le Globe Céleste*. Each hemisphere was 11.8 cm in diameter and was centered on an ecliptic pole using a polar stereographic projection with geocentric orientation. The south pole contains a square of four stars that Warner suggests is derived from the observations of Amerigo Vespucci. (W)

I. Gobille (active ca. 1677). French cartographer who in 1677 produced two circular celestial maps covering the sky from the north to the south equatorial pole, each of which was 42.8 cm in diameter using an equatorial stereographic projection with extenal orientation. One map covered right ascension 0–180 degrees and was labeled *Uranographie, seu globi caelestis mappa* ... The other covered right ascension 180–360 degrees and was labeled *I.G. Pardies societatis iesu presbiteri* ..., reflecting the fact that both maps were derived from Pardies. (W)

Theodore Graminaeus (active ca. 1573). Professor of Mathematics at Cologne who in 1573 wrote a book entitled *Erklärung oder Auszlegung eines Cometen ...* that contained 11 constellation maps derived from Dürer in polar stereographic projection with external orientation. It also contained an 11.7 cm diameter hemisphere centered on the north ecliptic pole in polar projection with external orientation that had no figures but showed the path of the comet of 1556. Warner cites a reference suggesting that Graminaeus also produced a map of the path of the comet of 1577 that appeared in the 1578 *Weltspiegel oder, Algemeiner widerwertigkeit ...* (W)

Orazio Grassi (1582–1654). Italian mathematician at the Collegio Romano who authored an anonymous treatise in 1619 on the comets of 1618 entitled *De Tribus Cometis Anni M.DC.XVIII ...* It contained three charts of comets against the background of constellations. The first comet is shown in Ursa Major on Aug. 29 and Sept. 2. The second comet is shown in Hydra, Corvus, and Crater on Nov. 18 and 30. The third comet is shown from Draco to Virgo from Nov. 29 to Dec. 29. All three maps used a gnomonic projection with geocentric orientation. (W)

J. Green (active early 1800s). Author of *Astronomical Recollections* in 1824, which contained 19 plates, 17 of which showed the constellations taken from Bode. The stars were shown down to magnitude 4. (B)

Christoph Grienberger (1561–1636). Jesuit astronomer and mathematician at the Collegio Romano who in 1612 published a book entitled *Catalogus veteres affixarum Logitudines ac Latitudines conferens cum novis*. This book contained both a catalog of 1,225 stars and a celestial atlas of 24 maps and 2 hemispheres. Each map focused on one or a few constellations, was quarto size, and used a gnomonic projection with geocentric orientation. The hemispheres were labeled *Hemisphaerium Boreale* or *Australe* and were each centered on an equatorial pole using a polar stereographic projection with external orientation. (W)

P. Grimaldi (or in Chinese, **Min Ming-Wo**). French Jesuit in China who in 1711 published in Peking a document entitled *Fang Sing-tou-kiai* that contained 6 star maps and 3 coordinate diagrams. The maps were around 22 cm square and used a gnomonic projection with geocentric orientation. The star positions and coordinates were copied from Pardies. The constellations were Chinese and were shown by lines joining the appropriate stars. Although the names of the stars and constellations were in Chinese, these maps were exported and used in Europe as well. (W)

Chr. Louis Joseph de Guignes (1759–1849). French orientalist who in 1785 published a document entitled *Mémoires de Mathématique et de Physique, Présentés à l'Académie Royal des Sciences*. This work contained two celestial hemispheres entitled *Planisphère Céleste Chinois. Partie Septentrionale or Meridionale*. Each was 42 cm in diameter and was centered on an ecliptic pole using a polar equidistant projection with geocentric orientation. Chinese constellations were superimposed on the western constellations, with the former taken from Grimaldi and the latter taken from de la Hire. Both French and Chinese were used to name the stars and constellations. (W)

Amédée Guillemin (active mid-1800s). Author of *Le Ciel*, which was later translated into English as *The Heavens* in 1867 and consisted of 5 plates of the stars as seen from Paris and 1 plate of the southern skies. (B)

Isaac Habrecht II (1589–1633). Physician and professor of mathematics and astronomy at the University of Strasbourg, who in 1628 produced two celestial hemispheres that appeared in his *Planiglobium Coeleste, et Terrestre* ... Each measured 23 cm in diameter and was centered on an equatorial pole using a polar stereographic projection with geocentric orientation. Further editions of the book appeared in 1650, 1662, and 1666. In addition, in 1621 Habrecht produced a 20.4 cm celestial globe that Warner says was influenced by Plancius and was the first to depict the southern constellation of Rhombus, now known as Reticulum. (W)

Thaddaeus Hagecius ab Hagek (1525–1600). Czech Imperial physician and astronomer who produced a map of the nova of 1572 in Cassiopeia labeled *Imago Cassiopeae* that appeared in his 1574 *Dialexis de Novae et Prius Incognitae Stellae* that was 21 × 15 cm with external orientation. He also produced a map of the path of the comet of 1577 that appeared in his 1578 treatise *Descriptio cometae, qui apparvit* ... that was 12 × 17 cm with geocentric orientation. The style of both of these maps was derived from Dürer. (W)

Johann George Hagen (active late 1800s and early 1900). Director, Specola Vaticana, who in 1899–1908 published the 6-part *Atlas Stellarum Variabilium*, which focused on variable stars and contained 311 star maps, along with catalogs and comparison stars. (B)

Sidney Hall (active 1800s). Author of *Thirty-two Plates of the Constellations*, where the stars were perforated in size according to their magnitudes, allowing the star patterns to be visible when held up against a light. (B)

C.L. Harding (active early 1800s). Friend of Frederick Wilhelm Argelander (see Section 10.1.4.1) who produced an atlas in 1822 entitled *Atlas Novus Coelestis* ... that showed some 40,000 stars down to magnitude 9 and went through several editions through 1856. (B)

E. Heis (active mid-1800s). Distinguished German astronomer who in 1872 published a star atlas entitled *Coelestis Novus*, along with an accompanying catalog. In 1878 he published the *Atlas Coelestis Eclipticus*, which contained zodiacal stars in 8 charts. (B)

John Hill (active mid-1700s). English physician who wrote a book entitled *Urania*, in which he added 14 new constellations. According to Brown, the book was published in 1754 and 1768, and it was a kind of astronomical dictionary, with topics arranged alphabetically. (B)

John Russell Hind (active mid-1800s). Author of *Atlas of Astronomy* in 1855, which contained 18 plates, including 6 star maps that showed white stars against a blue background. This atlas went through several editions and was used mainly in schools. (B)

Jodocus Hondius, Jr. (ca. 1594–1629). Dutch cartographer and son of cartographer Jodocus Hondius, Sr., who produced two 6 cm diameter celestial hemispheres entitled *Globus Coelestris* ..., which appeared in Petrus Bertius' 1616 *Tabularum Geographicarum Contractarum*. Each was centered on an ecliptic pole using a polar projection and exterior orientation. It was also reproduced in Henricus Regius' *Philosophia Naturalis*. (W)

Jakob Honold (died 1727). Clergyman and teacher in Wurttemberg who in 1677 published a document entitled *Sidereus Dei Clarigator, Das ist, Kurtzer Bericht von dem Neuen Cometen* ..., whose title page had a map labeled *Situs Cometae Novi A. C. M.DC.LXXVII*. This pictured the path of the comet of 1677 against a constellation backdrop using a geocentric orientation. Honold published another celestial map with a cometary path in 1681 (also in a geocentric orientation) entitled *Cometae Vespertini Situs*, which appeared in his *Monitor Hominum Novissimus. Das ist: Kurtzer Bericht von dem Ungewöhnlich Grossen Cometen* ... (W)

Robert Hooke (1635–1703). English experimental philosopher and member of the Royal Society who on August 26, 1663 presented a map of the Pleiades to the Society in geocentric orientation that contained over 80 stars (far more than earlier representations). A map based on this representation was published in 1665 in Hooke's *Micrographia*. (W)

Christiaan Huygens (1629–1695). Famous Dutch natural philosopher and astronomer who constructed his own telescopes and defined the ring around Saturn, discovered the satellite Titan, and was a co-discoverer of the deep-sky object M42. In 1659 he published the *Systema Saturnium*, which contained an often-reproduced 1656 drawing of the Orion nebula in geocentric orientation that contained 7 stars within the nebula and 5 stars outside of it. Huygens continued to study and define this well-known deep-sky object. (W)

Alexis Hubert Jaillot (active late 1600s). French cartographer who produced ca. 1690 the engraving *Four Systems of Cosmology*, depicting the planetary systems of Ptolemy, Brahe, Descartes, and a fourth image representing the Sun and fixed stars, with a curving figure representing the symmetry of the heavens. (M)

T. Jamard (active in the 1750s). French ecclesiastic who published two geocentric celestial maps dealing with the path of the first predicted return of Halley's Comet. The first map was produced in 1757, was 18×68 cm, and showed previous appearances and the predicted path of the upcoming appearance. It was entitled *Routes de la Comète qui a été observée pendant les Années 1531, 1607, 1682, et qui doit reparoitre en 1757, ou 1758, au plutard*. It was presented to the King and approved by the Académie Royale des Sciences. The second map predicted the path of the comet in May 1759 after it went behind the Sun. It was entitled *Route que doit tenir pendant le mois de May 1759, La Comète qui a été observée en 1531, 1607, et 1682, et qui a passé à son Perihelie* ... The constellations were derived from Flamsteed. (W)

Jan Januszowski (1550–1623). Polish jurist, publisher, and amateur astronomer who in 1585 edited a book entitled *Jan Kochanowski* that contained the latter's Polish translation of Aratus' *Phaenomena*, along with a pair of woodblock celestial prints labeled *Phaenomena Albo Wyraz Znakow Polnocnych*. Each print was 26 cm square and contained a 22 cm diameter celestial hemisphere centered on an ecliptic pole in polar stereographic projection with geocentric orientation. Aside each map was a list of astronomical terms, and the maps were derived from Honter. The book was reissued in 1604. (W)

Edme-Sebastien Jeaurat (1724–1803). French astronomer and Professor of Mathematics at the Ecole Royale Militaire in Paris who in 1782 published in *Mémoires, Académie Royale des Sciences* a catalog and a 20×24 cm geocentric map of the Pleiades labeled *Carte des 64. principales Etoiles des Playades*. (W)

T. Jefferys (active mid-1700s). Chart published by the British Parliament in early 1748 informing its citizens of the appearance of the solar eclipse of July 14. It was labeled *The Geography of the Great Solar Eclipse of July 14, 1748*. Around a central map of the path of totality are 24 eclipse images, as they would be seen from various cities. (M)

Vasilii O. Kipriyanoviji (died ca. 1725). Russian publisher who in 1707 produced a 94.1×130 cm print with two celestial hemispheres with the translated label of *Celestial globe or heavenly sphere in which are depicted the constellations of the fixed stars* ... Each hemisphere is centered on an ecliptic pole using a polar equidistant projection with geocentric orientation. The constellations are patterned after Coronelli. (W)

Gottfried Kirch (1639–1710). German publisher, member of the Berlin Academy of Sciences, and Director of the Berlin Observatory who produced a number of beautiful maps of individual constellations for *Acta Eruditorum* and introduced three new constellations, illustrated in Warner. One was published in 1684 as a 12×8.8 cm print with an external view of Gladii Electorales Saxonici (near Bootes) adjacent to an image of Scutum Sobiescianum. Two other maps were produced in 1688 and showed geocentric views of a 16.5×8 cm print of Pomum Imperiale (near Aquilla) and a 16.5×6 cm print of Sceptrum Brandeburgicum (near Eridanus). The first two honored Kirch's partron, Emperor Leopold I. The last honored the Brandenburg province of Prussia, where Kirch lived, and it was reintroduced by Bode a century later (see also Section 4.4). (W)

Hermann Joseph Klein (active late 1800s). Author of *Star Atlas*, which contained 18 celestial maps of stars and deep-sky objects down to 34 degrees S. dec., but no constellation figures. This book was translated into English in 1888. (B)

Georg Friedrich Kordenbusch (1731–1802). Professor of Mathematics and Physics at Aegidien-Gymnasium in Nuremberg who in 1789 produced a series of 6 celestial maps designed for schoolchildren that appeared bound in *Atlas Scholasticus et Itinerarius*. They were labeled *Pater Ignaz Gaston Pardies, der Gesellschaft Jesu ehemahligen Grossenlehrers Himmelskugel in Sechs Karten abgebildet* ... Each was 32 cm square (with overlap) and used a gnomonic projection with geocentric orientation. Two were centered on the equatorial poles, and the others were centered on the two equinoxes and two solstices. The constellations were derived from Pardies but included some constellations from Hevelius. (W)

Francis Lamb (active ca. 1675). London engraver and cartographer who produced two celestial hemispheres that were listed in the 1673 Term Catalogue for Trinity as *Two large Hemispheres of the Heavens projected upon the Poles of the World* ..., accompanied by a book describing their use entitled *Astroscopium*. Each unlabeled hemisphere was 35.5 cm in diameter and was centered on an equatorial pole using a polar stereographic projection with external orientation. The constellations were derived from Blaeu's globes and in turn were copied by Philip Lea (see below). In 1679, Lamb published another document with two hemispheres entitled *Two large Hemispheres in Plano, of above Thirty Inches diameter* ... The hemispheres were labeled *Planispherium Boreale* (or *Australe*) *Stellarum*. Each measured around 76 cm in diameter and was centered on an equatorial pole using a polar stereographic projection with geocentric orientation. Insets of different cosmological systems, maps of the Moon and Earth, and the comparative sizes of the planets were included in the corners. (W)

E. Lapie (1756–1844). French geographer who in 1838 authored the *Atlas Universel*, which contained the engraving *Systèmes Planétaires*. It showed the systems of Ptolemy, Copernicus, and Tycho Brahe, along with a depiction of the universe as it was known in the early 1800s. Also shown were lunar and solar eclipses, the phases of the Moon, and a depiction of the seasons. In the center was an armillary sphere. (J, M)

Joseph-Jerôme Lefrançais de Lalande (1732–1807). Famous French astronomer who published a celestial map of the path of a comet entitled *Comète de 1762. Calculée par M. De La Lande* in the 1762 issue of *Mémoires, Académie Royale des Sciences*. The map was 19.2 × 14.9 cm, and the constellations around Lynx were shown in a geocentric orientation (see also Section 4.4). (W)

Philip Lea (active 1666–1700). London cartographer and globe and instrument maker who around 1686 produced two untitled celestial hemispheres that were sold at the "Atlas and Hercules in Cheapside London". Each was essentially identical to Lamb's 1673 hemispheres in size and orientation, although Lea added a bold border design and included additional data from Halley in his southern map. (W)

Pierre Le Clerc (active ca. 1770). Deacon at the church of Rouen who in 1772 produced a celestial planisphere published in Amsterdam labeled *Planisphère Céleste* . . . (in French) and *Nieuwe Hemel-Kaart* . . . (in Dutch). It was 58 cm in diameter, centered on the north equatorial pole down to 55 degrees S. dec., used a polar stereographic projection down to the equator, and had an external orientation. The constellations were derived from Senex, and there was a small inset map of the remaining southern skies. (W)

Jean Le Febure (active ca. 1700). Parisian instrument maker who around 1700 produced a celestial planisphere labeled *Planisphère nouvellement mis en pratique par LeFebure* . . . It was around 60 cm in diameter, centered on the north equatorial pole down to 30 degrees S. dec., used a polar stereographic projection, and had a geocentric orientation. The constellations were derived from Royer. (W)

Pierre Le Lorrain (see **Abbé de Vallemont**).

Georges-Louis Le Rouge (ca. 1741–ca. 1779). French military engineer, hydrographer, and cartographer who was appointed Ingénieur Géographe du Roi to King Louis XV and published many atlases of fortifications, North American maps, hydrographic charts, and other things. In his 1756 *Atlas Nouveau Portatif* is bound a double celestial hemisphere labeled *Le Globe Céleste en Deux Plans Hemisphères*. The print is 20.2 × 26.9 cm, with each hemisphere measuring 9.5 cm in diameter. Each is centered on an ecliptic pole using a polar projection and geocentric orientation. The maps are derived from de la Hire. Inset around the hemispheres are images of the Sun, Moon, and planets according to famous astronomers; diagrams of lunar phases and the great circles on the Earth, and two small hemispheric terrestrial maps centered on the poles. (W)

Andreas van Luchtenburg (active 1664–1706). Instructor of astronomy, mathematics, geography, and navigation in Rotterdam who published several celestial maps. For example, in 1684 he produced a planisphere labeled *Platte Globus* . . . that was centered on the north equatorial pole down to 38 degrees S dec. in a polar stereographic projection with external orientation. Similar to a map by Voogdio (see below), it contained an overlaid rete that represented a projection of the sky over the Netherlands that served as a star finder. Around the same time, Luchtenburg produced a map of the Earth labeled *Nieuwe Werelt* that was centered on the north equatorial pole down to 60 degrees S dec. It was encircled by a map of the stars from the north to south pole using a polar equidistant projection with geocentric orientation (but no constellation figures). The print measured around 126 × 105 cm, and in the corners were drawings of great men and allegorical figures. According to Warner, the same map was also issued under the name of Jacobus Robijn. A similar map was issued around 1697 under the label *Nieuwe Wereld*. Finally, around 1688 Luchtenburg issued a planisphere labeled *Nieuwe Hemels Spiegel waer door den hemel* . . . that was centered on the north equatorial pole down to 40 degrees S dec. in a polar equidistant projection with external orientation. It was 49 cm in diameter, and the print itself was 60 × 53 cm. The corners were decorated with scientific instruments and diagrams of the Copernican universe and the Earth's orbit around the Sun. This piece was reprinted with faces of the winds replacing instruments at the end of the 17th Century and appeared ca. 1703 in Danckert's *Atlas*. (W)

Friedrich Madeweis (1648–1705). German astronomer, linguist, and postman who in 1681 published two treatises and a broadside showing the path of a comet entitled *Cometa A. MDC LXXX et LXXXI* . . . The background constellations were geocentric. In the right lower corner was a drawing of an egg on which was a picture of a comet, and the adjacent text recapitulates a myth of a Roman hen laying such an egg when the comet was visible. (W)

Michael Maestlin (1550–1630). Professor of Astronomy at the University of Tubingen who according to Warner produced the oldest known map of the Pleiades in 1579; a map of the path of the comet of 1577/8 entitled *Observatio et Demonstratio Cometae Aetherae Qui Anno 1577 et 1578* . . . in 1578; and a map of the comet of 1580 entitled *Consideratio & Observatio Cometae Aetherei Astronomica* . . . in 1581. The latter used an external orientation but no figures, and it also showed the nova of 1572 and the path of the comet of 1577. (W)

Lucilio Maggi (active ca. 1565). Italian philologist (a.k.a. Lucillus Philalthaeus) who in the 1565 *L. Philalthaei in 1111 libros Aristoteles* . . . included two celestial hemispheres labeled *Septentrionales Imagines* and *Meridionales Effigies* that were centered around the north and south ecliptic poles, respectively, and were each 12 cm in diameter with a polar stereographic projection and geocentric orientation. The Ptolemaic constellations were patterned after Honter. (W)

Greg Mariette (active ca. 1690). Parisian cartographer who produced two sets of celestial hemispheres labeled *Hemisphère Septentrional* (or *Meridional*) *du Globe du Firmament*. The hemispheres were 26.5 cm in diameter and were centered on an ecliptic pole using a polar projection. One set (produced in 1697 and labeled *Le Globe Concave* . . .) used a geocentric orientation, and the other set (produced around the same time and labeled *Le Globe Convèxe* . . .) used an external orientation. (W)

Pierre Mariette (died 1657). Book and print dealer in Paris who around 1650 produced a print with two celestial hemispheres labeled *Planisphère du Globe Céleste*. Each was 21 cm in diameter and was centered on an ecliptic pole using a polar projection with external orientation. The model was likely an early Blaeu globe. (W)

Johann Mayer (active ca. 1680). German clergyman who produced a hemisphere labeled *Observabat et delineabat ... Anno 1680*, which appeared in Ussn's 1681 *Vorstellung Dess jungst-erscheinen Cometen ...* The planisphere was 31 cm in diameter, centered on the north ecliptic pole, used a polar equidistant projection, and had a geocentric orientation. (B, W)

Pedro de Medina (1493–1576). Spanish historian and author of an influential 1545 book on navigation labeled *Arte de Navegar* (or *L'Art de Naviguer ...*). According to Warner, de Medina's claim to fame as a celestial cartographer stems from a 4-star diagram in his book of the Southern Cross and how to use it in determining latitude, and as a result he has been cited as an authority on the southern skies by such people as Petrus Plancius and Johannes Bayer. (W)

J. Messer (active late 1800s). Author of *Stern Atlas für Himmels-beobachtungen*, the second edition of which was published in 1902 and contained 28 star charts and a description of the chief objects in each constellation. (B)

Conrad Meyer (1618–1689). Artist and engraver in Zurich who in 1681 produced a planisphere labeled *Planisphaerium Coeleste Cura Philomusi ...*, which was centered on the north equatorial pole down to 28 degrees S dec., was about 30 cm in diameter, and used a polar stereographic projection with external orientation. Included was an overlaid rule with one-degree gradations. (W)

Charles Theodore Middleton (active mid-1700s). Author of the *Complete System of Geography*, published in London in 1750, which went through many editions. One of the images depicted matching terrestrial and celestial globes side by side. (M)

J. Middleton (active mid-1800s). Author of *Celestial Atlas* in 1842, which showed all of the constellations visible from Great Britain. This was followed up by his *Celestial Atlas and Student's Companion* in 1846, which had a large circulation in schools and showed double maps, with and without (white on black background) the constellation figures. (B)

Isiberandius Middoch (Frisius) (active 1558). Cartographer who produced two large woodcut maps in 1558 labeled *Sydera coeli utriusque hemisphaerii iam recens edita* for the students at the University of Wittenberg that according to a source cited in Warner were copied from the Dürer planispheres. (W)

Min Ming-Wo (see **P. Grimaldi**).

Otto Mollinger (active 1851). German cartographer who in 1851 published *Himmels Atlas mit transparenten Sternen*. The star maps were printed on cards with different-sized holes punched out at the star positions which, when held up to the light, revealed the star patterns. The card maps had no constellation images but had bright stars connected by lines and constellation boundaries. (A)

Germiniano Montanari (active 1660s). Italian cartographer who in 1665 produced a finely engraved star map on which was traced the path of the comet of 1664–5 entitled *Cometes Bononiae observatus, Anno 1664 et 1665* ... (B)

Henri Montignot (active ca. 1780). French ecclesiastic who in 1786 published a document labeled *Etat des Etoiles Fixes* ... that discussed the relationship of the positions of the stars in his day to the positions recorded by Ptolemy. Two star maps were included that were labeled *Position des Etoiles du Zodiaque ... Par Ptolemée* ...: one from 0–180 degrees longitude along the zodiac, and the other from 180 to 360 degrees. Each was 23.5 × 70 cm and showed the Ptolemaic constellations in a geocentric projection. (W)

Jonas Moore (1617–1679). Influential mathematician and surveyor who helped bring about the founding of the Royal Observatory in Greenwich, for which he provided astronomical instruments, and the Royal Mathematical School in London, at which he taught and wrote a navigation text entitled *A New System of Methematicks* ... Written in 1681, this book contained a set of 6 celestial maps, 2 covering the equatorial polar regions and 4 centered on the areas around the equinoxes and solstices. Each was 20 cm square and used a gnomonic projection with geocentric orientation. The maps were smaller versions of those drawn by Pardies. (W)

Joseph Moxon (1627–1700). Distinguished globe and instrument maker and cartographer in London who was the Hydrographer to Charles II and a Fellow of the Royal Society. He described two celestial hemispheres in a catalog from his 1674 book *A Tutor to Astronomy and Geography as: 2 Caelestial Hemispheres of all the Stars and Constellations in Heaven: Projected upon the Poles of the Equator. 17 inches Diameter* ... In his 1679 document *The English Globe* ..., he described a 23 cm diameter celestial planisphere in the base supporting a terrestrial globe which was centered on the north equatorial pole using a polar stereographic projection with external orientation. Finally, there is a print attributed to him ca. 1691 and dedicated to the Rev. John, Archbishop of Canterbury, which depicted side-by-side celestial and terrestrial planispheres surrounded by biblical scenes. The celestial planisphere was centered on the north equatorial pole to below the Tropic of Capricorn, was 22 cm in diameter, and used a polar equidistant projection with external orientation. (W)

Muggletonians (see **Isaac Frost**).

Frantisek Noël (1651–1729). Czech Jesuit astronomer who produced a planisphere labeled *In huius mappae projectione oculus ponitur* ..., which appeared in his 1710 *Observations Mathematicae, et Physicae in India et China* ... It was centered on the south equatorial pole to 40 degrees S dec. using a polar equidistant projection with geocentric orientation. (W)

Reginald Outhier (1694–1774). Astronomer, cartographer, and canon at the Bayeux Cathedral who corresponded with the Académie Royale des Sciences in Paris. In 1755, a celestial map of his appeared in the Académie's *Mémoires Scavans Etrangers* labeled *Carte des Etoiles des Pleiades ...*, which was 20.2 × 16.3 cm and used a geocentric orientation. Warner states that this was simply an elaboration of Le Monnier's Pleiades map. (W)

Samuel Parker (active 1717–1728). British engraver who around 1728 produced two celestial hemispheres labeled *N.* (or *S.*) *Coelestial Hemisphere*. Each was 34 cm in diameter and was centered on an ecliptic pole using polar stereographic projection with geocentric orientation. The maps resembled those of John Seller. (W)

W.E. Parry (active early 1800s). Author of an 1816 book on navigation entitled *Nautical Astronomy by Night*, which consisted of 6 tables and 5 plates showing the principal fixed stars. (B)

Sir William Peck (active late 1800s). Director of the City Observatory of Edinburgh who in 1890 published the *Popular Handbook and Atlas of Astronomy*, with 44 plates and a number of text illustrations. This prolific astronomer also wrote *The Constellations and How to Find Them* for beginners, and the *Observer's Atlas of the Heavens* for more advanced students, which contained star and deep-sky catalogs and 30 star charts. (B)

C.H.F. Peters (active late 1800s). American astronomer at the Lichfield Observatory in New York who in 1882 published *Celestial Charts*, which contained 20 star maps of the zodiacal heavens. (B)

Lucillus Philalthaeus (see **Lucilio Maggi**).

Noel-Antoine Pluche (1688–1761). French naturalist who wrote a popular book in 1732 that went through many editions and translations that was labeled *Spectacle de la Nature*. It contained 4 celestial half-planispheres. Two were labeled *Première (*or *Seconde) Partie de l'Hemisphère Céleste Septentrional* and were centered on the north equatorial pole from 180–360 degrees and 0–180 degrees right ascension, respectively. The other two were labeled *Première* (or *Seconde*) *Partie de l'Hemisphère Céleste Austral* and were centered on the south equatorial pole from 180–360 degrees and 0–180 degrees right ascension, respectively. Each was 28 cm in diameter using a polar stereographic projection with geocentric orientation. According to Warner, the maps were copied from the work of Senex. (W)

Johannes Mauricius Polzius (active ca. 1680). Author of a celestial map of the path of a comet entitled *Com. Mens Augus Anno 1682*, which appeared in Rostock's *Eine neue grosse Himmels-Lampe ...* The map was 10.7 × 14.8 cm, and the constellations around Virgo were poorly drawn and in a geocentric orientation. (W)

Guillaume Postel (1505–1581). French savant who was a Professor of Mathematics and of Oriental Languages at the College Royal in Paris who included four woodcut celestial maps in his 1553 *Signorum Coelestium Vera Configuratio aut Asterismus ...*

Two were hemispheres centered around an ecliptic pole that were each 23 cm in diameter and used a polar stereographic projection and geocentric orientation, without figures. The other two were the northern and southern ecliptic hemispheres of Honter. (W)

Etienne-André Philippe de Pretot (ca. 1708–1787). History professor and member of the Académie Royale des Sciences et Belles-Lettres d'Angers who in 1768 published a book entitled *Cosmographie Universelle, Physique et Astronomique* ... The book contained two celestial hemispheres labeled *Hemisphère Céleste Septentrional* (or *Meridional*), each of which was 23.3 cm in diameter and centered on an equatorial pole using a polar stereographic projection with external orientation. (W)

W. H. Rosser (active mid-1800s). Author of a book on navigation in 1867 entitled *The Stars—How to Know Them and How to Use Them* which consisted of 4 large maps showing white stars on a black background and figures of the chief constellations. (B)

Alexandre Ruelle (born 1756). French army deserter who took refuge with a relative at the Paris Observatory. He subsequently became a student there, and from 1793 to 1795 he was appointed as one of the astronomers. In 1785 he published a 3-part celestial map and instruction book labeled *Nouvelle Uranographie* ... Two parts were hemispheres entitled *Hemisphère Septentrional ou Boréal* (or *Meridional ou Austral*), each of which was 37 cm in diameter and centered on an equatorial pole using a polar equidistant projection with geocentric orientation. The third part consisted of constellations above and below the equatorial belt using a geocentric orientation, and the print was 23.2 × 93.5 cm. There were no constellation figures, just lines connecting the stars, and paths of the planets also appeared. A second edition was published in 1786, and a third edition followed. (W)

Giovanni Antonio Rusconi (active 1590). Architect in Venice who published a book in 1590 entitled *Dell'Architettura Secundo i Precetti di Vitruvio* ... that included two celestial hemispheres to illustrate Vitruvius' discussion of constellations. Each was 12.2 cm in diameter and centered on an ecliptic pole using a polar stereographic projection with external orientation. They were patterned after Dürer. (W)

Johann Adam Schall von Bell (1591–1666). German Jesuit who collaborated with the Chinese Imperial Bureau of Astronomy. He published a treatise on the telescope in 1626, a calendar for the Emperor in 1634, and a book on astronomy in collaboration with other Chinese and Jesuit scholars. In 1645 he published a document entitled *Hsi Yang Hsin Fa Li Shu*, in which 24 celestial maps appeared. Two were hemispheres centered on an equatorial pole and using a polar stereographic projection; two were hemispheres centered on an ecliptic pole and using a polar stereographic projection; and two were planispheres centered on an ecliptic pole down to 65 degrees lat. and using a polar projection. The rest of the sky was covered by the remaining 18 maps, all of which used a trapezoidal projection. The constellations were Chinese and were viewed from an external orientation. (W)

Wilhelm Schickard (1592–1635). Astronomer, mathematician, and professor at Tubingen who corresponded with Kepler and supplied some of the Arabic names for Schiller's Judeo-Christianized atlas (see Section 6.2.3.1). In 1623, Schickard produced a book entitled *Astroscopium pro facillima Stellarum cognitione noviter excogitatum* that contained two conical celestial maps of 13.5 cm in diameter and centered on an equatorial pole with geocentric orientation. Biblical as well as Greco-Arabic names were provided for the constellations. Other editions were published in 1655, 1659, and 1665. In 1687, Schickard published another book with the title *Astroscopium Pro facillima Stellarum cognitione excogitatum & Commentariolo illustratum*, some issues of which contained two similar conical maps, but much larger (33 cm in diameter). Warner mentions three important aspects of Schickard's work: one of the first uses of a conical format in a celestial map; the use of a geocentric orientation based on his interpretation of Ptolemy, which later influenced Flamsteed's work; and a reflection of the interest at the time in Germanic countries to apply Judeo-Christian concepts to the constellations. (W)

R. Schurig (active 1886). German producer of a popular star atlas entitled *Himmels Atlas*, published in 1886. A recent edition came out in 1960 by Schurig and Goetz and was entitled *Tabulae Caelestis*. (B)

Edward Sherburne (1618–1702). British poet and member of a Catholic royalist family who in 1675 published a book entitled *The Sphere of Marcus Manilius made an English Poem …* that contained two celestial hemispheres centered on an equatorial pole using a polar stereographic projection with geocentric orientation. The constellation figures were dressed in contemporary rather than classical clothes. The book also contained Moon maps and other cosmological diagrams illustrating updated astronomical information, such as the new constellation Cor Caroli Regis Martyris, invented in 1673 by Charles Scarborough to honor Charles I. (A, W)

Admiral William Henry Smyth (active mid-1800s). English creator of *A Cycle of Celestial Objects* in 1844, which contained a number of woodcuts and the Bedford Catalogue of double and multiple stars and deep-sky objects. (B)

Edward James Stone (active late 1800s). Cape Astronomer Royal who in 1881 published his *Catalogue of 12,441 Stars from the Observations made at the Cape of Good Hope, 1871–9*, which was accompanied by a very fine folding star map on linen. This work was awarded the Lalande prize by the French Academy. (B)

Melchior Tavernier (ca. 1594–1665). Parisian cartographer who around 1650 produced a pair of celestial hemispheres with the label *A Paris Chez Melchior Tavernier Graveur …* Each was 26.5 cm in diameter and was centered on an ecliptic pole using a polar stereographic projection with external orientation. According to Warner, they were very similar to the hemispheres produced by his neighbor, Antoine de Fer. (W)

Jan Thesing (active ca. 1700). Dutch printer and publisher who in 1699 was granted a 15-year privilege to publish maps and books in Russia by Czar Peter I two years after his visit to Amsterdam. In that same year, Thesing produced a celestial planisphere with the label *To the most serene and most enlightened Lord, the Blizhnyi Boiarin ...*, which was bound in Kopijewicz's *Ougotovanie I tolkovanie ...* The map was 56 × 48 cm, and the planisphere was centered on the north equatorial pole down to 40 degrees S dec. using a polar equidistant projection with external orientation. It was overlaid by a rete projecting the sky as seen from 52 degrees N lat. that served as a star finder. Warner cites it as the first printed Russian star map and views it as a Russian translation of Voogdio's popular 1680 celestial map (see below). (W)

Leonhard Thurneysser (1531–1596). German adventurer who according to Warner was an alchemist, astrologer, soldier, merchant, engineer, and physician. She attributes 7 complex astronomical diagrams to him that appeared in the 1575 book *Des Menschen Circkel und Lauff*, and she describes one hemisphere labeled *Der Sonnen Circkel und Lauff* that was 42 × 58 cm and centered on the south ecliptic pole using a polar stereographic projection and geocentric orientation. This map included a number of concentric circles, some of which were printed on the page and others of which were volvelles. Some of the items featured included a perpetual calendar for the dates of the new Moon, the names and latitudes of places on Earth, the days and parts of the year in Hebrew and Latin, a ring with astrological phrases, a rete with northern stars attached, and a rotating arm in the shape of a dragon tracing the solar path. Warner also cites a reference attributing to him a tract on the comet of 1577 labeled *Ein kurtzer und einfältiger Bericht ...* that contains two celestial maps. (W)

Abbé de Vallemont (1649–1721). French author and cleric also known as Pierre Le Lorrain who in 1696 published a book in Paris entitled *Les Elemens de l'Histoire*. It contained a pair of celestial hemispheres entitled *Le Globe Céleste*, each of which was 9.9 cm in diameter and was centered on an ecliptic pole using a polar projection with external orientation. It was repeated in the various editions and translations of Vallemont's book, such as the *Atlante Portatile*, published in Venice in 1748. (W)

Didier Robert de Vaugondy (1723–1786). French cartographer and globe maker who was descended from the well-known cartographers Nicolas Sanson and Gilles Robert de Vaugondy. In 1764 he produced two celestial hemispheres described in *Uranographie ou Description du Ciel en Deux Hemisphères* that were labeled *Hemisphère Céleste Arctique* (or *Antarctique*) ... Each was 58 cm in diameter and was centered on an equatorial pole using a polar stereographic projection with external orientation. Although the views of his constellations were influenced by Bayer, he chose to use an external orientation, as befit his globe-making background. In 1779 he produced another pair of hemispheres that were identical in name, size, and orientation to the 1764 version, except that he included the term *Nouvelle Edition* in the titles, added a few constellations, and omitted some of the his previous labeling. (W)

Ludovico Vlasblom (active ca. 1675). Physician and mathematician who around 1675 produced a print of a pair of celestial hemispheres labeled *Stellatum Planisphaerium* that was bound in a number of sea atlases by Van Keulen. Each hemisphere was 25.4 cm in diameter and was centered on an ecliptic pole using a polar stereographic projection with external orientation. They were nearly identical to those appearing in Joan Blaeu's great world map and to those in the double-hemisphere print of de Wit (see below). Two inserts were included showing the cosmologies of Ptolemy and Copernicus. (W)

Johann Heinrich Voigt (active 1640–1685). German astronomer and author of almanacs and cometary prognostications. Two of the latter are described by Warner. The first was labeled *Nord: Uben. Cometa Ao 1677 ...* and appeared in his 1677 *Christmassige Betrachtung Des Cometen Im Aprili Anno 1677 ...* It was 16.2 cm square and showed the cometary path through stars without constellation figures using a geocentric orientation. The paths of two earlier comets were also shown. The second was labeled *Im Gebrauch mus das Angesicht gegen Sud-Westen gekehret ...* and appeared in his 1681 *Cometa Matutinus & Vespertinus ...* It was 11.7×28.5 cm and showed the cometary path in 1680–1681 through stars without constellation figures using a geocentric orientation. (B, W)

Eufrosino Della Volpaia (ca. 1494–1554 or later). Well-known cartographer and member of a distinguished Florentine family of instrument makers who in 1530 produced a hemisphere centered on the south ecliptic pole labeled *Imagines Coeli Meridionales ...* that was 35 cm in diameter (on a 44 cm-square print) using a polar stereographic projection with external orientation. The style was copied from Dürer but also incorporated the 18 southern stars and Magellanic Clouds from Corsali's map (see above). At the top corners were images of his father and the astronomer Gauricus. (W)

N. Voogdio (active ca. 1680). Probable Dutch cartographer from Amsterdam who around 1680 produced a celestial planisphere that was 32 cm in diameter and centered on the north equatorial pole down to 40 degrees S dec. using a polar equidistant projection with external orientation. It included an overlaid rete representing the skies over Amsterdam that served as a star finder. There was explanatory text on the side labeled *Hemels Pleyn ...*, and *Auct: N. Voogdio ...* According to Warner, this map was printed several times, with variations in signatures or explanatory text, and it was translated into Russian by Jan Thesing. (W)

Caspar Vopel (1511–1561). Cartographer and mathematician living in Cologne who produced a cordiform world map labeled *Typo de la Carta Cosmographica de Gaspar Vopellio Medeburgense* that appeared in the 1556 book by de Girava entitled *Dos Libros de Cosmographia*. In the lower corners were two celestial hemispheres, each of which was centered on an ecliptic pole using a polar stereographic projection with external orientation. Warner states that this is probably a reduced copy of a now-lost larger wall map that Vopel produced in 1545. She cites another copy of this map that appeared in Vavassore's 1558 book *Nova et Integra Universalisque* ... These maps were all derived from Dürer. (W)

Lucas Janszoon Waghenaer (ca. 1534–ca. 1605). Dutch pilot and author who produced a volvelle map used for polar navigational purposes originally labeled *Tamme dragende compasse oft Instrumenten metten Ghesternte* that was 22 cm in diameter and centered on the north equatorial pole down to 40 degrees S dec. using a polar equidistant projection and external orientation, but no figures. Appearing originally in his 1584 book *Spiegel der Zeevaerdt* ..., it was later copied and appeared in the many editions of his popular navigation book *Speculum Nauticum* (or *The Mariners Mirror* in the English translation). (W)

Rev. T.W. Webb (active late 1800s). Author of a popular observer's handbook for amateur astronomers entitled *Celestial Objects for Common Telescopes* in 1859, which went through at least 6 editions into the 1900s. (B)

Erhard Weigel (ca. 1625–ca. 1699). Astronomer, inventor, architect, and Professor of Mathematics at the University of Jena who published a star map around 1688, which according to Brown has been called *Coelum Heraldicum*. In this map the ancient figures of the non-zodiacal constellations were replaced by the heraldic arms of many princes of Europe. (B)

Johann Christoph Weigel (died 1746). Cartographer in Nuremberg who in 1720 produced a print of a pair of celestial hemispheres labeled *Planiglobium Coeleste* that appeared bound in the 1720 and 1745 editions of his book *Atlas Portalis*. Each hemisphere was 10.6 cm in diameter and was centered on an equatorial pole using a polar projection with external orientation. Warner calls this a "sloppy, practically useless map" (p. 264). (W)

Frederic de Wit (1630–1706). Dutch engraver and map dealer who around 1700 produced a print of a pair of celestial hemispheres labeled *Planisphaerium Coeleste* that appeared bound in several atlases, such as Allard's 1705 *Atlas Major* and Sanson's *Atlas Nouveau*. Each hemisphere was 34 cm in diameter and was centered on an ecliptic pole using a polar stereographic projection with external orientation. The hemispheres were nearly identical to those appearing in Joan Blaeu's great world map and to those in the double-hemisphere print of Vlasblom (see above). (W)

R. Woosley (active 1800). Author of *Celestial Companion*, 1802, consisting of 5 plates, a portrait, and a treatise on astronomy. (B)

Thomas Wright (active mid-1700s). English amateur astronomer, instrument maker, lecturer, and writer who authored *Clavis Coelestis* in 1742, which contained a number of drawings of celestial bodies that integrated natural phenomena with ideas of the divine in the physical order. (J)

Thomas Young (1773–1829). British physician, physicist, and Egyptologist who was a Professor of Natural Philosophy at the Royal Institution in London who in 1807 published a book entitled *A Course of Lectures on Natural Philosophy and the Mechanical Arts*. This work included two celestial hemispheres that were labeled *Joseph Skelton sculp.*, each of which was 16.6 cm in diameter and was centered on an equatorial pole using a polar stereographic projection with geocentric orientation. They included a number of now-extinct constellations, including the Mockingbird perched on the tail of Hydra, and the Battery of Volta near Equuleus that reflected the interest in science at the time. (W)

Johann Jacob Zimmermann (1644–1693). German astronomer who in his 1681 treatise *Cometo-Scopia* included a map of the path of the comet of 1680–1681 that was 11.8 × 26.2 cm and used a geocentric orientation. In 1692 he produced two conical celestial hemispheres that were often bound in his *Coniglobium Nocturnale Stelligerum* ... that were each 29 cm in diameter and were centered on an equatorial pole using a polar projection. These conical celestial maps were printed several times from 1704 to 1770. The constellations were taken from Hevelius, and the conical format was taken from Schickard (see above). (W)

Appendix C

Indices of major constellation atlases

This appendix contains a list of the plates numbered in order as found in eight classical constellation atlases that are mentioned in the text. The intention of such a list is twofold: to give the reader an idea of the content of these works, and to help the collector locate the source of a plate from its title. In addition, the internal image dimension or the diameter of the hemisphere of one or two representatives is given as an indication of the size of the plates from the listed edition, which should also help in identification (keeping in mind that the exact sizes vary from plate to plate in some cases). The dimensions of plates from other editions are given in the main text. Although many of the titles are in a non-English language, I have listed them using regular script rather than italics in order to simplify the presentation. Text found on the plates is underlined; otherwise, the plates are identified by the major constellations depicted. Notes are in brackets. The spelling reflects that used in the atlas.

The listing for two of the atlases (Fortin and Piccolomini) came from the Nick and Carolynn Kanas collection. The other six atlases came from the British Library, and I have personally examined them. I am grateful to Peter Barber, the Head of Map Collections at the British Library, and his staff for helping me to locate these atlases. I am also grateful to the British Library for giving me permission to record this list based on their holdings (as well as the contents of the collection described in Appendix D). Bob Gordon, my friend and fellow celestial map collector, provided verification of some of the information on these lists. For more information, the reader is referred to the Internet, to the British Library's website, and to Rodney Shirley's excellent *Maps in the Atlases of the British Library* (London: The British Library, 2004).

C.1   Bayer, Johann—*Uranometria*, 1603

[There are no labels on the plates, just letters in the lower right corner. In the 1603 edition only, the constellations are named and described in catalog entries on the verso.]

1. Ursa Minor, A
2. Ursa Major, B ... 27.3 × 37.3 cm
3. Draco, C
4. Cepheus, D
5. Bootes, E
6. Corona, F
7. Hercules, G
8. Lyra, H
9. Cygnes, I
10. Cassiepeia, K
11. Perseus, L
12. Auriga, M
13. Serpentarius, N
14. Serpens, O
15. Sagitta, P
16. Aquila, Q
17. Delphinus, R
18. Equus Minor, S
19. Pegasus, T
20. Andromeda, V
21. Triangulum, VV
22. Aries, X
23. Taurus, Y
24. Gemini, Z
25. Cancer, Aa
26. Leo, Bb
27. Virgo, Cc
28. Libra, Dd
29. Scorpio, Ee
30. Sagittarius, Ff
31. Capricornus, Gg
32. Aquarius, Hh
33. Pisces, Ii
34. Cetus, Kk
35. Orion, Ll
36. Eridanus, Mmx
37. Lepus, Nn
38. Canis Major, Oo
39. Canis Minor, Pp
40. Navis, Qq
41. Centaurus, Rr
42. Crater, Ss
43. Corvus, Tt
44. Hydra, Rr
45. Lupus, Ww
46. Ara, Xx
47. Corona Meridionalis, Yy
48. Piscis Notius, Zz
49. Pavo, Toucan ... [12 new southern constellations], Aaa
50. Synopsis Coeli Superioris Borea, Bbb ... 22.7 cm dia.
51. Synopsis Coeli Inferioris Austrina,

C.2   Bode, Johann—*Uranographia*, 1801

[Labels only appear on the first two plates. Roman numerals appear on the upper right corner of the plates.]

1. Coelium Stellatum Hemisphaerium Arietis, Uranographia Tab I ... 56.7 cm dia.

2. Coelium Stellatum Hemisphaerium Librae, Uranographia Tab II

3. Draco, Cepheus, Cauda Draconis, Ursa Minor, III

4. Cassiopeja, Perseus, Caput Medusa, Triangulum Majus, Andromeda, IV

5. Auriga, Camelopardus, Lynx, Telescopium Herschelii, V

6. Ursa Major, Leo Minor, VI ... 52.5 × 73 cm

7. Bootes, Coma Berenices, Corona Borealis, Quadrans Muralis, Mons Maenalus, Canes Venatici, VII

8. Cygnus, Vultur et Lyra, Hercules, Sagitta, Vulpecula, VIII

9. Ophiuchus, Serpens Ophiuchi, Scutum Sobiesii, Taurus Poniatorii, Aquila, Antinous, IX

10. Pegasus, Delphinus, Equuleus, X

11. Aries, Musca, Pisces Borealis ..., XI

12. Gemini, Taurus, Canis Minor, Orion, Harpa Georgii, XII

13. Leo, Cancer, Sextans Uraniae, XIII

14. Virgo, Libra, Turdus Solitarius, Cauda Hydrae, XIV

15. Sagittarius, Corona Australis, Scorpius, Tubus Astronomicus, Ara, Norma et Regula, Lupus, XV

16. Aquarius, Capricornus, Piscis Notius, Globus Aerostaticus, Microscopium, XVI

17. Cetus seu Monstrum Marinum, Eridanus, Apparatus Chemicus, Machina Electrica, Apparatus Sculptoris, XVII

18. Canis Major, Caput Hydrae, Monocerus, Lepus, Officina Typographica, Argo Navis, Columbus ..., XVIII

19. Centaurus, Hydra seu Serpens Aquaticus, Corvus, Crater, Felis, Antilia Pneumatica, Argo Navis, XIX

20. Phoenix, Grus, Toucan, Indus, Pavo, Apus seu Avis Indica, Apis ..., XX

C.3   Cellarius, Andreas—*Harmonia Macrocosmica*, 1660

[All titles are plate labels, some abbreviated. The second printing of the first edition and later editions have ordinal numbers in the lower right corner corresponding to the list numbers.]

1. Planisphaerium Ptolemaicum ...

2. Scenographia Systematis Mundani Ptolemaici

3. Orbium Planetarum ...

4. Planisphaerium Copernicanum ...

5. Scenographia Systematis Copernicani

6. Planisphaerium Braheum ...

7. Scenographia Compagis Mundanae Brahea

8. Planisphaerium Arateum ...

9. Tychonis Brahe Calculus, Planetarum ...

10. Corporum Coelestium Magnitudines

11. Situs Terrae Circulis Coelestibus Circundate

12. Haemisphaeria Sphaerarum Rectae et Obliquae ...

13. Hemisphaerium Orbis Antiqui ...

14. Hypothesis Ptolemaica, sive Communis Planetarum ...

15. Typus Aspectuum, Oppositionum et Coniunctionum ...

16. Theoria Solis per Eccentricum sine Epicyclo ... 42.1 × 50.4 cm, 38.5 cm dia.

17. Solis Circa Orbem Terrarum ...

18. Theoria Lunae, Eius Motum per Eccentricum et Epicyclum ... 42 × 50.4 cm, 36.8 cm dia.

19. Typus Selenographicus Lunae Phases ...

20. Theoria Trium Superiorum Planetarum

21. Theoria Veneris et Mercurii

22. Coeli Stellati Christiani Haemisphaerium Prius

23. Coeli Stellati Christiani Haemisphaerium Posterius

24. Haemisphaerium Stellatum Boreale Antiquum

25. Haemisphaerii Borealis Coeli et Terrae Sphaerica Scenographia

26. Haemisphaerium Stellatum Boreale cum Subiecto Haemisphaerio Terrestri

27. Haemisphaerium Stellatum Australe Antiquum

28. Haemisphaerium Scenographicum Australe Coeli Stellati et Terrae

29. Haemisphaerium Stellatum Australe Aequali Sphaerarum Proportione

C.4 Doppelmayr, Johann—*Atlas Coelestis*, 1742

[All titles are plate labels, some abbreviated. Ordinal numbers appear in upper right corner corresponding to the list numbers.]

1. Sphaera Mundi ...
2. Systema Solarae et Planetarium ... 48.2 × 56.8 cm, 43.6 cm dia.
3. Systema Mundi Tychonicum ...
4. Theoria Planetarum Primariorum ...
5. Phaenomena in Planetis Primariis
6. Phaenomena ...
7. Phaenomena Motuum Irregularium ...
8. Ephererides Motuum Coelestium Geometricae ...
9. Motus in Coelo Spirales ...
10. Motus Planetarum Superiorum, qui, Secundum Tychonis Hypothesin ... 48 × 57.2 cm, 43.8 cm dia.
11. Tabula Selenographica ... tam Hevelii quam Riccioli ... 47.9 × 57.2 cm, each hemisphere 27.7 cm dia.
12. Theoria Lunae ...
13. Theoria Eclipsium ...
14. Theoria Satellitum Iovis et Saturni ...
15. Basis Geographiae Recentioris Astronomica ...
16. Hemisphaerium Coeli Boreale in qua loca stellarum fixarum ...
17. Hemisphaerium Coeli Australe in qua loca stellarum fixarum ...
18. Hemisphaerium Coeli Boreale in quo Fixarum ...
19. Hemisphaerium Coeli Australe in quo Fixarum ...
20. Globi Coelestis in Tabulas Planas Redacti Pars I ... 48 × 58.8 cm
21. Globi Coelestis in Tabulas Planas Redacti Pars II ...
22. Globi Coelestis in Tabulas Planas Redacti Pars III ...
23. Globi Coelestis in Tabulas Planas Redacti Pars IV ...
24. Globi Coelestis in Tabulas Planas Redacti Pars V ... 48.5 × 57.4 cm
25. Globi Coelestis in Tabulas Planas Redacti Pars VI ...
26. Theoria Cometarum ...
27. Motus Cometarum In Hemisphaerio Boreali ...
28. Motus Cometarum In Hemisphaerio Australi ...
29. Astronomia Comparativa ... e Sole Mercurio et Luna exhibentur ...
30. Astronomia Comparativa ... Marte, Jove, et Saturno siftuntur ...

C.5   Flamsteed, John—*Atlas Coelestis*, 1729

[All plates are labeled except for 3, 9, 11, 25, 26, and 27.]

1. Aries
2. Taurus
3. Gemini
4. Cancer
5. Leo
6. Virgo
7. Libra
8. Sagittarius
9. Aquarius
10. Pisces
11. Cetus
12. Eridanus Orion Lepus
13. Monoceros, Canis Major & Minor, Navus, Lepus ... 47.3 × 58.2 cm
14. Hydra Crater Corvus Sextans Virgo ... 44.5 × 97.1 cm
15. Cassiopea Cepheus Ursa Minor Draco
16. Andromeda Perseus Triangulum
17. Camelopardal & Auriga
18. Lynx & Leo Minor
19. Ursa Major ... 44.9 × 58.9 cm
20. Comae Berenices Bootes Canes Venatici
21. Hercules Corona & Lyra
22. Ophiunchus & Serpens
23. Aquila Sagitt Vulpecula & Anser Delphinus
24. Lyra Cygnus Lacerta, Vulpec. & Anser Sagitta
25. Pegasus
26. Southern celestial hemisphere ... 50.6 cm dia.
27. Northern celestial hemisphere ... 50.6 cm dia.

C.6   Fortin, Jean—*Atlas Céleste de Flamsteed*, 1776

[All titles are plate labels and are taken from the index listing. Ordinal numbers appear in upper right corner corresponding to the list numbers.]

1. Hémisphere Boréal
2. Cassiopée, Cephée, le Réene, la Petite Ourse, le Dragon
3. Andromede, Persée, le Triangle
4. La Giraffe, le Cocher
5. Le Linx, le Petit Lion
6. La Grande Ourse
7. Le Bouvier, les Levriers, la Chevelure, de Berénice
8. Hercule, la Couronne
9. Le Serpentaire & le Serpent
10. L'Aigle, Antinoüs, la Flêche, le Renard, le Dauphin
11. La Lyre, le Cigne, le Lézard, le Renard
12. Pegase, le Petit-Cheval, le Dauphin
13. Le Bélier
14. Le Taureau
15. Les Gemeaux
16. Le Cancer
17. Le Lion
18. La Vierge
19. La Balance & le Scorpion
20. Le Sagittaire ... 15.8×20.7 cm
21. Le Capricorne & le Verseu
22. Les Poissons
23. La Baleine
24. L'Eridan, Orion & le Lievre
25. La Licorne, le Grand Chien
26. L'Hydre, le Sextant
27. L'Hydre, la Coupe, le Corbeau
28. Hémisphere Austral
29. Hémisphere Austral, suivant M. l'Abbé de la Caille ... 19.4×23.9 cm image, 18.4 dia. hemisphere

C.7   Hevelius, Johannes—*Firmamentum Sobiescianum*, 1687, appearing in *Prodromus Astronomiae*, 1690

[Except for the first two plates, there are no labels, but "Figure" letters appear on the plates.]

1. Hocce Hemisphaerium Firmamenti Sobiescianai Boreale ... 46.5 cm dia.
2. Hocce Hemisphaerium Australe Serenissimo Principe ...
3. Ursa Minor, A
4. Draco, B
5. Cepheus, C
6. Ursa Major, D ... 29.1 × 36.3 cm
7. Canes Venatici, E
8. Bootes, Coma Berenices, Mons Maenalus, F
9. Corona, G
10. Hercules, H
11. Lyra, I
12. Cygnus, K
13. Anser, Vulpecula, L
14. Lacerta sive Stellio, M
15. Cassiopeia, N
16. Camelopardalus, O
17. Serpentarius, Serpens, P
18. Scutum Sobiescianum, Q
19. Aquila, Antinous, R
20. Sagitta, Delphinus, Equuleus, S
21. Pegasus, T
22. Andromeda, V
23. Perseus, W
24. Auriga, X
25. Lynx, Y
26. Leo Minor, Z
27. Triangulum Majus, Triangulum Minus, Musca, AA

28. Aries, BB
29. Taurus, CC
30. Gemini, DD
31. Cancer, EE
32. Leo, FF
33. Virgo, GG
34. Libra, HH
35. Scorpius, II
36. Sagittarius, KK
37. Capricornus, LL
38. Aquarius, MM
39. Pisces, NN
40. Cetus, OO
41. Phoenix, Toucan, Eridanus, PP
42. Orion, QQ
43. Monoceros, RR
44. Canis Minor, SS
45. Hydra, TT
46. Sextans Uraniae, VV
47. Crater, Corvus, WW
48. Centaurus, Crux, XX
49. Lupus, YY
50. Ara, Pavo, Triangulum Australis, ZZ
51. Corona Australis, AAa
52. Piscis Notius, Grus, BBb
53. Lepus, Columba, CCc
54. Canis Major, DDd
55. Argo Navis, EEe
56. Polus Antarcticus, FFf

C.8   Piccolomini, Alessandro—*De le Stelle Fisse*, 1579 edition

[All plates are labeled with titles and "Figura" Roman numerals.]

1. De l'Orsa Minore, I ...
   17.5 × 13.9 cm
2. De l'Orsa Maggiore, II ...
   17.7 × 14.1 cm
3. Del Drago, III
4. Di Cefeo, IIII
5. Di Boote, o vero Arturo, V
6. Della Corona d'Arianna, VI
7. Di Hercole, VII
8. Della Libra, VIII [probably Lyra]
9. Del Cigno, IX
10. Di Cassiopea, X
11. Di Perseo, XI
12. De l'Inventor del Carro, XII
13. Quel che Tiene il Serpe, XIII
14. Del Serpente d'Esculapio, XIIII
15. Della Saetta, XV
16. [Missing chart, probably Equuleus]
17. Dell'Aquila, XVII
18. Del Delfino, XVIII
19. Del Cavallo Alato, XIX
20. D'Andromeda, XX
21. Del Triangolo, XXI
22. Del Montone, XXII
23. Del Tauro, XXIII
24. De I Gemelli, XXIIII

25. Del Cancro, XXV
26. Del Leone, XXVI
27. Della Virgine, XXVII
28. Della Libra, XXVIII
29. Dello Scorpione, XXIX
30. Del Sagittario, XXX
31. Del Capricorno, XXXI
32. Dell'Aquario, XXXII
33. De I Pesci, XXXIII
34. Del Ceto, o ver Balena, XXXIIII
35. Di Orione, XXXV
36. Del Fiume, XXXVI
37. Della Lepre, XXXVII
38. Del Can Sirio, o ver Maggiore,
    XXXVIII
39. Della Canicula, XXXIX
40. Della Nave, XL
41. Del l'Hidra, XLI
42. Della Tazza, o ver Vaso, XLII
43. Del Corvo, XLIII
44. Del Centauro, XLIIII
45. Del Lupo, XLV
46. Dell'Altare, XLVI
47. Della Corona Astrale, XLVII
48. Del Pesce Australe, XLVIII

Appendix D

The British Library "King's" edition

In the British Library is a series of collected works under the listing of "King's Topographical Collection" (Maps K.Top.1.58). Volume I has two parts devoted to astronomical maps and prints: Part 1 includes numbered entries 50–69, and Part 2 includes numbered entries 70–91. Some of these entries are from well-known sources, others are not.

A list of items in Part 1 includes: celestial prints by Barlow, Watson, Nicholas de Fer, de la Fosse, None, Senex, de Vaugondy, Chrysologue, and Wollason; two unusual German celestial planispheres in conic projection, with unidentified authorship; and two other unidentified German planispheres. A list of items in Part 2 includes celestial prints by de l'Isle, Lalande, Ferguson, and others (both named and unnamed) showing: planetary orbits; transits of Mercury and Venus; the paths of comets in the sky; solar eclipse paths on the Earth; and volvelles, one with an accompanying booklet.

The researcher of celestial maps would do well to investigate this collection in addition to the better-known celestial atlases and books in the British Library, which are readily accessed by title or author in the computerized catalog.

Appendix E

Glossary

alidade   the sighting bar on the back of an **astrolabe**

armillary sphere   an observational and calculating instrument developed by the ancient Greeks that is made up of several graduated rings that represent circles in the celestial sphere

astrolabe   an observational and calculating device developed by the ancient Greeks and refined by Islamic astronomers that allowed for a physical representation of the heavens and movements of celestial bodies to be projected on flat plates, usually made out of metal

autumnal (fall) equinox   *see* **equinox**

azimuth   a horizontal direction in the sky usually measured in degrees clockwise from the north

azimuthal map projection   a map projection where the surface of a globe is projected onto a flat surface that touches it at a single point

back-staff   a navigation instrument that allowed the user to measure the altitude of the Sun by facing away from its glare

celestial equator (or equinoctial)   the great circle in the sky that is a projection of the Earth's equator

celestial latitude   the angular distance of a celestial body north or south of the **ecliptic**

celestial longitude   the angular distance of a celestial body measured eastward along the **ecliptic** from a reference point (usually the **First Point of Aries**)

charge-coupled device (CCD)   an electronic detector sensitive across a range of wavelengths that has replaced photographic plates in imaging faint heavenly bodies

clepsydra   an ancient water clock

colure   the great circle in the sky passing through the north and south celestial poles and the location in the **zodiac** of the Sun during the two **equinoxes** (equinoctial colure) or the two **solstices** (solstitial colure)

conic map projection   a map projection created as if a cone was placed around the corresponding globe and touched it along a line of latitude

cosmography   the description and representation of the universe's structure

cosmology   the study of the structure and evolution of the universe

cross-staff   an instrument used for measuring the angle between two celestial objects

cylindrical map projection   a map projection created as if the map was rolled around the corresponding globe along its equator

decan   in ancient Egyptian astronomy, one of 36 stars or star groups used for telling the time at night

declination   the angular distance of a celestial body north or south of the **celestial equator**

deferent   in ancient geocentric Greek astronomy, a circle going around the Earth that carries the Sun or an **epicycle** of the Moon or a planet

dominical letter   the letter representing Sundays in old calendars where the seven weekdays were each assigned a different letter of the alphabet

eccentric circle   in ancient geocentric Greek astronomy, a circle going around the Earth whose center is not the Earth

ecliptic   the great circle in the sky carrying the yearly path of the Sun as seen from the Earth (and along which also move the Moon and planets)

ecliptic pole (north or **south)**   the north or south extension of the axis of the celestial sphere that has the **ecliptic** as its equator

Empyrean   the highest level of heaven, inhabited by the blessed, angels, and God

engraving   *see* **intaglio printing**

ephemeris (pl. ephemerides)   an astronomical table giving the location and other information regarding heavenly bodies over a sequence of dates (compare with **zij**)

epicycle   in ancient geocentric Greek astronomy, a small circle carrying the Moon or a planet; the center of the epicycle is in turn carried on a **deferent** around the Earth

equant   in ancient geocentric Greek astronomy, the point which is an equal and opposite distance from the Earth with respect to the center of a **deferent** circle; the motion of the **epicycle** being carried on the deferent is set so that the Moon or planet it carries revolves around the equant point at a uniform angular speed (even though its speed around the deferent is variable)

equatorial pole (north or **south)**   the north or south extension of the axis of the celestial sphere that has the **celestial equator** as its equator

equatorium   a device with moving disks (or **volvelles**) that is used to demonstrate (or sometimes calculate) the motions of planetary bodies in space

equidistant azimuthal map projection   a type of **azimuthal map projection** where the meridians are equally spaced straight lines and the parallels are equally spaced circles; deformation increases away from the center point

equinoctial   *see* **celestial equator**

equinoctial colure   *see* **colure**

equinox (vernal or **autumnal)**   the time when the Sun crosses the **celestial equator** going north or south (currently around March 21 and September 23, respectively); in both cases, the hours of day and night are equal

etching   *see* **intaglio printing**

ether   in ancient Greek astronomy, the substance out of which the heavenly bodies and spheres are made

external orientation   the pattern of stars in a constellation as seen from outside the heavenly sphere or on a celestial globe; a left-to-right, mirror-image reversal of the **geocentric orientation**

First Point of Aries   the location in the sky where the Sun crosses the celestial equator at the time of the **vernal equinox**; historically, this was in the constellation Aries, but due to **precession** it is now in the constellation Pisces

geocentric orientation   (1) the pattern of stars in a constellation as seen from the Earth looking up at the heavens (contrast with **external orientation**); (2) the cosmological view of the heavens where the Earth is in the center of the universe (contrast with **heliocentric orientation**)

gnomonic map projection   a type of **azimuthal map projection** where the meridians are equally spaced straight lines and the parallels are unequally spaced circles, with spacing (and deformation) increasing away from the center point; all great circles are straight lines

heliacal rising   the first reappearance at dawn in the eastern sky of a star or planet that previously had been lost in the Sun's glare

heliocentric orientation   the cosmological view of the heavens where the Sun is in the center of the universe (contrast with **geocentric orientation**)

hemisphere (celestial)   on a star map, a circular area of the sky showing the **ecliptic** or **equatorial pole** in the center and the **ecliptic** or **celestial equator** around the rim, 90 degrees away

hippopede   the figure-of-eight curve that could be generated by two nested concentric spheres; this model helped explain a planet's motion in the sky based on Eudoxus' theory

horoscope   a record of the computed locations of the Sun, Moon, and planets, as well as certain astrologically important points on the **ecliptic**, for a given date (usually the date of birth)

intaglio printing   a type of printing where the lines to be printed are cut into the printing plate, which is usually made out of metal; this is done using a sharp engraving tool (in an engraving) or acid (in an etching)

lithographic printing   a type of **planographic printing** where ink adheres to an oily drawing on a printing plate that classically is made out of stone

lunar mansion   in ancient astronomy, one of 28 areas of the sky occupied by the Moon during its monthly revolution around the Earth (compare with **naksatra**)

lunation   the period of the Moon's synodic revolution (e.g., new Moon to new Moon), or one lunar month, lasting approximately $29\frac{1}{2}$ days

macrocosm   in astrological lore, the universe as a whole (contrast with **microcosm**)

meridian   the great circle in the sky passing through the celestial poles and the observer's **zenith**

microcosm   in astrological lore, the living body, which is seen as being analogously organized to the **macrocosm**

micrometer   a telescope accessory used to measure small distances or angles

mural quadrant   a **quadrant** that is mounted on a north–south wall that is used in measuring the altitude of a heavenly body as it crosses the observer's **meridian**

naksatra (sometimes written **nakshatra**)   (1) in ancient Hindu astronomy, one of 27 or 28 areas of the sky occupied by the Moon during its monthly revolution around the Earth (compare with **lunar mansion**); (2) the name of the representative star or star group within each of these areas

nocturnal   an instrument used for telling the time at night from stellar positions

octant   an instrument with a graduated arc of 45 degrees (i.e., an eighth of a circle) that is used for measuring the angles between celestial bodies

orthographic map projection   a type of **azimuthal map projection** where the meridians are equally spaced straight lines and the parallels are unequally spaced circles, with spacing decreasing (and deformation increasing) away from the center point; the image is like a real globe seen from a far distance

planisphere (celestial)   on a star map, a circular area of the sky showing the **ecliptic** or **equatorial pole** in the center and a perimeter that is usually greater that 90 degrees away

planographic printing   a type of printing using images drawn onto the surface of the printing plate

pneuma   the basic celestial substance according to ancient Greek cosmology (compare with **prana** and **qi**)

polar map projection (celestial)   a type of **azimuthal map projection** where the center is the north or south **ecliptic** or **equatorial pole**

prana   the basic celestial substance according to ancient Hindu cosmology (compare with **pneuma** and **qi**)

precession   the motion of the Earth's axis around an **ecliptic pole** in a period of nearly 26,000 years, which is caused by the gravitational pull of the Sun and Moon

Prime Mover   based on Aristotle, the name given to the entity or force responsible for keeping the heavenly spheres in motion

qi   the basic celestial substance according to ancient Chinese cosmology (compare with **pneuma** and **prana**)

quadrant   an instrument with a graduated arc of 90 degrees (i.e., a quarter of a circle) that is used for measuring the angles between celestial bodies

relief printing   a type of printing where the non-printing area has been removed from the printing block or plate and the part to be printed is flat and raised

rete   the rotating cutaway plate on the face of an **astrolabe** that contains the relative positions of important stars

retrograde motion   as seen from the Earth, the apparent reversal of a planet's movement in the sky from east to west instead of its normal west-to-east direction

right ascension   the angular distance of a celestial body measured eastward along the **celestial equator** from a reference point (usually the **First Point of Aries**)

Sanson–Flamsteed map projection   *see* **sinusoidal map projection**

save the phenomena (or appearances)   the characteristic of ancient Greek astronomy whereby a notion was accepted if it mathematically accounted for the observed location of a heavenly body in the sky, even though it may not necessarily represent physical reality

sexagesimal system   in ancient Babylonian mathematics, a place–value notation based on the number 60

sextant   an instrument with a graduated arc of 60 degrees (i.e., a sixth of a circle) that is used for measuring the angles between celestial bodies

shadow square   a right angle coordinate system on the back of an **astrolabe** that could be used to trigonometrically calculate the height of a building or some other object on Earth

sinusoidal map projection   a modified **conic map projection** where the meridian lines on each side of a central straight vertical meridian line are projected as sine curves

solstice (summer or **winter)**   the time when the Sun reaches its maximum distance in the sky north and south from the celestial equator (currently, around June 21 and December 22, respectively)

solstitial colure   *see* **colure**

stereographic map projection   a type of **azimuthal map projection** where the meridians are equally spaced straight lines and the parallels are unequally spaced circles, with spacing increasing away from the center point; this projection is conformal, in that it is free of angular deformation

syzygy   an alignment of three (or more) celestial objects, such as the Sun, Earth, and Moon during full Moon or eclipses

trapezoidal map projection   a type of **cylindrical map projection** with straight, equally spaced parallels and straight, converging meridians

tropics (celestial)   the circles in the sky that mark the extreme positions of the Sun north (Tropic of Cancer) and south (Tropic of Capricorn) of the **celestial equator**

tympan   a plate on the face of an **astrolabe** beneath the **rete** that is inscribed with altitude and **azimuth** coordinates

vellum   parchment made out of animal skin (usually lamb) that has been treated for use as a writing surface

vernal (spring) equinox   *see* **equinox**

volvelle   an instrument composed of one or more rotating circular disks or pointers, usually made out of paper, which are attached to a page of an old book (e.g., 16th Century books on astronomy); this device typically is used for calculating time and the location of heavenly bodies, but it may also be used for educational demonstrations

wandering stars   in the **geocentric** universe, the term given to the planets (and sometimes the Sun and Moon as well) due to their differing motions from the background stars

woodblock printing   a type of relief printing where the printing block is made out of wood

zenith   the point in the sky directly overhead

zij   an Islamic astronomical table with numbers that could be used to calculate the positions of heavenly bodies in the sky (compare with **ephemeris**)

zodiac   the circular belt in the sky through which move the Sun, Moon, and planets

Index

Printed by Publishers' Graphics LLC